AF880498

Mechanisms and Machine Science

Volume 85

Series Editor

Marco Ceccarelli, Department of Industrial Engineering, University of Rome Tor Vergata, Roma, Italy

Advisory Editors

Sunil K. Agrawal, Department of Mechanical Engineering, Columbia University, New York, USA

Burkhard Corves, RWTH Aachen University, Aachen, Germany

Victor Glazunov, Mechanical Engineering Research Institute, Moscow, Russia

Alfonso Hernández, University of the Basque Country, Bilbao, Spain

Tian Huang, Tianjin University, Tianjin, China

Juan Carlos Jauregui Correa, Universidad Autonoma de Queretaro, Queretaro, Mexico

Yukio Takeda, Tokyo Institute of Technology, Tokyo, Japan

This book series establishes a well-defined forum for monographs, edited Books, and proceedings on mechanical engineering with particular emphasis on MMS (Mechanism and Machine Science). The final goal is the publication of research that shows the development of mechanical engineering and particularly MMS in all technical aspects, even in very recent assessments. Published works share an approach by which technical details and formulation are discussed, and discuss modern formalisms with the aim to circulate research and technical achievements for use in professional, research, academic, and teaching activities.

This technical approach is an essential characteristic of the series. By discussing technical details and formulations in terms of modern formalisms, the possibility is created not only to show technical developments but also to explain achievements for technical teaching and research activity today and for the future.

The book series is intended to collect technical views on developments of the broad field of MMS in a unique frame that can be seen in its totality as an Encyclopaedia of MMS but with the additional purpose of archiving and teaching MMS achievements. Therefore, the book series will be of use not only for researchers and teachers in Mechanical Engineering but also for professionals and students for their formation and future work.

The series is promoted under the auspices of International Federation for the Promotion of Mechanism and Machine Science (IFToMM).

Prospective authors and editors can contact Mr. Pierpaolo Riva (publishing editor, Springer) at: pierpaolo.riva@springer.com

Indexed by SCOPUS and Google Scholar.

More information about this series at http://www.springer.com/series/8779

Jaroslav Beran · Martin Bílek ·
Miroslav Václavík · Petr Žabka
Editors

Advances in Mechanism Design III

Proceedings of TMM 2020

Springer

Editors
Jaroslav Beran
Faculty of Mechanical Engineering
Technical University of Liberec
Liberec, Czech Republic

Martin Bílek
Faculty of Mechanical Engineering
Technical University of Liberec
Liberec, Czech Republic

Miroslav Václavík
VÚTS a.s.
Liberec, Czech Republic

Petr Žabka
Faculty of Mechanical Engineering
Technical University of Liberec
Liberec, Czech Republic

ISSN 2211-0984 ISSN 2211-0992 (electronic)
Mechanisms and Machine Science
ISBN 978-3-030-83593-4 ISBN 978-3-030-83594-1 (eBook)
https://doi.org/10.1007/978-3-030-83594-1

© The Editor(s) (if applicable) and The Author(s), under exclusive license
to Springer Nature Switzerland AG 2022
This work is subject to copyright. All rights are solely and exclusively licensed by the Publisher, whether the whole or part of the material is concerned, specifically the rights of translation, reprinting, reuse of illustrations, recitation, broadcasting, reproduction on microfilms or in any other physical way, and transmission or information storage and retrieval, electronic adaptation, computer software, or by similar or dissimilar methodology now known or hereafter developed.
The use of general descriptive names, registered names, trademarks, service marks, etc. in this publication does not imply, even in the absence of a specific statement, that such names are exempt from the relevant protective laws and regulations and therefore free for general use.
The publisher, the authors and the editors are safe to assume that the advice and information in this book are believed to be true and accurate at the date of publication. Neither the publisher nor the authors or the editors give a warranty, expressed or implied, with respect to the material contained herein or for any errors or omissions that may have been made. The publisher remains neutral with regard to jurisdictional claims in published maps and institutional affiliations.

This Springer imprint is published by the registered company Springer Nature Switzerland AG
The registered company address is: Gewerbestrasse 11, 6330 Cham, Switzerland

Preface

The International Conference on the Theory of Machines and Mechanisms has been held since 1973 in regular four-year intervals. The expert agenda of the conference focuses on a wide range of problems, including theoretical and practical findings related to theories pertaining to machines and mechanisms. The person involved in the creation of this conference was an important personality and leading expert in mechanics and mechanism theories doc. Ing. Jaroslav Charvát, CSc. He was not only a renowned expert regarding mechanism theories, but he was also an excellent teacher and promoter of mechanics for the professional and lay public. He impressed students with his lectures and boosted their enthusiasm for the study field. His scientific and research activities were focused to a great extent on the mechanisms of textile machines. He regularly contributed to expert magazines. He was the author of many monographs and scripts, and he actively attended conferences and seminars. He was responsible to a significant extent for the development of young technical intelligence and contributed to the development of the field of textile machine mechanics.

The 13th International Conference on the Theory of Machines and Mechanisms was held from September 7–9, 2021, at the Technical University of Liberec, Liberec, Czech Republic. The conference was organized under the auspices of the Czech National Committee for Theories of Machines and Mechanisms IFToMM, the Czech Society for Mechanics and the VUTS a.s. The lectures focused particularly on the areas of analysis, synthesis of linkages, cam, planar and spatial mechanisms.

A major part included a focus on problems related to the dynamics of machines and mechanisms as well as mechatronic and biomechanical systems. Robotic systems were another major focus of the conference.

In this book, contributions from the conference have been included into five main sections.

- General theory of machines and mechanisms, analysis and synthesis of planar and spatial mechanisms, linkages and cams.

- Dynamics of machines and mechanisms, computational mechanics, rotor dynamics, biomechanics, vibration and noise in machines.
- Robots and manipulators.
- Mechatronics, control and monitoring systems of machines, accuracy and reliability of machines and mechanisms
- The mechanisms of textile machines, optimization of mechanisms and machines.

As it has for more than 48 years, this volume of proceedings offers both a broad perspective on the state of the art in the field and an in-depth look at its leading-edge research. It is our privilege to be able to offer this collection and we express our sincere thanks to the contributing authors for making this series a continuing success.

We appreciate the interest in this conference and believe it will bring many suggestions for further extension of knowledge in the field of machines and mechanisms theory and will provide new ideas for international cooperation in this field.

Jaroslav Beran

Organization

Program Chairs

Jaroslav Beran	Technical University of Liberec, Czech Republic
Martin Bilek	Technical University of Liberec, Czech Republic
Miroslav Václavík	VÚTS, a.s., Czech Republic
Petr Zabka	Technical University of Liberec, Czech Republic

Scientific Committee

J. Beran (Conference Chair), Czech Republic
M. Václavík (Honorary Chair), Czech Republic
M. Bílek (Secretary), Czech Republic
V. Arakelyan, France
M. Ceccarelli, Italy
B. Corves, Germany
J. Horáček, Czech Republic
G. Kiper, Turkey
M. Lima, Portugal
E. CH. Lovasz, Romania
V. Natarajan, India
V. Petuya, Spain
J. Petruška, Czech Republic
L. Pešek, Czech Republic
Š. Segľa, Slovakia
J. Stadnicki, Poland
J. Szrek, Poland
I. Tempea, Romania
H. Terada, Japan
J. Vimmr, Czech Republic
J. Zapomĕl, Czech Republic

The Conference is Organized Under the Patronage of:

TECHNICAL
UNIVERSITY
OF LIBEREC
www.tul.cz

Technical University of Liberec
Studentská 1402/2, 461 17 Liberec 1, Czech Republic
http://www.tul.cz

Czech Society of Mechanics
Dolejškova 5, 182 00 Praha 8, Czech Republic
http://www.csm.cz

IFToMM

IFToMM
International Federation for the Promotion of Mechanism and Machine Science
http://www.iftomm.net/
Czech National Commitee of IFToMM
Institute of Thermomechanics AS CR, v. v. i., Dolejškova 5, 182 00 Praha 8
Czech Republic

VÚTS
LIBEREC

VÚTS, a.s.
Svárovská 619 Liberec XI- Růžodol I 460 01 Liberec Czech Republic
http://www.vuts.cz

Contents

Dynamics of Machines and Mechanisms, Computational Mechanics, Rotor Dynamics, Biomechanics, Vibration and Noise in Machines

Mechanisms of Textile Machines, Optimization of Mechanisms and Machines

General Theory of Machines and Mechanisms, Analysis and Synthesis of Planar and Spatial Mechanisms, Linkages and Cams

Experiences in Leadership IFToMM: Achievements and Challenges

Marco Ceccarelli(✉)

Department of Industrial Engineering, University of Rome Tor Vergata, Laboratory of Robot Mechatronics, 00133 Rome, Italy
marco.ceccarelli@uniroma2.it

Abstract. The paper presents a recent history of IFToMM from personal experiences in leadership positions of IFToMM bodies during the last twenty years. A community is characterized by its activities, but also by the leaders guiding /supervising/stimulating it. The peculiarities of IFToMM mission are discussed with the challenges attached during leadership positions up to reach results indicating new goals and trends in MMS developments, without ignoring open issues for future attention. In the last years it is possible to recognize an evolution of IFToMM community with activities with more multidisciplinary aspects enforcing also the international collaborations. Limits are experienced in the organization structure that has not always been able to react properly and in due time. Based on the past experiences a challenging future of IFToMM is also outlined with main targets to be worked out by next leaders and whole IFToMM community.

Keywords: History of IFToMM · History of MMS · IFToMM leaderships · IFToMM Presidents · Marco Ceccarelli

1 Introduction

A scientific community is generally recognized in the history of the community's evolution [1], but also by the personality of its leaders within the professional and scientific evolution of the leaders with their activities in addressing the challenges for the community and in achieving planned objectives.

Significance of IFToMM can be stressed by its history and the challenges that have been successfully experienced facing the future, as related to the worldwide community working in the broad areas of disciplines of Mechanism and Machine Science (MMS) for theoretical aspects up to practical implementations for service and benefits in the society [1–7].

IFToMM is an emblematic result of needs and convenience in aggregation of people with common interests and activities to strengthen their visibility and impacts both in technical-scientific collaboration aims and contractual purposes within the society [2–4]. IFToMM as federation is an aggregation of national/territory communities of scientists and engineers working in MMS with a vision a world community servicing for the welfare of the mankind [5, 7, 8]. Aggregation is a natural action of humans with

© The Author(s), under exclusive license to Springer Nature Switzerland AG 2022
J. Beran et al. (Eds.): TMM 2020, MMS 85, pp. 3–16, 2022.
https://doi.org/10.1007/978-3-030-83594-1_1

common interests and with the aims of achieving strong impacts in the surrounding frames. A society is an aggregation that is motivated by common cultural views and professional interests with the aim both to have a community within which is possible to share successfully needs and interests and to have full understanding of the activity results with good visibility and influence with future trends of developments. The above aspects can be recognized in IFToMM community along its history since its foundation 50 years ago [9].

This work aims to summarize the author's experiences within the IFToMM international community in his leadership activity with a gradual evolution from a simple congress participant up to reaching the maximum representative authority of president, as partially presented in [10]. The paper presents the challenges that the author has faced especially in leadership positions and the results achieved by not renouncing to indicate the aspects still to be addressed and which can be considered future challenges for the next generations of the IFToMM community, as reported in IFToMM documents like [11–15]. The work therefore presents a historical survey regarding the author's experiences but with a perspective referring to the historical development in the IFToMM community with its peculiarities as an international community present in all continents within a wide spectrum of disciplines that are synthetically grouped in the domain of Mechanism and Machine Science with aspects of theory, training, design, research, and application.

2 A Timeline of History of IFToMM

My life in IFToMM frames started with a participation at the 1985 CNIM national Spanish conference in Gijon, at the 1987 IFToMM World Congress in Seville and then at SYROM in 1989 organized by prof Manolescu, who showed me enthusiasm in IFToMM confirming the passion that prof Vinciguerra transmitted me during my PhD formation [10]. A timeline of my experience in IFToMM is shown in Fig. 1 within a historical outline of IFToMM with main events [9].

IFToMM was founded as the International Federation for the Theory of Mechanisms and Machines in Zakopane, Poland on September 29, 1969 during the Second World Congress on TMM (Theory of Mechanisms and Machines). IFToMM is the International Federation of a world community working in the broad area of Mechanism and Machine Science including not only aspect of Mechanical Engineering. Its mission is clearly stated in the articles 2.1–2.8 of IFToMM Constitution [8] as: 'To promote research and development in the field of Machines and Mechanisms by theoretical and experimental methods, along with their practical application'. The bodies of IFToMM of IFToMM as per the Constitution are General Assembly (GA), Executive Council (EC), 3 Commissions of the General Assembly (GACs), 14 Technical Committees (TCs), and 4 Permanent Commissions (PCs). IFToMM activity is characterized by the main aspects on collaboration and dissemination in research, application, and formation through several initiatives like meetings, conferences, editorial works, teaching technological transfers and so on, Main conference event is the World Congress (WC) and main student-oriented event is the Student International Olympiad on MMS (SIOMMS).

The history of IFToMM has been outlined from several perspectives mainly by the past IFToMM Presidents, as in the reports [3–6], very often with an eye to future trends

[9]. The history of IFToMM was also outlined in a poster exhibition during the IFToMM 2019 World Congress in Krakow, Poland, celebrating the 50th anniversary of IFToMM. The History of IFToMM can be outlined looking at the generation that can be identified as follows, Fig. 1:

- 1950's –'79: First generation with founding fathers and their friendly colleagues up to the 4-th IFToMM World Congress in New Castle upon Tyne in 1975 with prof Leonard Maunder as Congress Chair
- 1980–95: Second Generation with pupils and educated people by founding fathers and their friendly colleagues up to the 9-th World Congress in Milan in 1995 with prof Alberto Rovetta as Congress Chair
- 1996–2011: Third Generation with educated people in the frames of IFToMM and within IFToMM activity with Prof. Carlos Lopez-Cajùn as General Chair for 2011 Congress
- 2012 – today: Fourth Generation with educated people in local MMS frames that are linked to IFToMM and within IFToMM activity with 44 organizations as IFToMM member organizations.

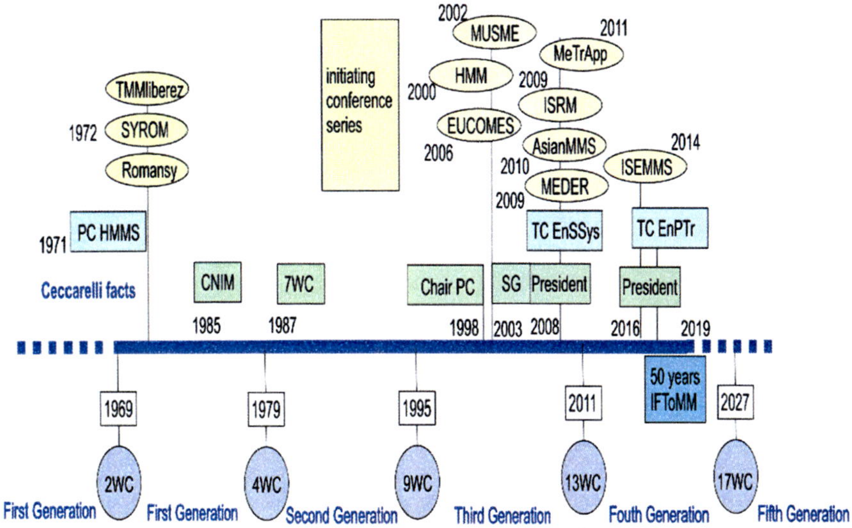

Fig. 1. A historical timeline of IFToMM with Ceccarelli experiences in light-green rectangles, starting dates of PCs-TCs in light-blue rectangles and main IFToMM conferences in light-yellow ellipses

The IFToMM community has grown continually and TMM has evolved to include even new emerging discipline leading in the year 2000 to an update of the name of the IFToMM Federation as IFToMM International Federation for the Promotion of Mechanism and Machine Science [7]. In particular, Presidents and Secretaries General had significant roles in guiding the growth and success of IFToMM. Their personalities are also representative of the IFToMM community in terms of reputation and visibility worldwide. The Presidents were Ivan I. Artobolevsky (1969–1971 and 1972–1975) (USSR), Leonard Maunder (1976–1979) (UK), Bernard Roth (1980–1983) (USA), Giovanni Bianchi (1984–1987 and 1988–1991) (Italy), Adam Morecki (1992–1995) (Poland), Jorge Angeles (1996–1999) (Canada), Kenneth J. Waldron (2000–2003) and 2004–2007) (USA), Marco Ceccarelli (2008–2011) and (2016–2019) (Italy), Yoshihiko Nakamura (2012–2015) (Japan), and today Andres Kecskeméthy (Germany).

The first generation (1960–1975) was characterized by efforts in starting and advertising the activities of the new Federation making clear its mission, with great hopes for future success. Several EC meetings discussed long agendas with many details in even one-week of duration with reports that are stored in the IFToMM Archives. Significant in 1972 is the start of the conference series of Romansy, CISM-IFToMM Symposium on Theory and Practice of Robots and Manipulators as the first conference event on Robotics in the world, SYROM: IFToMM International Symposium on Linkages and Computer Aided Design Methods, and TMM conference in Liberez as a continental conference. Those conference series are still very successful IFToMM events. The Journal of Mechanisms by Elsevier was affiliated to IFToMM in 1972 and was renamed as Mechanism and Machine Theory to link it clearly to IFToMM. Most of the first IFToMMists were active in IFToMM for several decades and up to their last days.

The activity of the second generation (1976–1995) was characterized by activity with a similar enthusiasm and vision as being the pupils the first generation to enlarge the interests and participation to the federation. The initiatives were enlarged both in number and participation. Other international conferences were started with several TCs, like TC for Computational Kinematics in 1991, TC for Gearing in 1976, C for Rotordynamics in 1977, TC for Human-Machine Systems in 1986, TC for Mechatronics in 1994, and TC for Micromachines in 1994. The participation in WCs grew continuously and one with the highest number of papers was the event in Milan in 1995 and one of the most socially participated ones was the one in Seville in 1987.

The third generation (1996–2011) was characterized by a worldwide presence of the community with 48 MOs in IFToMM in 2003. This growth is reflected both in renewed and revitalized activities for the already existing TCs and PCs that have culminated in a period of relevant results in the 2008–2011 term. One characteristic operation of the third generation was an extensive use of informatics means as typical of Information Age. New TCs were established in new areas of MMS, like TC for Biomechanical Engineering and TC for Energy Sustainable Systems in 2010, and a TC on Gearing and Transmissions has been re-established with a reinvigorated group of colleagues.

The Fourth generation (2012 – todays) is characterized by an intense international activity with collaboration in teaching, research and technological transfer of MMS results with challenging trends in confirming the significance of traditional subjects in emerging mechatronic approaches for new and update solutions of systems with less and less mechanical parts.

The next Fifth generation is expected with interests and no barriers in looking and developing systems that, although with more and more aspects in non-mechanical traditional aspects, will be conceived with innovations but still for helps and benefits of human users, thanks still to solutions from MMS.

3 Leadership Experiences

The author's leadership experience can be summarized in the following periods with the relative roles, Fig. 1:

- 1998–2004: chairman of the permanent committee for history of MMS [11, 16]
- 2004–2007: IFToMM general secretary [12]
- 2008–2011 and 2016–2019: IFToMM president [13, 14]
- 2012–2015 and 2020–2023: chairman of IFToMM nominating committee as past president

The leadership activity has been carried out by interaction with individuals (IFToMMists or not) through initiatives with achievements and challenges that are presented in the following short accounts whereas details are available in the reports [11–15].

The activities of Chair of the PC for History of MMS were centered on a program agreed also with the previous Chair Professor Teun Koetsier to increase the number of members of the commission and to develop an activity related to the analysis of the history of technological and cultural development of machines from a technical point of view by MMS researchers in the IFToMM fields. The main effort in the early years of chairmanship was to recruit new members for the commission with the indication of sharing and developing a vision of the history of machines from a specifically technical point of view in the sense of understanding and interpreting the historical developments of the machines from a technical point of view even if endured from a cultural point of view typical of the history of science and technology. These activities have given rise to a rapid growth of the PC also following a periodic meeting activity between the members which began at the beginning of 1998 with only four people and with 48 members at the beginning of 2004, as shown in the example in Fig. 2. Figure 3 briefly shows the result of the activities of increased interest in historical analysis from a technical-scientific point of view with the start of the HMM symposium in 2000 which also produced conference proceedings with publication for international dissemination and a more informal meeting with the HMMS workshop to facilitate collaboration and the beginning of collaboration between the members of the commission. Figure 4 emphasizes the results of these activities even after the period of chairmanship with the periodic organization of the HMM symposium and the corresponding publication of the proceedings, [18], which then have also determined the beginning of the Springer book series on the history of machines (http://www.springer.com/series/7481) as well as the beginning of an encyclopedia of distinguished figures in MMS, Fig. 4a) and finally the success of being able to obtain an adequate number of works from which to extract also Journal special issues such as the recent one reported in Fig. 4b).

In summary, the period of the chairmanship of the PC was characterized by an intense activity of initiatives to attract the attention of people interested in the historical developments of the machinery of mechanisms in all its aspects referring to both inventors and inventions as well as theoretical works and to the schools that led to the development of MMS. Although this activity was essentially of leadership, it also had aspects of sharing especially with the past chair prof Koetsier and collaboration with all the members of the PC who from time to time were recruited by sharing programs, expectations and initiatives as obtained also in the following years after chairmanship thanks to a continuous collaboration with the next chairs of the commission that still characterizes the success of the PC, despite the diminished interest on the past by the new generations.

Fig. 2. Examples of leadership activity as Chair of PC for History of MMS in chairing PC meeting in 2004.

Fig. 3. Examples of leadership activity as Chair of PC for History of MMS in conference initiative HMM symposium series since 2000 [17]

As Secretary General during 2004–07, [12], the activity has been focused in dealing with the handling of IFToMM business in coordination with the President, prof. Kenneth Waldron referring mainly to:

- forwarding information on IFToMM activity and stimulating opinion exchanges among IFToMM officers and with contributions of IFToMMists

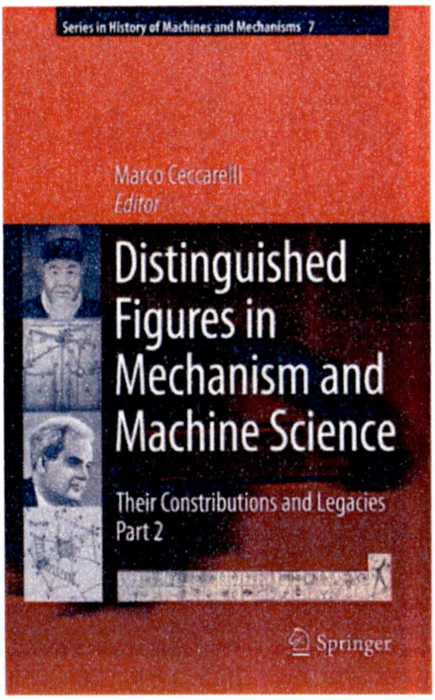

a)

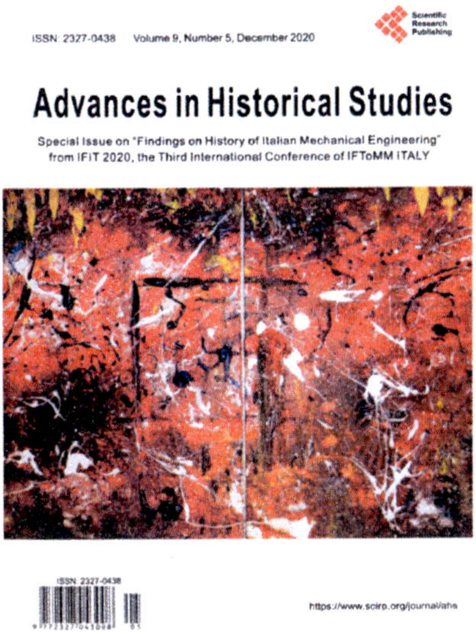

b)

Fig. 4. Examples of later achievements following leadership activity as Chair of PC for History of MMS in publications initiatives: a) volumes on Distinguished Figures in MMS [19]; b) a special journal issue [20].

- assisting and collecting information for EC meeting, Fig. 5a), and GA and preparing minutes for distribution and archiving in the reinforce IFToMM archive
- working to finalize the actions that have been decided at EC meetings
- updating information on the organization of IFToMM MOs for a correct visibility of the Federation nature of IFToMM
- asking nominations for representatives of IFToMM MOs in the PCs and TCs
- contributing in the IFToMM Taskforce for Journals for plans of better publications frames with correct liaisons
- collecting material for IFToMM Archive
- contributing in IFToMM Working Group for a campaign of information with proposals on posters and flyer, Fig. 5b) [21], and on improve of the clarity of the IFToMM webpage
- cooperating for regular payment of annual fee from IFToMM MOs
- working out email postal ballots in EC on matters identified with the President
- promoting actions for new IFToMM MOs
- representing IFToMM in conference events and meetings with presentations on general MMS topics and IFToMM mission.

During the term 2008–2011 the presidency activity has been focused on guiding IFToMM activity within president planned candidature program for Visibility-Activity of IFToMM by also improving the functioning of the IFToMM bodies. The Visibility-Activity plan has been intended to increase the visibility of IFToMM and its activity

a)

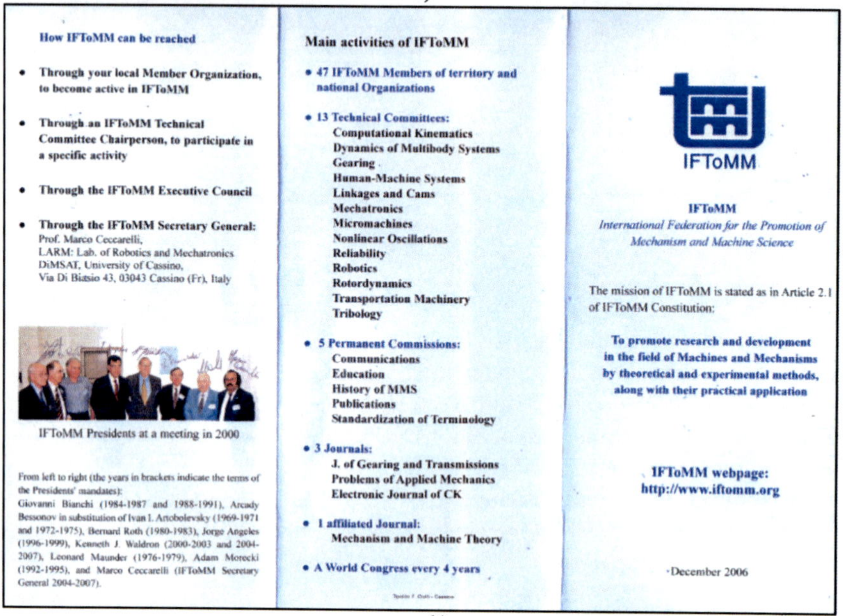

How IFToMM can be reached

- **Through your local Member Organization, to become active in IFToMM**
- **Through an IFToMM Technical Committee Chairperson, to participate in a specific activity**
- **Through the IFToMM Executive Council**
- **Through the IFToMM Secretary General:**
  Prof. Marco Ceccarelli,
  LARM: Lab. of Robotics and Mechatronics
  DiMSAT, University of Cassino,
  Via Di Biasio 43, 03043 Cassino (Fr), Italy

IFToMM Presidents at a meeting in 2000

From left to right (the years in brackets indicate the terms of the Presidents' mandates):
Giovanni Bianchi (1984-1987 and 1988-1991), Arcady Bessonov in substitution of Ivan I. Artobolevsky (1969-1971 and 1972-1975), Bernard Roth (1980-1983), Jorge Angeles (1996-1999), Kenneth J. Waldron (2000-2003 and 2004-2007), Leonard Maunder (1976-1979), Adam Morecki (1992-1995), and Marco Ceccarelli (IFToMM Secretary General 2004-2007).

Main activities of IFToMM

- **47 IFToMM Members of territory and national Organizations**
- **13 Technical Committees:**
  Computational Kinematics
  Dynamics of Multibody Systems
  Gearing
  Human-Machine Systems
  Linkages and Cams
  Mechatronics
  Micromachines
  Nonlinear Oscillations
  Reliability
  Robotics
  Rotordynamics
  Transportation Machinery
  Tribology
- **5 Permanent Commissions:**
  Communications
  Education
  History of MMS
  Publications
  Standardization of Terminology
- **3 Journals:**
  J. of Gearing and Transmissions
  Problems of Applied Mechanics
  Electronic Journal of CK
- **1 affiliated Journal:**
  Mechanism and Machine Theory
- **A World Congress every 4 years**

IFToMM

IFToMM
International Federation for the Promotion of Mechanism and Machine Science

The mission of IFToMM is stated as in Article 2.1 of IFToMM Constitution:

To promote research and development in the field of Machines and Mechanisms by theoretical and experimental methods, along with their practical application

IFToMM webpage:
http://www.iftomm.org

December 2006

b)

Fig. 5. Examples of leadership activity as IFToMM Secretary General: a) assisting EC meeting in 2005; b) (first) IFToMM flyer in 2006 [21]

by promoting new and existing activities with an explicit mention of IFToMM and to facilitate new and existing initiatives under the umbrella of IFToMM.

During the term 2008–2011, results are achieved in increasing main aspects of meetings, exchanges, publications, teaching, and international collaborations, as prescribed for the mission of IFToMM in the IFToMM constitution.

New significant initiatives were the start of SIOMMS: IFToMM Olympiad of MMS (thanks to the late prof Veniamin Goldfarb) and affiliation of more journals and book series on MMS. Functioning of the IFToMM has been improved with better functioning the Executive Council also through specific activity of newly established EC Working

Groups in attaching specific matters with active role of EC members. New MOs have been accepted from Turkey, Portugal, Egypt, Denmark, after preliminary negotiations for proper candidature submissions reaching s 48 MOs with presence in all continents.

A new TC has been started on Sustainable Energy Systems and a TC on Gearing and Transmissions has been re-established with a reinvigorated group of colleagues, although still few TCs were with a weak activity. Particular mention for a continuous dedication to IFToMM is deserved to the Secretary General late prof. Carlos Lopez-Cajùn, the Treasurer Joseph Rooney, but also to late prof. Veniamin Goldfarb working continuously also thanks to telemeetings that have been started in 2010. The important new means of EC work has been successfully experienced with Working Groups (WGs) following procedures in new-established EC regulations. The 40-th year anniversary of IFToMM was organized and celebrated with a ceremony in foundation city Zakopane, Poland.

The second Presidency period in 2016–19 addressed attention not only referring to the plans as proposed in the candidature in terms of Visibility-Activity-Benefits but also problems and trends that were brought to the attention to the President and EC during this term, even coming from the previous terms. In particular, one of the essential services as President has been spent with almost a diary dedication in planning, coordinating, soliciting, summarizing, email exchanges and communications, and chairing problem discussions and solution proposals with indications coming not only within the EC but also from the IFToMM community, even at the level of individual IFToMMists. A measure of this dedication (also for most of the EC members) can be indicated by the number of EC tele-meetings and face-to-face meetings (in total 16) that have been worked out during the term with a plan to have EC discussions every three-four months and the number of Working Groups (WGs) (in total 21) that were established to attach specific problems to have well studied proposals for the EC also in EC tele-meetings. Special care and time have been devoted to assist IFToMM Member Organizations (MOs), Technical Committees (TCs), and Permanent Commissions (PCs) in critical situation (with weak activity and irregular status).

Matters for Activity were discussed to increase the quantity and quality of the initiatives for the IFToMM aim of promoting MMS in technological-scientific frames for the benefits of mankind and peace (as per the 1969 founding principles still valid).

Results were obtained in reinvigorated action of TC/PC also with increased TC/PC members representing MOs; start of a new TC on Engines and Powertrains; new IFToMM MOs; new affiliations of publications frames; support (also with presence) to old and new initiatives (conferences, summer schools, Olympiads, meeting); considerations and solutions for better activity and new matters (like procedures for IFToMM functioning as in the Manual for IFToMM procedures, and ethical issues); improved participation of individuals and MOs in IFToMM activity and its planning. Open issues were recognized in coordination of activities to avoid overlapping of time schedule and topics; influent action of EC on IFToMM bodies, influent action of PCs and TCs in the worldwide community; influence and representative of IFToMM in funding programs; coordination with other federations in engineering; efficient response and interaction from IFToMM bodies with respect to EC indications.

Matters for Visibility were attached to make IFToMM well known and well reputed in the scientific community and even in the general public. Results were achieved with increased number of IFToMM events as conferences, meetings, tutorials and schools at local and international levels; improvement of visibility means (webpage, call for papers), preparation of posters and flyers. But it was noted that IFToMM is still unknown in several (scientific and professional) frames and countries; IFToMM is not fully understandable to the young generations; newsletter readability is not accepted; publicity of IFToMM is weak in technical-scientific and public frames.

Matters for Benefits were worked out to increase quantity and quality of benefits for communities (MOs, TCs, PCs) and individuals, especially for young generations; to reinvigorate the Student MMS Olympiad; to identify new benefits of interest and attraction towards IFToMM. Results were obtained in terms of increase of publication frames and conference events with desired aspects and benefits for individuals; reduced fees for conference participation and for publication of papers and proceedings; attention to ethical issues; conference Best paper Awards in 3 categories (research, application, student) and possibly in 3 levels (gold, silver, bronze); establishment of IFToMM fellow grade. However, understanding and planning proper benefits with significant values for young and senior IFToMMists and MOs are not yet at satisfactory levels requiring new and update proper benefits for individuals and MOs; attraction of individuals from professional and industrial worlds; continuous attention to new needs and benefits for the IFToMM community; participation of individuals and TC/PC representatives to WGs.

Figure 6 show examples of President acts during the two presidency terms.

In addition, the President was involved as chair of the Committee for Honors and Awards with activity in increasing interest and visibility of IFToMM awards with establishing BPA and fellow grade; soliciting nominations; chairing the evaluation of the nominees; participating personally at the award ceremonies; working and writing on the updates of the manuals. Several new honorary members were elected thanks to proper nominations and several recipients received Award of Merit and Dedicated Service Award.

As Past President, although still in the EC as member without vote right, the leadership role is influential as reference person for the recent past and experience in handling the IFToMM business, that however not very often is fully exploited, like in the current term. Additionally, as Past president the role of Chair of IFToMM Nominating Committee still requires dedication to IFToMM community in searching and stimulating proper nominations from IFToMMists in the roles of future leaders in the IFToMM bodies, for which is still experienced lack of interest and proper awareness.

Summarizing, the commitment as IFToMM leader in the various roles was to fulfill with the obligations as established by constitution, but also to improve and increase the activities and impact of IFToMM first in its IFToMM community as well as to greater dissemination and influence also in neighboring communities and more generally for public visibility in society, not neglecting to emphasize aspects that, despite having been carefully addressed, have not worked out with expected results and therefore remain as future challenges and targets for future leaders.

a)

b)

c)

Fig. 6. Examples of leadership activity as IFToMM President in chairing: a) EC meeting at Sousse, Tunisia, in 2010; b) EC meeting discussion in 2017; c) General Assembly in Krakow, Poland, in 2019

4 Challenges for Today and Tomorrow

The main challenges for IFToMM can be summarized in Fig. 7 for the following aspects, as coming from my experiences in the IFToMM leadership (as also presented at the 2019 IFToMM GA), [15]:

- Attraction and interest to IFToMM and its activity from young and senior MMS scientists
- Aggregation and activity of more and more MOs from all continents
- Increase of domains of interests, with more interdisciplinary activity
- Collaboration with other federations and communities, not only in engineering
- Improvement of benefits for communities and individuals in IFToMM
- Increase visibility and influence of IFToMM in scientific, professional, industrial frames at national and international levels
- Increase of interest and quality in leadership of IFToMM
- Share of IFToMM initiatives and challenges not only within the IFToMM bodies but even more and more with individual IFToMMists.

Even if the above aspects can be well understood as challenges and plans for future activities by IFToMM leadership in MMS activities, the practical implementation of solutions and initiatives as specific issues for short-medium programs requires flexibility in the leader actions as to try to satisfy all the expectations from a large variety of conditions and communities in IFToMM that is a world community with different cultural backgrounds. One key point is an understanding and appreciation of the benefits that IFToMM can offer or can start avoiding that those benefits can be either impossible to reach for someone or even to be considered constrains or obstacles for personal promotion. An emblematic example of such a benefit differentiation can be considered in the area of publications where in some countries there is a need or even an obligation to have indexed publications in pre-defined publication frames both for career promotion and project funding, whereas in other national communities is already asked to have the dissemination open-source free for the public and not linked to those mentioned indexed frames. Therefore, although a mission of IFToMM is to guide towards the future for new publication frames, IFToMM should provide yet the possibilities to communities and individuals all the variety of conditions for their promotion, impact, and influence with their publications in all kind of frames.

One another general challenge for IFToMM even in short time can be considered in the fact that in general, for the new generations an aggregation in societies or entities with common interests and targets is not felt useful like in the past and the fragmentation in individuals or in small groups, even only in social media, seems to give sufficient benefits for a required short horizon. This seems to happen also towards IFToMM community since both ignorance and not-understanding of the motivations and mission of IFToMM give limited or not properly attractive benefits of being involved in IFToMM. These two last aspects can be reflected also in the considerably variability of situations in which IFToMM will have to act more and more in the future.

A peculiar aspect for aggregations and communities to which the IFToMM community is particularly sensitive, is related to communication and information within the community and mainly from the leaders. It is strategic that the leaders such as especially the IFToMM president, as official representative of IFToMM towards outside of the community, and the members of the EC maintain continuous and constant communication and information with the community through direct channels, as well as with the periodic meetings of the EC, and characteristic ways of IFToMM through the Chairs of

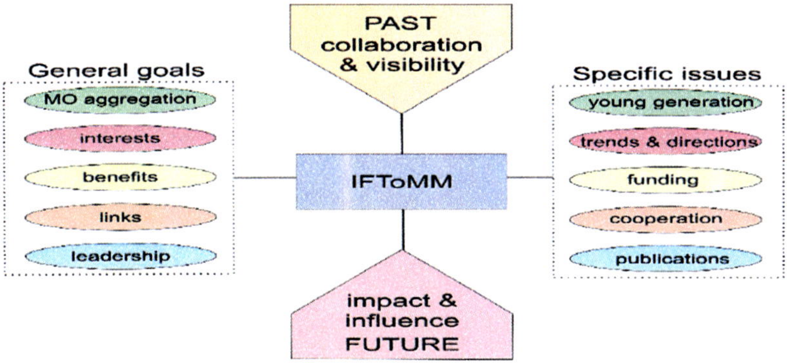

Fig. 7. A summary of issues for IFToMM future challenges

member organizations, not neglecting the possibility of direct contacts with IFToMMist individuals.

5 Conclusions

This paper presents a brief historical account of the activity of IFToMM from a personal perspective of the author in his leadership roles that he has lived in the last twenty years allowing not only an analysis of the historical development but also a critical analysis of the evolution of the IFToMM community and its activities. In particular, the positive character of IFToMM is emphasized as a structure that allows an aggregation and collaboration of communities in common interests for research, training, and technology transfer in the broad domain of the science of machines and mechanisms. The author has had the honor and responsibility of contributing to the growth of the IFToMM community over the past 20 years in the role of permanent committee chair, general secretary, president, and then past president.

References

1. Koetsier, T.: Mechanism and machine science: its history and its identity. In: Ceccarelli, M. (ed.) International Symposium on History of Machines and Mechanisms Proceedings HMM 2000, pp. 5–24. Springer Netherlands, Dordrecht (2000). https://doi.org/10.1007/978-94-015-9554-4_2
2. Ceccarelli, M.: Activity and trends in MMS from IFToMM community. In: Ceccarelli, M. (ed.) Technology Developments: the Role of Mechanism and Machine Science and IFToMM, pp. 3–24. Springer Netherlands, Dordrecht (2011). https://doi.org/10.1007/978-94-007-1300-0_1
3. Ceccarelli, M.: A short account of history of IFToMM and its role in MMS. Mech. Mach. Theory **89**, 75–91 (2015). https://doi.org/10.1016/j.mechmachtheory.2014.09.007
4. Ceccarelli, M. (Editor): Chapter 2: History of IFToMM. In: International Symposium on History of Machines and Mechanisms – Proc. of HMM2000. Kluwer, Dordrecht, pp. 27–86 (2000)
5. IFToMM webpage: History of IFToMM. www.iftomm.net (2021)

6. Ceccarelli, M.: A short introduction on IFToMM officers over time. In: Ceccarelli, M. (ed.) International Symposium on History of Machines and Mechanisms, pp. 3–10. Springer Netherlands, Dordrecht (2004). https://doi.org/10.1007/1-4020-2204-2_1
7. Ceccarelli, M.: From TMM to MMS: a vision of IFToMM, Bull. IFToMM Newslett. 10(1). http://www.iftomm.org (2001)
8. IFToMM: IFToMM Constitution and By-Laws – 2019. Available at IFToMM Archives and in the IFToMM webpage (2019)
9. Ceccarelli, M: IFToMM: yesterday, today, and tomorrow. In: T. Uhl (Eds.), Advances in Mechanism and Machine Science. Proceedings of the 15th IFToMM World Congress on Mechanism and Machine Science, vol. 73, Springer Nature Switzerland AG, pp. ix–xxi. https://doi.org/10.1007/978-3-030-20131-9_96 (2019)
10. Ceccarelli, M.: Twenty-five years of activity in IFToMM, J. Theory Mech. Machine, St Petersburg State University, 2014, 11(2), 3–14. htpp//tmm.spbstu.ru (2014)
11. Ceccarelli, M.: Chair Reports 1998–2003 of IFToMM PC for Histoiry of MMS. Available at IFToMM archive in CISM of Udine (Italy) (2003)
12. Ceccarelli, M.: Reports 2004–2007 of IFToMM Secretary General. Available at IFToMM archive in CISM of Udine (Italy) (2007)
13. Ceccarelli, M.: Reports 2008–2011 of IFToMM President. Available at IFToMM archive in CISM of Udine (Italy) (2011)
14. Ceccarelli, M.: Reports 2015–2019 of IFToMM President. Available at IFToMM archive in CISM of Udine (Italy) (2019)
15. Ceccarelli, M.: End-term message from the IFToMM President. J. Vibration Eng. Technol. **8**(2), 381–389 (2020). https://doi.org/10.1007/s42417-020-00200-x
16. Ceccarelli, M., Koetsier, T.: On the IFToMM permanent commission for history of MMS. In: Proceedings of International Symposium on History of Machines and Mechanisms HMM2004. Kluwer, Dordrecht, pp. 10–25 (2004)
17. Ceccarelli, M. (ed.): Proceedings of International Symposium on History of Machines and Mechanisms HMM2000. Kluwer, Dordrecht (2000)
18. Ceccarelli, M. (Ed.): Explorations in the history of machines and mechanisms, Book series on History of Machines and Machine Science. Springer, Dordrecht (2008–2018).
19. Ceccarelli M. (Ed.): Distinguished figures in mechanism and machine science: their contributions and legacies – Part 1–4, Book series on History of Machines and Machine Science. Springer, Dordrecht (2007–2020)
20. Ceccarelli, M.: Niola, V. (Guest Editors): Special Issue on 'Findings on history of Italian mechanical engineering', J. Adv. Hist. Stud. 9(5) https://www.scirp.org/journal/ahs/ (2020)
21. IFToMM: IFToMM flyer. Available at IFToMM archive in CISM of Udine (Italy) (December 2006)

Minimization of Shaking Moment in Fully Force Balanced Planar Four-Bar Linkages

Vigen Arakelian[1,2(✉)]

[1] INSA Rennes / Mecaproce, 20 Avenue des Buttes de Coesmes, CS 70839, 35708 Rennes, France
vigen.arakelyan@insa-rennes.fr

[2] LS2N-ECN UMR 6004, 1 rue de la Noë, BP 92101, 44321 Nantes, France

Abstract. This paper deals with a solution of shaking force and shaking moment balancing of planar four-bar linkages. The shaking moment balancing is realized by displacement of the axis of rotation of the counterweight connected with the input link. The conditions for balancing are formulated by the minimization of the root-mean-square value of the shaking moment. This approach is well known. However, the paper describes another of its properties. It is about the choice of the shaking force balancing solution, which significantly affects the minimization of the shaking moment. It is well known that the shaking force in four-bar linkages can be balanced in various ways. The aim of this paper is to show that the choice of the balancing scheme of shaking forces can influence the minimization of shaking moment. To show this difference, two balancing schemes are compared: by two and three counterweights. It is shown that the application of the mentioned balancing technique for minimization of the shaking moment is more efficient for shaking forces balancing by three counterweights. Numerical simulations carried out via ADAMS software illustrate the mentioned observations.

Keywords: Shaking force · Shaking moment · Dynamic balancing · Planar four-bar linkage · Inertia forces · Minimization · Root-mean-square value

1 Introduction

The balancing of mechanisms is a well-known problem in the field of high-speed machinery because the variable dynamic loads cause vibration and noise of the machines. The resolution of this problem consists in the balancing of the shaking force and shaking moment, fully or partially, by internal mass redistribution or by adding auxiliary links [1].

A reliable and simple way to balance shaking forces is to redistribute the mass of the moving links of the mechanism by adding counterweights. It is widespread and quite attractive for industrial applications.

However, balancing of the shaking moment is more challenging and can only be reached by a considerably complicated design of the initial mechanism or by unavoidable increase of the total mass.

© The Author(s), under exclusive license to Springer Nature Switzerland AG 2022
J. Beran et al. (Eds.): TMM 2020, MMS 85, pp. 17–23, 2022.
https://doi.org/10.1007/978-3-030-83594-1_2

R. Berkof [2], Ye and Smith [3], Arakelian and Smith [4], Feng [5] have proposed methods for complete shaking moment balancing by planetary gear trains. Esat and Bahai [6] used a toothed-belt transmission to cancel the shaking moment in four-bar linkages. Kochev [7] proposed to balance shaking moment by a prescribed input speed fluctuation achieved by non-circular gears or by a microprocessor speed-controlled motor.

Moore, Schicho and Gosselin have proposed all possible sets of design parameters for which a planar four-bar linkage is balanced: both shaking force and shaking moment [8]. Briot and Arakelian [9] used this approach for complete shaking force and shaking moment balancing of four-bar linkages.

The complete shaking force and shaking moment balancing of four-bar linkages via copying properties of pantograph systems formed by gears was also considered [10].

A comparison of various shaking moment balancing principles has been carried out by van der Wijk, Herder and Demeulenaere [11]. This overview summarizes, compares and evaluates the existing principles of complete shaking force and shaking moment balancing regarding the addition of mass and the addition of inertia.

As was mentioned above, the complete shaking moment balancing can often be achieved by a considerably complicated design of the initial mechanism and by unavoidable increase of the total mass. This is the reason why methods of partial dynamic balancing of mechanisms have also been developed.

Freudenstein, J.P. Macey, E.R. Maki [12] derive the equations for minimizing any order of combined pitching and yawing moments by counterweighting the driveshaft or a shaft geared to the driveshaft. The equations are given directly as a function of the harmonic coefficients of pitch and yaw and apply to any plane machine configuration. J.L. Wiederrich and B. Roth [13] proposed simple and general conditions for determination of the inertial properties of a four-bar linkage that allow partial momentum balancing. Dresig and Schönfeld [14, 15] examined the optimum balancing conditions for various structural forms of planar six and eight-bar linkages. A last-square theory for the optimization of the shaking moment of fully force-balanced inline four-bar linkages, running at constant input angular velocity, is developed in the studies of J.L. Elliot and D. Tesar [16] and R.S. Haines [17].

V.A. Shchepetilnikov [18] suggested the minimization of the unbalance of shaking moment by transferring the rotation axis of the counterweight mounted on the input crank. In his works the first harmonic of the shaking moment is eliminated by attaching the required input link counterweight, not to the input shaft itself, but to a suitable offset one which rotates with the same angular velocity. This approach is original in that, while maintaining the shaking force balance of the mechanism, it is possible to create an additional moment, reducing thereby the shaking moment. The similar studies have been developed in [19, 20].

This paper represents the further development of shaking moment balancing technique based on the last mentioned principle, i.e. by parallel displacement of the rotation axis of the counterweight mounted on the input crank. The improvement of the known approach resides in the fact that the choice of the scheme of the shaking force balancing essentially influences at the level of the shaking moment minimization.

2 Shaking Moment Minimization

Let us consider an in-line four-bar linkage with constant input angular velocity: $\dot{\varphi}_1 = d\varphi_1/dt$. Two schemes of the shaking force balancing of the linkage will be considered: by two and three counterweights (Fig. 1a and Fig. 1b).

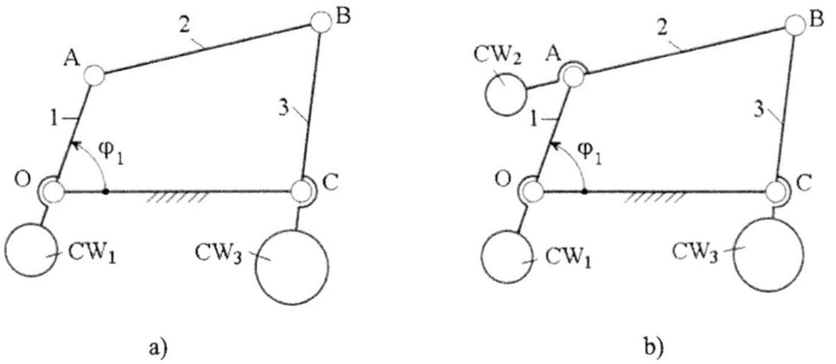

Fig. 1. Force-balanced in-line four-bar linkage: a) by two counterweights and b) by three counterweights.

After shaking force balancing of the in-line four-bar linkage by two counterweights connected to links 1 and 3 (Fig. 1a), the shaking moment can be expressed as [21]:

$$M^{sh} = K_2\ddot{\varphi}_2 + K_3\ddot{\varphi}_3 \tag{1}$$

with $K_2 = -m_2\left(k_2^2 + r_2^2 - l_2r_2\right)$ and $K_3 = -\left(m_3 + m_{CW_3}\right)\left(k_3^2 + r_3^2 + l_3r_3\right)$, where, m_2 is the mass of link 2, m_3 is the mass of link 3, m_{CW_3} is the mass of the counterweight mounter on the link 3, k_2 is the radius of gyration of link 2, k_3 is the radius of gyration of link 3, $l_2 = l_{AB}$ is the length of link 2, $l_3 = l_{BC}$ is the length of link 3, $r_2 = l_{AS_2}$ is the distance of the joint center A from the center of mass S_2 of link 2, $r_3 = l_{CS_3}$ is the distance of the joint center C from the center of mass S_3 of link 3.

In the case of the shaking force balancing by three counterweights (Fig. 1b), considering that the center of mass of the rocker 3 is on the axis of the joint C and the center of mass of the connecting rod 2 is on the axis of the joint B, the shaking moment may be expressed as: $K_2 = -\left(m_2 + m_{CW_2}\right)k_2^2$ and $K_3 = -\left(m_3 + m_{CW_3}\right)k_3^2$, where, m_{CW_2} is the mass of the counterweight mounter on the link 2.

By parallel displacement of the axis of rotation of the counterweight CW_1 (Fig. 2) from center O to the center $O_1(x_1, y_1)$, the balancing of the shaking force of the mechanisms can be maintained, but, in addition to the unbalanced shaking moment, a supplementary moment M_1^{bal} will be created:

$$M_1^{bal} = F_1(x_1 \sin\varphi_1 - y_1 cos\varphi_1) \tag{2}$$

with $F_1 = m_{CW_1} r_{CW_1} \dot{\varphi}_1^2$, where, φ_1 is the angle of rotation of the input link, m_{CW_1} is the mass of the counterweight mounted on the input link, $r_{CW_1} = l_{O_1S_{CW1}}$ is the rotation radius of the center of mass of the counterweight with respect to center O_1.

The counterweight with mass m_{CW_1} moved in parallel is driven (by gears or toothed belts for example) at the same rotational speed as the input link, i.e. $\dot{\varphi}_1$. For clarity, the driving mechanisms are not shown here.

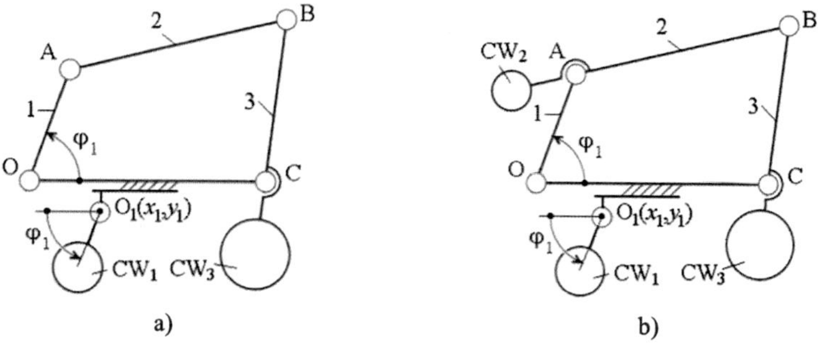

Fig. 2. Shaking moment balancing of a force-balanced four-bar linkage.

For minimization of the root-mean-square value (rms) of the shaking moment of the modified mechanism, it is necessary to minimize the sum:

$$\Delta_{rms} = \sum_{i=1}^{N} \left(M_1^{bal} + M^{sh}\right)^2 \rightarrow \min_{x_1, y_1} \tag{3}$$

where, N is the number of calculated positions of the linkage.
For this purpose, the following conditions must be fulfilled:

$$\partial\Delta_{rms}/\partial x_1 = 0 \text{ and } \partial\Delta_{rms}/\partial y_1 = 0 \tag{4}$$

Conditions (4), taking into account that $\sum_{i=1}^{N} \sin\varphi_1 \cos\varphi_1 = 0$ for $\varphi_1 \in [0; 2\pi]$, lead to a system of linear equations, from which the following expressions are obtained:

$$x_1 = \sum_{i=1}^{N} M^{sh} \sin\varphi_{1i}/F_1 \sum_{i=1}^{N} \sin^2\varphi_{1i} \text{ and } y_1 = -\sum_{i=1}^{N} M^{sh} \cos\varphi_{1i}/F_1 \sum_{i=1}^{N} \cos^2\varphi_{1i} \tag{5}$$

Observations showed that the choice of the shaking force balancing scheme influences the minimization of the shaking moment. In order to demonstrate this for an arbitrarily four-bar linkage, a numerical comparison has been carried out.

3 Illustrative Example with Numerical Simulations

The in-line four-bar linkage used for numerical simulations has the following parameters: the lengths of links: $l_{OA} = 0.2\ m$; $l_{AB} = 0.45\,m$; $l_{BC} = 0.45\,m$; $l_{OC} = 0.6\,m$, the location of the centers of mass: $l_{OS_1} = 0.1\,m$; $l_{AS_2} = 0.225\,m$; $l_{CS_3} = 0.225\,m$, the masses:$m_1 = 2\,kg$; $m_2 = m_3 = 4\,kg$, the axial inertia moments: $I_{S_2} = I_{S_3} = 0.08\,kg\,m^2$.

The shaking force of the four-bar linkage has been balanced via two mentioned methods.

a) By two counterweights (Fig. 1a) with following parameters: the location of the counterweights' centers of mass: $r_{CW_1} = r_{OS_{CW1}} = 0.1\,m$; $r_{CW_3} = r_{OS_{CW3}} = 0.225\,m$, the masses of counterweights: $m_{CW_1} = 6\,kg$; $m_{CW_3} = 8\,kg$, the axial inertia moments after shaking force balancing: $I_{S_2} = 0.08\,kg\,m^2$; $I_{S_3} = 0.5\,kg\,m^2$.

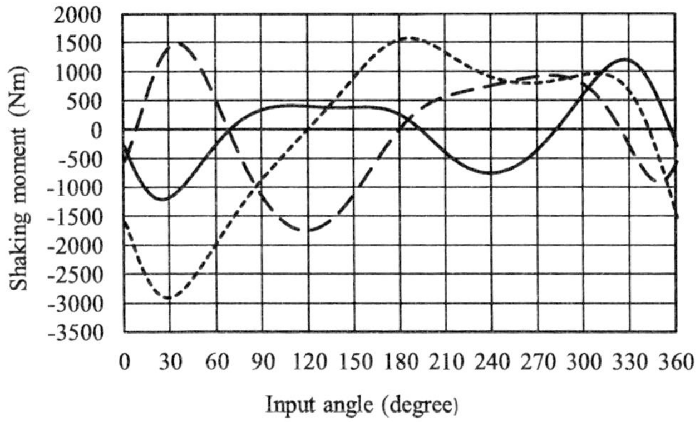

Fig. 3. Shaking moments of the four-bar linkage balanced by two counterweights: unbalanced (dash line), force-balanced (dot line) and with minimized shaking moment (solid line).

b) By three counterweights (Fig. 1b) with following parameters: the location of the counterweights' centers of mass: $r_{CW_1} = r_{OS_{CW1}} = 0.15\,m$; $r_{CW_2} = r_{BS_{CW2}} = 0.225\,m$, $r_{CW_3} = r_{OS_{CW3}} = 0.225\,m$, the masses of counterweights: $m_{CW_1} = 12\,kg$; $m_{CW_2} = 4\,kg$; $m_{CW_3} = 4\,kg$, the axial inertia moments after shaking force balancing: $I_{S_2} = 0.3\,kg\,m^2$; $I_{S_3} = 0.3\,kg\,m^2$.

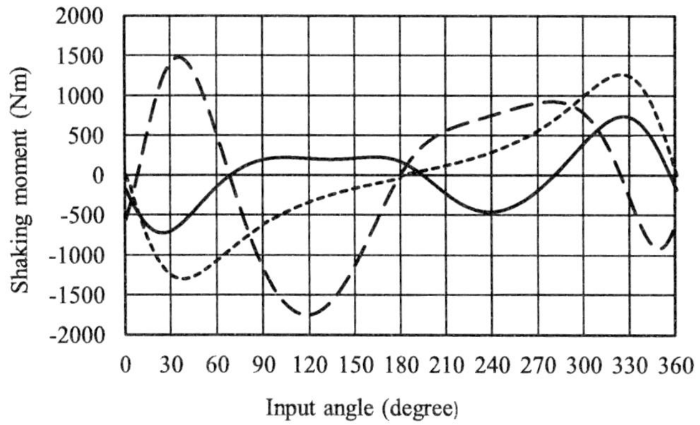

Fig. 4. Shaking moments of the four-bar linkage balanced by three counterweights: unbalanced (dash line), force-balanced (dot line) and with minimized shaking moment (solid line).

According to expressions (5) the following values of the coordinates of the axis O_1 have been obtained: a) for the linkage balanced by two counterweights (Fig. 1a): $x_1 = -0.48\,m$ and $y_1 = 0.56\,m$; b) for the linkage balanced by three counterweights (Fig. 1b): $x_1 = -0.11\ m$ and $y_1 = 0$.

The obtained results (Fig. 3) show that for the force-balanced mechanism given in Fig. 2a, a 31% reduction in the shaking moment is achieved. With regard to the force-balanced mechanism given in Fig. 2b, a 50% reduction in the shaking moment is achieved (Fig. 4). This comparison was made according to the maximum values of the shaking moments of the unbalanced and the moment-balanced mechanisms.

4 Conclusions

In the paper, it is shown that when applying the method of shaking moment minimization of four-bar linkages by transferring the rotation axis of the counterweight mounted on the input crank, the choice of the shaking force balancing approach influence on the moment minimization. To evaluate the efficiency of the shaking moment balancing of four-bar linkages, two force-balanced linkages are numerically compared: by two and three counterweights. At first sight, the balancing approach carried out by two counterweights seems more attractive as it leads to a smaller increase in the total mass of the mechanism. Moreover, from the point of view of the design, the shaking force balancing of the four-bar linkage by two counterweights mounted on the crank and the rocker is easier to implement. However, as shown in the paper, the application of the mentioned balancing technique for minimization of the shaking moment is more efficient for shaking forces balancing by three counterweights.

One should not get the impression that a solution with three counterweights is always more optimal from the point of view of minimizing the shaking moment according to the described method. The conclusion that should be retained is that different shaking force balancing schemes lead to different shaking moment minimization results. Therefore, when applying the described method to minimize the shaking moment, it is important to choose an optimal shaking force balancing scheme, as it affects the results of minimization.

References

1. Arakelian, V., Briot, S.: Balancing of linkages and robot manipulators. Advanced Methods with Illustrative Examples, Springer, Switzerland (2015)
2. Berkof, R.S.: Complete force and moment balancing inline four-bar linkages. Mech. Mach. Theory **8**, 397–410 (1973)
3. Ye, Z., Smith, M.R.: Complete balancing of planar linkages by an equivalence method. Mech. Mach. Theory **29**(5), 701–712 (1994)
4. Arakelian, V., Smith, M.R.: Complete shaking force and shaking moment balancing of linkages. Mech. Mach. Theory **34**(8), 1141–1153 (1999)
5. Feng, G.: Complete shaking force and shaking moment balancing of four types of six-bar linkages. Mech. Mach. Theory **24**(4), 275–287 (1989)
6. Esat, I., Bahai, H.: A theory of complete force and moment balancing of planar linkage mechanisms. Mech. Mach. Theory **34**, 903–922 (1999)

7. Kochev, I.S.: Active balancing of the frame shaking moment in high-speed planar machines. Mech. Mach. Theory **27**(1), 53–58 (1992)
8. Moore, B., Schicho, J., Gosselin, C.M.: Determination of the complete set of shaking force and shaking moment balanced planar four-bar linkages. Mech. Mach. Theory **44**, 1338–1347 (2009)
9. Briot, S., Arakelian, V.: Complete shaking force and shaking moment balancing of in-line four-bar linkages by adding a class-two RRR or RRP Assur group. Mech. Mach. Theory **57**, 13–26 (2012)
10. Arakelian, V.H., Smith, M.R.: Shaking force and shaking moment balancing of mechanisms: a historical review with new examples. J. Mech. Des. **127**(2), 334–339 (2005)
11. van der Wijk, V., Herder, J.L., Demeulenaere, B.: Comparison of Various Dynamic Balancing Principles Regarding Additional Mass and Additional Inertia. J. Mech. Robot. **1**(4), 041006 (2009)
12. Freudenstein, F., Macey, J.P., Make, E.R.: Optimum balancing of combined pitching and yawing moments in high-speed machinery. J. Mech. Des. **103**(3), 571–577 (1981)
13. Wiederrich, J.L., Roth, B.: Momentum balancing of four-bar linkages. J. Eng. Ind. **98B**(4), 1289–1295 (1976)
14. Dresig, H., Schönfeld, S.: Rechnergestützte Optimierung der Antribs-und Gestellkraftgrössen ebene. Koppelgetriebe-Teil 1. Mech. Mach. Theory **11**, 363–370 (1976)
15. Dresig, H., Schönfeld, S.: Rechnergestützte Optimierung der Antribs-und Gestellkraftgrössen ebene, Koppelgetriebe-Teil 2. Mech. Mach. Theory **11**, 371–379 (1976)
16. Elliot, J.L., Tesar, D.: The theory of torque, shaking force and shaking moment balancing of four link mechanisms. J. Eng. Ind. **99B**(3), 715–722 (1977)
17. Haines, R.S.: Minimum RMS shaking moment or driving torque of a force-balanced 4-bar linkage using feasible counterweights. Mech. Mach. Theory **16**, 185–190 (1981)
18. Shchepetilnikov, V.A.: Balancing of mechanisms. Mashinostroenie, Moscow (1982)
19. Arakelian, V., Dahan, M.: Partial shaking moment balancing of fully shaking force balanced linkages. J. Mech. Mach. Theory **36**(11–12), 1241–1252 (2001)
20. Arakelian, V., Smith, M.: Shaking moment minimization of fully force-balanced linkages. In: Proceedings of the 11th World Congress in Mechanism and Machine Science, vol. 2, pp. 633–636. Tianjin, China, 1–4 April 2004
21. Berkof, R., Lowen, G.: Theory of shaking moment optimization of force-balanced four-bar linkages. ASME J. Eng. Ind. **93**(1), 53–60 (1971)

MLP Neural Network for a Kinematic Control of a Redundant Planar Manipulator

Vladimír Hlaváč(✉)

Faculty of Mechanical Engineering, Czech Technical University in Prague, Prague, Czech Republic
hlavac@fs.cvut.cz

Abstract. A non-redundant manipulator inverted kinematics can be easily solved by a multilayer perceptron neural network. For redundant manipulators, the inverted function cannot exist. Many advanced types of neural networks have been used at least for kinematic and dynamic control. This article describes a solution, when the redundancy is compensated by a simple quality function, which serves at the same time as a solution of the obstacle avoidance problem. This additional function is not combined with the functions describing the manipulator forward kinematics, but is applied to the data, prepared for the network training. This makes the whole process much simpler to realize, although the preparation of data for the training is computationally demanding.

Keywords: Redundant manipulator · Inverted kinematics · Neural network · MLP neural network · Obstacle avoidance · Planar manipulator

1 Introduction

The primary task of the manipulator is to move the end-effector to follow a desired trajectory [1]. Many articles have been written about this problematic and both kinematics and dynamics have been solved, see Introduction in [1].

A planar manipulator with two segments (links) has the exact geometric solution. This type of manipulator has 2DOF (degrees of freedom). While the position has two coordinates, any possible position in the working space has exactly two solutions (one solution if the limit of the manipulator range is reached). Possible solution of inverted kinematics is to the use of neural networks, which can be easily and precisely trained [2, 3].

A planar manipulator with three and more segments (links) has infinity solutions. This type of manipulator is called redundant. The typical solution for the position control (problem of inverted kinematics) is to use pseudo-inverse of the manipulator's Jacobian (see for example [4, 5]). Advantage of this type of manipulator is a better flexibility of use; a typical task is to solve obstacle avoidance [4]. A neural network can be used instead of solving the pseudo-inverse of Jacobian [5], or as an adaptive controller. Special types of neural networks have been tested [6].

© The Author(s), under exclusive license to Springer Nature Switzerland AG 2022
J. Beran et al. (Eds.): TMM 2020, MMS 85, pp. 24–32, 2022.
https://doi.org/10.1007/978-3-030-83594-1_3

In this article, the MLP (multilayer perceptron) neural network with one hidden layer, trained by the Matlab nftool toolbox, is used. At first, the well-known use for 2DOF planar manipulator is presented. In the next chapter, the derivation of use of the same type of neural network for a planar manipulator with 5DOF and obstacle avoidance is explained.

2 Inverted Kinematics for a 2DOF Planar Manipulator

As is presented in Fig. 1, the end-effector position can be calculated using (1a) and (1b):

$$x = a\cos\beta_1 + a\cos(\beta_1 + \beta_2) \tag{1a}$$

$$y = a\sin\beta_1 + a\sin(\beta_1 + \beta_2) \tag{1b}$$

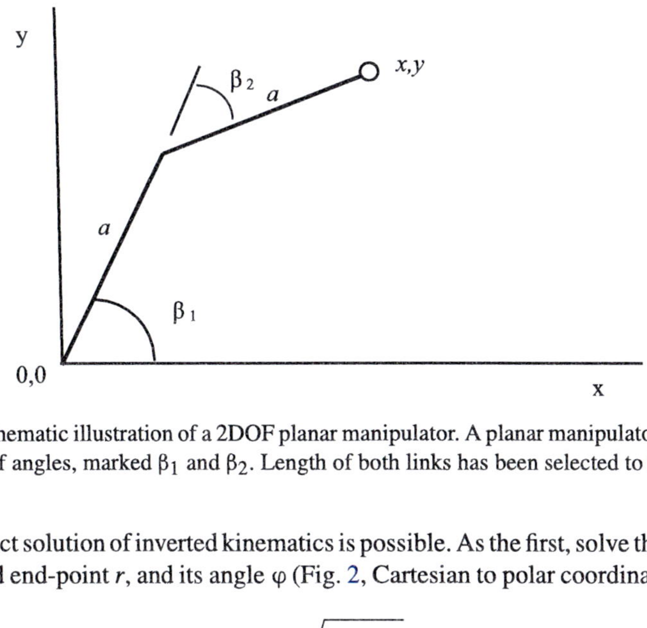

Fig. 1. A schematic illustration of a 2DOF planar manipulator. A planar manipulator is controlled by settings of angles, marked β_1 and β_2. Length of both links has been selected to be same, *a*.

The exact solution of inverted kinematics is possible. As the first, solve the distance of the required end-point *r*, and its angle φ (Fig. 2, Cartesian to polar coordinates transfer):

$$r = \sqrt{x^2 + y^2} \tag{2a}$$

$$\varphi = \arctan(x/y) \tag{2b}$$

From the required length of the arm *r*, the angle β_0 can be evaluated:

$$r = 2a\cos\beta_0 \tag{3a}$$

$$\beta_0 = \arccos(r/2a) \tag{3b}$$

As a result:

$$\beta_1 = \varphi + \beta_0 \tag{4a}$$

$$\beta_2 = -2\beta_0 \tag{4b}$$

The second angle is taken negative. This problem have two solutions, the second is similar, with mirrored shape of the arm:

$$\beta_1 = \varphi - \beta_0 \tag{5a}$$

$$\beta_2 = 2\beta_0 \tag{5b}$$

The exact solution for links of different length is possible too, but would be more complicated.

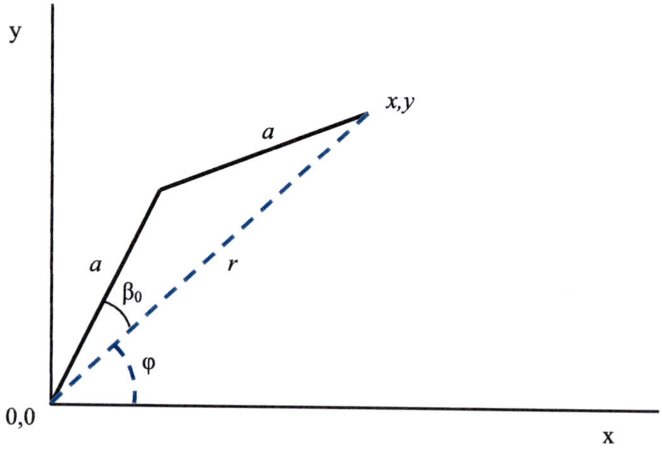

Fig. 2. The transformation into the polar coordinates.

The data for the neural network training has been prepared for a manipulator with the range of β_1 from 1 to 3 radians, in the step of 0.1 (0.2 radians for testing, Fig. 3). The length of both segments was 20 (units are not important; it could be cm or inches).

The β_2 has been generated in the range of [−2, 0] radians (with the same step), to avoid duplicate solutions, mentioned in Eqs. (5a) and (5b). The MLP network cannot have two solutions for the same value; in this case the training would be impossible.

For training, the Matlab nftool toolbox has been used. For two inputs and two outputs, 5 and 15 neurons in the hidden layer have been set. Bayesian regularization has been used as a batch training method. From the set, 10% of the samples have been used for testing and another 10% for validation (the default value of the nftool is 15%).

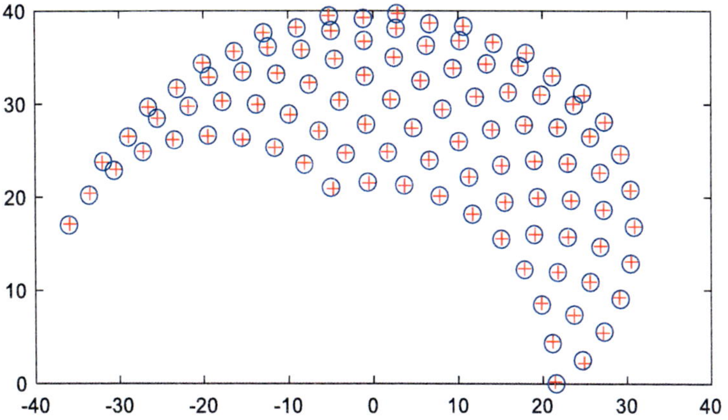

Fig. 3. Data for training (circles) and result (x,y) of the trained neural network, when x and y were transferred by the neural network into angles β_1 and β_2 and the angles have been transferred by Eqs. (1a) and (1b) back to the rectangular coordinates. Results are marked by the red crosses.

To test the trained neural network, the original x and y coordinates have been applied as the neural network input. Results of the neural network are angles β_1 and β_2. These angles have been substituted in Eqs. (1a) and (1b) to get the generated x and y. In Fig. 3, the circles represent the required positions and the cross symbols represent the calculated data. The presented result is for the neural network with 15 hidden neurons.

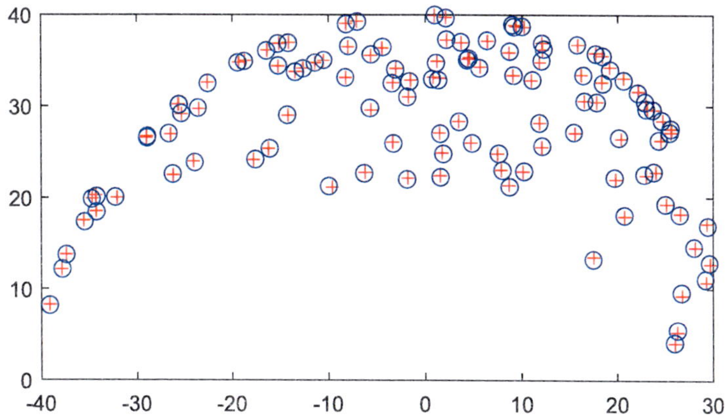

Fig. 4. Results of training of the same manipulator, as on Fig. 3, with randomly generated data.

The solution in Fig. 3 is very good for neural network training, generating points for training evenly through the training space. The task has been repeated and randomly generated positions have been used for training. The resulting graph (Fig. 4) is not so nice, but the accuracy is nearly the same (110 samples drawn). This test demonstrates the possibility of a random generator used for this task. Advantage is that the number of generated points can be set without solving any step or range of the manipulator.

3 Redundant Manipulator with Obstacle Avoidance

The goal of this work was to test a possibility to achieve similar precision (or at least possibility) of use of MLP NN for a planar redundant manipulator with obstacle avoidance. To solve this problem, a planar manipulator with 5 DOF and links of the same length has been selected. The required path and the obstacle are illustrated in Fig. 5. The transfer function from the set joint angles to coordinates is:

$$\alpha_i = \alpha_{i-1} + \beta_i \tag{6}$$

$$x_i = x_{i-1} + a\cos(\alpha_i)$$

$$y_i = y_{i-1} + a\sin(\alpha_i)$$

The values for the zero index are zero. To get data usable for MLP NN training, the function must be smooth and unique. The neural network will solve the inversion of kinematics, so not only for all combinations of inputs and outputs must be unique, but the opposite condition must be valid, all combinations of outputs must be generated by a unique combination of inputs. This does not comply with the definition of the redundant manipulator. The next paragraphs describe a possible workaround, not necessary only possible.

The problem is that the probability that the program will generate a combination, which will be part of the required path, is very low. For the training of the neural network, not exact points are required. Any point near the path is sufficient.

If we use some circumambience around the required point, it looks reasonable to prepare only set of these cells on the required path and in the surrounding and record only the points, which will be generated in these cells. For each of cells, an object can be created, containing an array of possible combinations of angles, resulting polar coordinates and quality describing values, explained in the next.

Here should be emphasized, that it is not important, which data will generate the final trained neural network outside the required path.

It is necessary to select, which of the randomly generated points will be used for the neural network training. For this, a quality function has been set. The requests to this function are to be continuous for any change of parameters and it must inherit all of them (in this case the joint angles). In this test, the quality function is defined as the sum of the distances of each of joints from the obstacle. Obstacle is defined here as a rectangle and the distance is a standard Euclidean distance.

3.1 Description of the Process

1. The first cycle – raw data generation:

   1.1. A random combination of the joint angles is generated.
   1.2. End-effector position for this combination is evaluated.

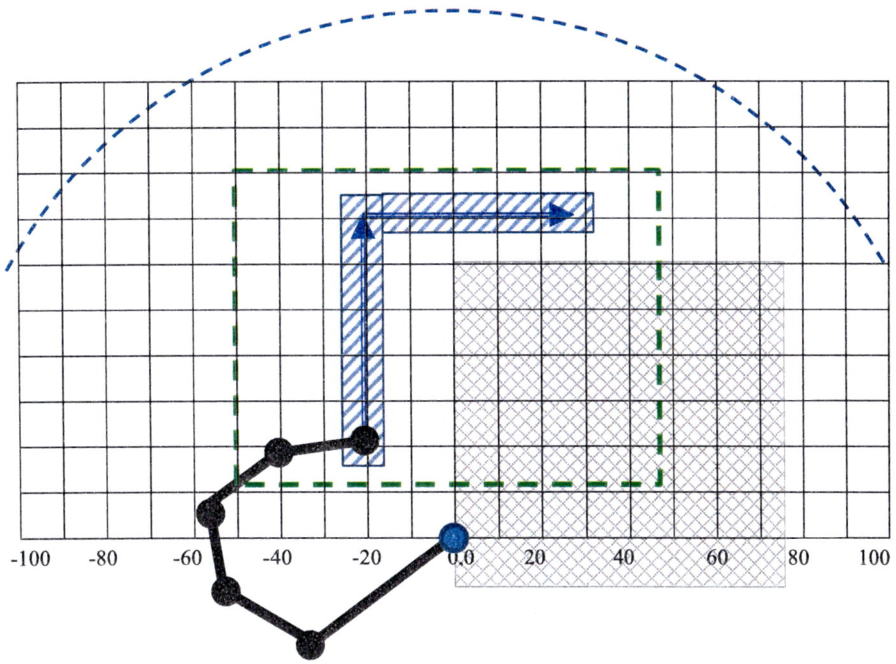

Fig. 5. The required path is drawn in blue, the surrounding area as a hatched blue rectangle. Thick black lines represent manipulator links, the dashed arc is the arm range. After a series of tests, the length of the first link has been extended to 40 to generate more data in the last point on the required path. The gray rectangle is the obstacle to be avoided. This is a similar task as to reach behind a cabinet. The empty green dashed rectangle represents the area of the graph in the next picture (Fig. 6).

   1.3. If the position of the end-effector falls into the working area, the quality function is evaluated and the data are recorded into the respective object (cell).
   1.4. The size of the array in the cell object is limited, if the cell is full, nothing is recorded. In this case, the counter of the lost solutions is incremented. The cycle ends, when it counts a preset value of the lost solutions.

2. The second cycle – smoothing:

   2.1 The cycle starts from a preselected cell. The reasonable solution is the cell, where is the worst access of the manipulator. In this case, this is the last point of the required path.
   2.2 In this selected cell, the quality function is used to select three or five best solutions. The others are abandoned.
   2.3 The process continues with the next cell on the path (if not started from the end, then in the both directions), and later with the surrounding cells.

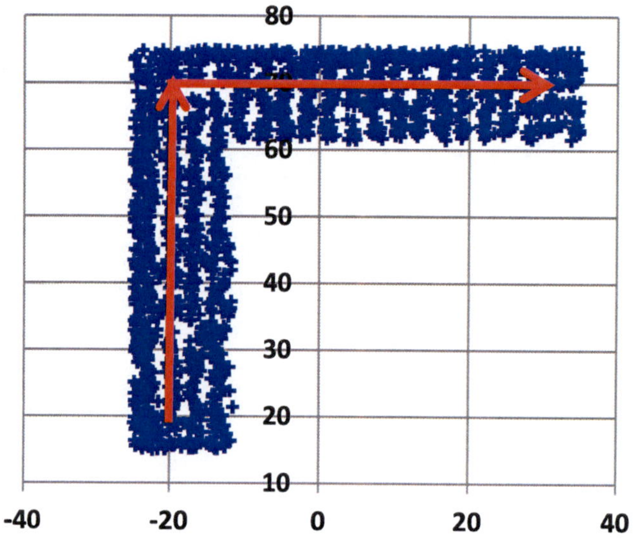

Fig. 6. Data generated for the neural network training, after the quality function and smoothing conditions applied. The desired path is marked by red arrows.

- as the first, a smoothing condition is applied: No one joint angle can differ more than a preset value from the best solution of the previous cell (in the same joint). In this test, the preset value is 0.2 radians.
- the second step is to select the best solutions from them using the quality function.

3. The third cycle – selection of the samples to be trained: After this cleaning, the best three or five solutions survive in each of the cells and they are used for the neural network training. It is obvious, that the number of abandoned solutions is huge. This is not a problem, because all data preparation works offline. The surprising fact from the testing is that a standard office PC generated billions of solutions in the order of minutes.

For this test a similar neural network like at Sect. 2 has been used. Because there are five neurons in the output layer, the number of neurons in the hidden layer has been increased to 25. This type of MLP neural network has $25 \times (2 + 1) + 5 \times (25 + 1) = 205$ settable parameters. Reasonable value of training points should be ten times bigger, so the path and its surroundings must be separated to 400 cells, while the five best points per cell will be recorded.

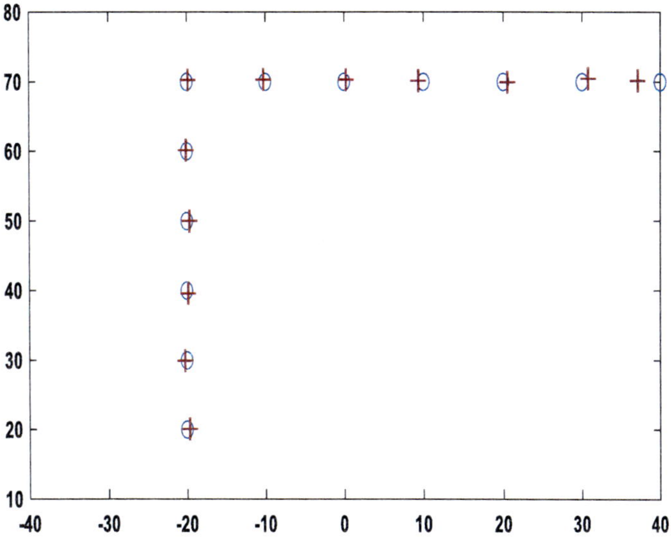

Fig. 7. Circles represent the required path; red crosses have been evaluated by insertion of the result of the trained neural network (joint angles) into Eq. (6). The last point (40, 70) is outside the prepared training data. Extrapolation is not a strong point of neural networks.

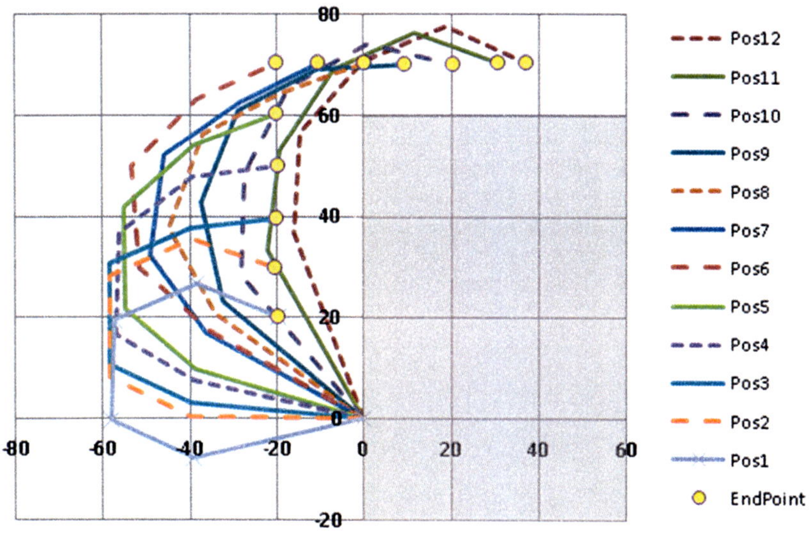

Fig. 8. Arm positions drawing. End effector marked as a circle. Gray rectangle represents the obstacle. The position marked "Pos11" proves the obstacle avoidance is solved. The last point is outside the trained data (extrapolation).

4 Conclusion

Feed-forward multilayer perceptron neural network, MLP neural network, is routinely used for system identification and as an inverted function substitution. This type of neural network can be easily used for inverted kinematics of non-redundant manipulators. This article demonstrates that a MLP neural network can be used for the inverted kinematics problem of redundant manipulator too, when the redundancy is solved by the same function, providing solution of the obstacle avoidance problem (Fig. 8).

The described solution is easy to implement and due to simplicity, many human errors can be avoided. The most demanding part, where really huge amounts of data must be processed, is very easy to implement. It makes this method very robust. For the solution in Fig. 7, more than 19 million positions have been evaluated, about 300 thousand saved and using the quality function, 1730 have been selected for the neural network training.

References

1. Chembulya, V.V., Satish, M.J., Vorugantia, H.K.: Trajectory planning of redundant manipulators moving along constrained path and avoiding obstacles. In: International Conference on Robotics and Smart Manufacturing. Procedia Computer Science, vol. 133, pp. 627–634 (2018)
2. Shah, J., Rattan, S.S., Nakra, B.C.: Kinematic analysis of 2-DOF planer robot using artificial neural network. Int. J. Mech. Mechatron. Eng. **5**(9), 1720–1723 (2011)
3. Shah, J., Rattan, S.S., Nakra, B.C.: Kinematic analysis of a planer robot using artificial neural network. Int. J. Robot. Autom. **1**(3), 145–151 (2012)
4. Zlajpah, L., Petric, T.: Obstacle avoidance for redundant manipulators as control problem. In: Küçük, S. (Ed.), Serial and Parallel Robot Manipulators – Kinematics, Dynamics, Control and Optimization, BoD – Books on Demand. InTech, pp. 203–230 (2012)
5. Mayorga, R.V., Sanongboon, P.: An artificial neural network approach for inverse kinematics computation and singularities prevention of redundant manipulators. J. Intell. Rob. Syst. **44**, 1–23 (2005)
6. Li, S., Zhang, Y., Jin, L.: Kinematic control of redundant manipulators using neural networks. IEEE Trans. Neural Netw. Learn. Syst. **28**(10), 2243–2254 (2017)

Determination of Tooth Profile Curves of Variable Rack and Pinion Gears

Yusuke Honjo(✉), Daisuke Matsuura, Tsune Kobayashi, and Yukio Takeda

Tokyo Institute of Technology, Meguro-ku, Tokyo 152-8552, Japan
{honjyo.y.aa,matsuura.d.aa,kobayashi.t.cs, takeda.y.aa}@m.titech.ac.jp

Abstract. Variable rack and pinion gears are one of the mechanical transmission systems with non-constant speed ratio. The design methodology for variable rack and pinion gears tooth profile based on the gear meshing theory has not yet been well established, and achievable speed ratio and corresponding essential conditions to be satisfied are not clear. In this paper, a design methodology of the tooth profile curves of the variable rack and pinion gears which are capable of achieving a given input/output relationship and are geometrically valid is derived. First, design variables were defined, and geometric conditions to be satisfied were shown. In addition, a calculation method of a tooth profile curve of a variable rack which meshes with a pinion and achieves target speed ratio was formulated from the meshing theory and instantaneous kinematics. Furthermore, conditions that must be satisfied in order for gears to be geometrically valid was formulated. The relationship between geometric conditions and design variables was clarified by these equations to represent the conditions. This paper contributes to designer's guideline for designing variable rack and pinion gears tooth profile curves that achieve target speed ratios.

Keywords: Variable rack and pinion gears · Tooth profile curve · Design methodology · Geometric conditions

1 Introduction

Variable rack and pinion gears are one of the mechanical transmission systems with non-constant speed ratio. Its composition is simple and capable of being used in an electric power steering mechanism of automobile to achieve both stability during high-speed driving and steerability during low-speed turning [1, 2]. In order to figure out a pair of variable rack and pinion gears with a wider range of speed ratio, there is a problem that the tooth profile may not be valid in some cases depending on the target speed ratio and design conditions. This is because the design methodology of the rack and pinion gears tooth profile based on the gear meshing theory has not yet been well established, and possible speed ratio and its design conditions are not clear. Furthermore, considering mass production, there are restrictions on the processing method, and at present,

© The Author(s), under exclusive license to Springer Nature Switzerland AG 2022
J. Beran et al. (Eds.): TMM 2020, MMS 85, pp. 33–43, 2022.
https://doi.org/10.1007/978-3-030-83594-1_4

generation of the profiles of variable rack is empirically performed by transferring the profiles of pinions [3, 4]. Therefore, a design methodology of variable rack and pinion gears tooth profile is demanded taking into account the meshing theory and geometric conditions such as undercut. For this reason, we are conducting research with the aim of deriving a valid tooth profile design methodology to obtain reasonable the rack and pinion gears tooth profile curves.

In this paper, a design methodology of the tooth profile curve of the variable rack and pinion gears which are capable of achieving a target input/output relationship and are geometrically valid is derived and demonstrated through numerical examples. First, design variables are defined, and geometric conditions to be satisfied are shown. In addition, a series of equations are derived to calculate the tooth profile curve of a variable rack which meshes with a pinion corresponding to given design variables, and is capable of achieving a target stroke ratio against each of pinion's rotation angles, which is the target input/output relationship. Furthermore, conditions that must be satisfied in order for the gears to be geometrically valid is formulated. The relationship between geometric conditions and design variables is clarified by these equations to represent the conditions. When the target stroke ratio and preliminarily determined variables are given, the variable rack and pinion gears tooth profile curves can be obtained. This paper contributes to show a design guideline that enables the design of variable rack and pinion gears tooth profile that realizes target stroke ratio.

2 Derivation of Variable Rack and Pinion Gears Tooth Profile Curve

2.1 Variable Rack and Pinion Gears Tooth Profile Curves on the Normal Plane

In deriving a design methodology of the tooth profile of variable rack and pinion gears, the tooth profile curves on the normal plane, which is the plane perpendicular to the tooth trace direction, is considered. Variable rack and pinion gears tooth profile curves are designed to satisfy an input/output relationship. In this paper, the pinion is considered as an involute pinion with constant module and pressure angle. A speed ratio is given as $i_{12} = v/\omega$, where v and ω are translational and angular velocities of the rack and pinion gear, respectively. This can also be expressed with respect to an instantaneous center as $i_{12} = r'_1$, where r'_1 is a radius of a working pitch circle of the pinion as shown in Fig. 1. The rack and pinion gear's input/output relationship is given as a target stroke ratio $S(= 2\pi r'_1)$. Once the base circle of the pinion is given, the meshing point is determined and the tooth profile curve (tangent at the point) of the rack at this point can be obtained. By tracking the tangent line with respect to each instantaneous meshing point and drawing its envelope, a rack tooth profile curve can be obtained as shown in Fig. 2. Here, it should be noted that there are parameters that can be freely selected, and these must be determined from the geometric conditions described later. Pinion tooth profile curve is determined by giving a module m_1, number of teeth z_1, pressure angle α_1, and addendum coefficient h^*_{a1}. Rack tooth profile curve can be calculated from the condition of meshing with pinion after defining addendum coefficient h_{a2}^* and working tooth depth h_{w2}. A pitch circle radius r_1, base circle radius r_{b1}, addendum h_{a1}, h_{a2} and tip circle radius r_{a1} are given as follows:

$$r_1 = \frac{m_1 z_1}{2},\ r_{b1} = \frac{m_1 z_1 \cos\alpha_1}{2},\ h_{a1} = h_{a1}^* m_1,\ r_{a1} = \left(\frac{z_1}{2} + h_{a1}^*\right) m_1,\ h_{a2} = h_{a2}^* m_1.$$

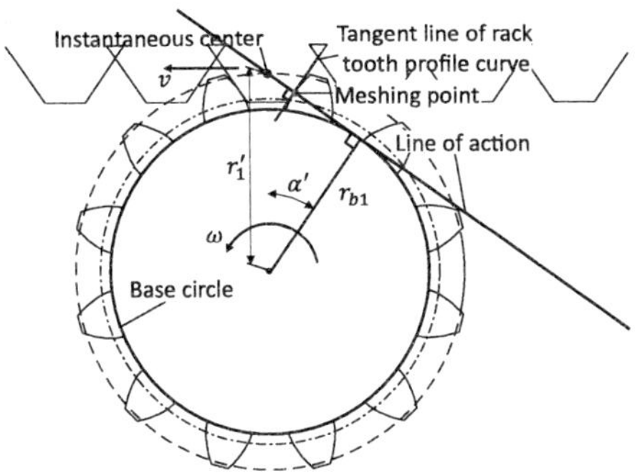

Fig. 1. Meshing of rack and pinion gear.

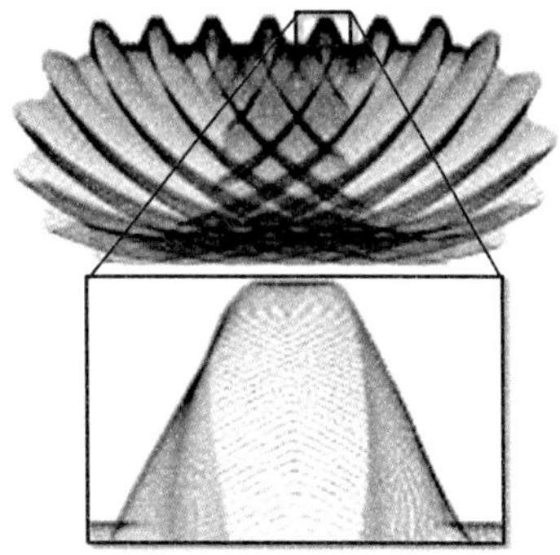

Fig. 2. Envelope of pinion with non-constant speed ratio.

2.2 Definition of Design Variables for Variable Rack and Pinion Gears

In this paper, design variables to determine the profiles of gears are a module m_1, number of teeth z_1, pressure angle α_1, addendum coefficient h_{a1}^* and h_{a2}^*, shift coefficient x, working tooth depth of rack h_{w2}, initial rotation angle θ_0 and fillet radii R_P and R_R.

2.3 Conditions that Variable Rack and Pinion Gears Must Satisfy to Be Geometrically Valid

Even when the profiles of rack and pinion gears satisfy a target stroke ratio, the tooth profile curve may not be geometrically valid. It is necessary to design so as to satisfy not only the target stroke ratio but also essential geometric conditions. In detail, determine all the design variables as follows:

1. Stroke ratio S satisfies $2r_{b1}\pi \leq \min(S(\theta)),\ \ \max(S(\theta)) \leq 2r_{a1}\pi$.
2. No interference of the tooth profile curves.
3. Tip of the pinion does not exceed the point limit.
4. Contact ratio should be greater than 1 (continuous meshing is realized).
5. No interference occurs at the fillet and root of tooth.

2.4 Formulation of Tooth Profile Curve Calculation Method for Variable Racks

From the target stroke ratio and the involute curve of the pinion, the rack tooth profile curve that meshes with the pinion is derived. In this paper, only the tooth curve on the left side is described, while that of the right side can be obtained in a similar way.

The coordinate system is set as shown in Fig. 3. $o_0 - x_0y_0$ and $O_0 - X_0Y_0$ represent the pinion and rack's positions at the beginning (the pinion has an initial rotation angle, θ_0), and $o - xy$ and $O - XY$ are with respect to them at θ. Coordinate (x_p, y_p) of the point P on the involute curve of the left side of the n-th ($n = 1 \ldots z_1$) tooth of a pinion on $o - xy$ frame are written as follows using $u(= \angle \mathrm{AoB})$ as shown in Fig. 4,

$$\begin{cases} x_p = r_{b1}\sin(\sigma_0 + \sigma_n + u) - r_{b1}u\cos(\sigma_0 + \sigma_n + u) \\ y_p = r_{b1}\cos(\sigma_0 + \sigma_n + u) + r_{b1}u\sin(\sigma_0 + \sigma_n + u) \end{cases} \tag{1}$$

σ_0, σ_n are expressed as follows, where inv is the involute function: inv $\varphi = \tan\varphi - \varphi$,

$$\sigma_0 = \frac{1}{2}\left(\frac{\pi}{z_1} - 2\mathrm{inv}\,\alpha_1\right), \quad \sigma_n = \frac{2\pi(n-1)}{z_1}. \quad (n = 1 \cdots z_1)$$

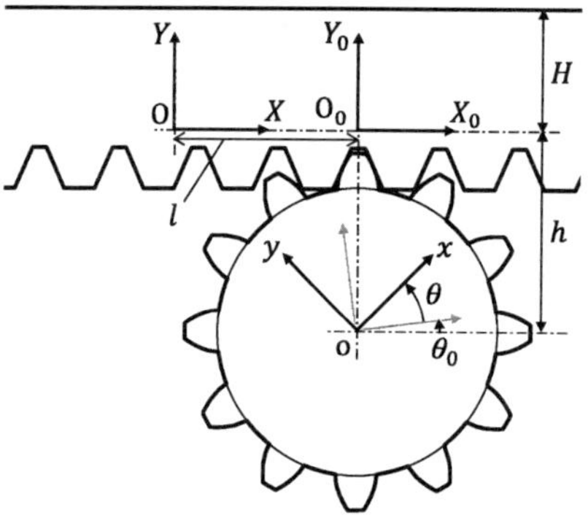

Fig. 3. Coordinate system of rack and pinion gear.

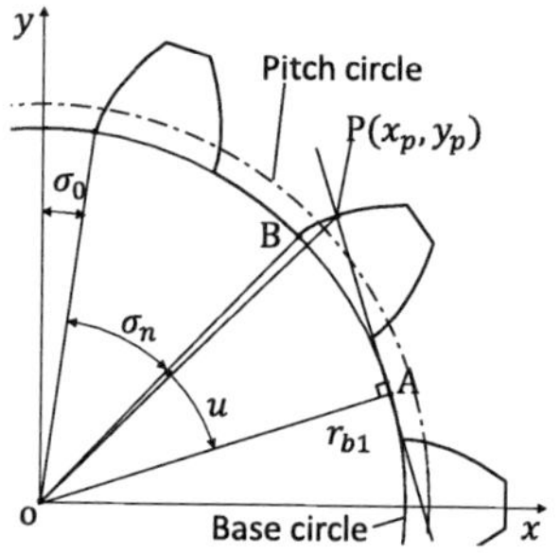

Fig. 4. Involute curve of a pinion.

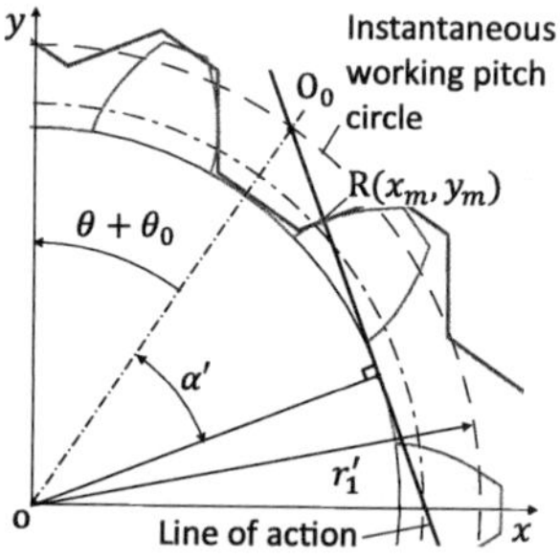

Fig. 5. Meshing point of rack and pinion gear.

As shown in Fig. 5, the meshing point of the rack and pinion gears is obtained as an intersection of the line of action and the involute curve. The working pressure angle α' and the instantaneous working pitch circle radius r_1' can be obtained with respect to the target stroke ratio S and the rotation angle θ,

$$\alpha'(\theta) = \cos^{-1}\left(\frac{m_1 z_1 \pi \cos\alpha_1}{S(\theta)}\right), \quad r_1'(\theta) = \frac{r_{b1}}{\cos\alpha'(\theta)}.$$

Equation of the line of action is obtained as

$$y = -\tan(\theta + \theta_0 + \alpha')x + \frac{r_{b1}}{\cos(\theta + \theta_0 + \alpha')}. \tag{2}$$

Coordinate (x_m, y_m) of the intersection point R is obtained by solving the following equation with respect to u and substituting it into the involute curve equation,

$$\begin{aligned} f(u) &= r_{b1}\left(-\tan(\theta + \theta_0 + \alpha') - u\right)\sin(\sigma_0 + \sigma_n + u) \\ -r_{b1}\left(1 - u\tan(\theta + \theta_0 + \alpha')\right)&\cos(\sigma_0 + \sigma_n + u) + \frac{r_{b1}}{\cos(\theta+\theta_0+\alpha')} = 0. \end{aligned} \tag{3}$$

Finally, the coordinate transformation of the meshing point is performed. For a given rotation angle θ, the rack stroke l is calculated by

$$l(\theta) = \int_0^\theta r_1'(\varphi)d\varphi.$$

Coordinate (X_m, Y_m) of the meshing point R on $O - XY$ frame are obtained as

$$\begin{bmatrix} X_m \\ Y_m \end{bmatrix} = \begin{bmatrix} \cos(\theta + \theta_0) & -\sin(\theta + \theta_0) \\ \sin(\theta + \theta_0) & \cos(\theta + \theta_0) \end{bmatrix} \begin{bmatrix} x_m \\ y_m \end{bmatrix} + \begin{bmatrix} l \\ -h \end{bmatrix}. \tag{4}$$

When the calculated meshing point is in between the tip line and the line passing through the end point of meshing ($-h_{a2} \le Y_m \le h_{w2} - h_{a2}$), the meshing point is a point on the tooth profile curve of the rack on $O - XY$ frame. The tooth profile curve of the rack meshing with the pinion is obtained as a path of all meshing points. Calculations for all pinion teeth ($n = 1 \cdots z_1$) and the full rotation angle ($\theta = \theta_i, i = 1 \cdots i_{end}$) give the entire rack tooth profile curve.

2.5 Derivation of the Geometric Conditions with a Constant Stroke Ratio

In this subsection, the geometric conditions 1 to 5 that the rack and pinion gears must satisfy are formulated for a case that the target stroke ratio is constant.

The condition 1 is expressed by the following equation,

$$2r_{b1}\pi = m_1 z_1 \pi \cos\alpha_1 \le S \cap S \le m_1\left(z_1 + 2\left(h_{a1}^* + x\right)\right)\pi = 2r_{a1}\pi. \tag{5}$$

For the condition 2, when the tooth profile curves are derived by the calculation method shown in Sect. 2.4, the pinion's involute curve and the rack's profile curve may interfere with each other as shown in Fig. 6. When such interference occurs, the desired meshing motion will be not realized and undesirable stress and deformation, fluctuation of motion and decrease of contact ratio will happen. Therefore, the interference should be avoided. When the stroke ratio is constant, the rack's profile simply becomes a straight line. The condition for not having the interference therefore becomes clear, under considering a geometric relationship, that "the tip line of the rack does not exceed the meshing limit point on the pinion's base circle". The condition 2 is therefore expressed by the following equation,

$$h_{a2} \le r_1' \sin^2\alpha' \Leftrightarrow \alpha' \ge \cos^{-1}\left(-\frac{h_{a2}^* + \sqrt{h_{a2}^{*2} + (z_1\cos\alpha_1)^2}}{z_1\cos\alpha_1}\right). \tag{6}$$

In addition, the interference may be prevented by introducing a profile shift. Figure 7 shows an example of shifted gears by preventing the tooth tip lines from exceeding the limit point by introducing a positive shift. For shifted gears, the base circle, pitch circle, and working pitch circle are not affected, except the addendum and the width of the tooth groove. The addendum of the rack and pinion gears and the spacewidth half-angle that undergoes profile shift are given as

$$h_{a1} = \left(h_{a1}^{*} + x\right)m_1, \quad h_{a2} = \left(h_{a2}^{*} - x\right)m_1, \quad \sigma_0 = \frac{\pi}{2z_1} - \frac{2x}{z_1}\tan\alpha'\frac{\cos\alpha'}{\cos\alpha_1} - \operatorname{inv}\alpha_1.$$

Gears with profile shift, condition 2 is expressed by the following equation,

$$\left(h_{a2}^{*} - x\right)m_1 \leq r_1' \sin^2\alpha' \Leftrightarrow x \geq h_{a2}^{*} - \frac{z_1 \cos\alpha_1}{2\cos\alpha'}\sin^2\alpha'. \tag{7}$$

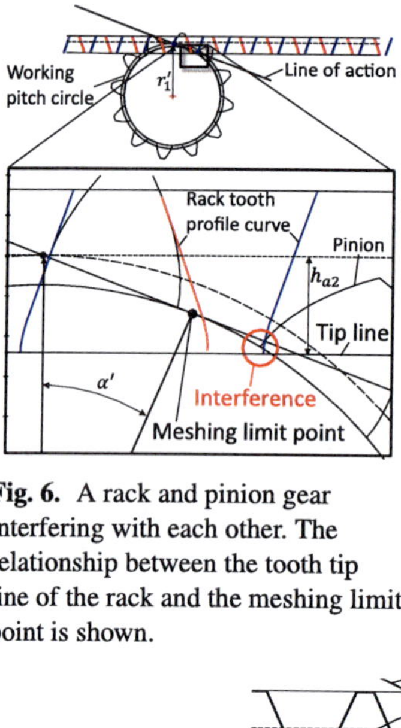

Fig. 6. A rack and pinion gear interfering with each other. The relationship between the tooth tip line of the rack and the meshing limit point is shown.

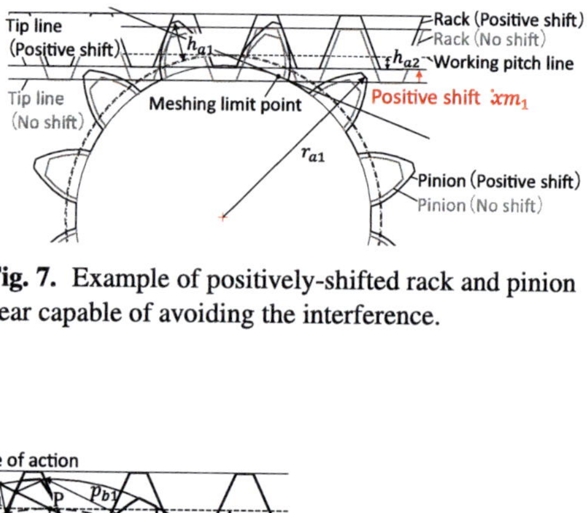

Fig. 7. Example of positively-shifted rack and pinion gear capable of avoiding the interference.

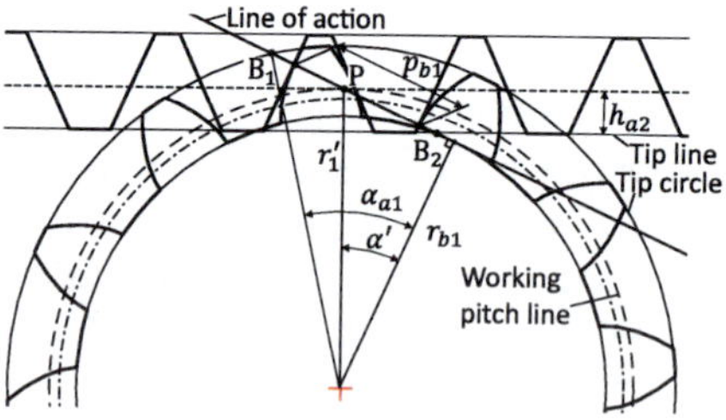

Fig. 8. Rack and pinion gear diagram about contact ratio.

For the condition 3, the pinion tooth tip may exceed the point limit and disappear depending on the tooth depth and the magnitude of the profile shift coefficient. The presence of pinion tooth tips can be indicated by the tooth thickness of the pinion on the tip circle being greater than zero. Therefore, condition 3 is expressed by the following equation with respect to the tooth thickness s_{a1} on the tip circle of the pinion,

$$\begin{aligned} s_{a1} &= 2r_{a1}\left(\frac{s'}{2r'_1} + \operatorname{inv}\alpha' - \operatorname{inv}\alpha_{a1}\right) \geq 0 \\ \Leftrightarrow \frac{\pi}{2z_1} &+ \operatorname{inv}\alpha_1 + \frac{2x\sin\alpha'}{z_1\cos\alpha_1} - \operatorname{inv}\left(\cos^{-1}\left(\frac{z_1\cos\alpha_1}{z_1+2(h^*_{a1}+x)}\right)\right) \geq 0. \end{aligned} \tag{8}$$

For the condition 4, the continuous meshing motion of the gears can be guaranteed by the contact ratio being greater than 1. The contact ratio can be obtained from the geometric relationship when the stroke ratio is constant. From Fig. 8, if working tooth depth is high enough ($h_{w2} \geq \overline{B_1B_2}\sin\alpha'$), the contact ratio ε is expressed by

$$\varepsilon = \frac{\overline{B_1B_2}}{p_{b1}} = \frac{\overline{B_1P} + \overline{B_2P}}{p_{b1}}, \tag{9}$$

where p_{b1} represents the pitch on the base circle. From the geometric relationship, $\overline{B_1P}$, $\overline{B_2P}$ and p_{b1} can be calculated, and condition 4 is expressed by

$$\varepsilon = \frac{z_1}{2\pi}\tan\left(\cos^{-1}\left(\frac{z_1\cos\alpha_1}{z_1+2(h^*_{a1}+x)}\right)\right) - \frac{z_1}{2\pi}\tan\alpha' + \frac{(h^*_{a2}-x)}{\pi\sin\alpha'\cos\alpha_1} \geq 1. \tag{10}$$

For the condition 5, a tooth tip of a gear may interfere with a tooth root of another gear, so the tooth root curve should have a fillet described as an arc to avoid this situation. First, the design method of the pinion tooth fillet curve is shown. In order to make a tooth root curve smoothly connecting the involute curve and the root circle, as shown in Fig. 9, the tooth fillet curve is an arc that touches to the start point of the involute curve on the base circle and the root circle. The fillet radius R_P therefore satisfy

$$\sigma_0 \geq \tan^{-1}\frac{r_{b1}}{R_P}. \tag{11}$$

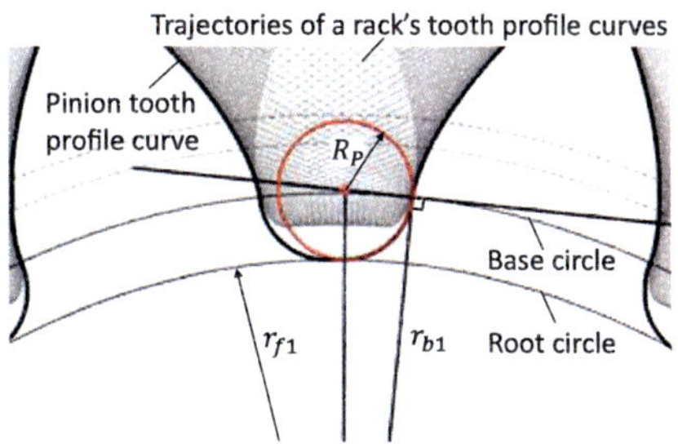

Fig. 9. Tooth root curve on the left side of the pinion to avoid interference at the root.

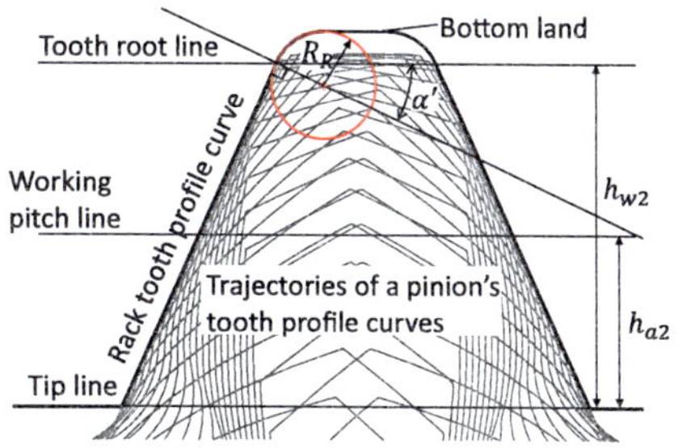

Fig. 10. Tooth root curve on the left side of the rack to avoid interference at the root.

Next, the design method of the rack tooth fillet curve is shown. In order to make a tooth root curve smoothly connecting a rack's tooth profile curve and a bottom land, a tooth fillet curve is defined as an arc touching to the rack tooth profile at the intersection with a tooth root line as shown in Fig. 10. The fillet radius R_R therefore satisfy

$$\frac{1}{2}\left[\frac{S}{z_1}\left(1-\frac{z_1}{\pi}\left(\sigma_0+\mathrm{inv}\,\alpha'\right)\right)-2(h_{w2}-h_{a2})\tan\alpha'\right]\geq R_R\cos\alpha'. \tag{12}$$

Here, the conditional Eqs. (5), (7), (8), (10), (11), and (12) that the design variables should satisfy have been explicitly derived with respect to the conditions 1 to 5. By giving design variables so as to satisfy all conditions and calculating the tooth profile curve from the given input/output relationship, the target constant stroke ratio can be achieved and the geometrically valid tooth profile curves can be derived.

2.6 Derivation of the Geometric Conditions with a Variable Stroke Ratio

In this subsection, the geometric conditions 1 to 5 that the rack and pinion gears must satisfy are formulated for a case that the target variable stroke ratio is given.

The condition 1 is expressed by the following equation,

$$m_1 z_1 \pi \cos\alpha_1 \leq \min(S(\theta)) \cap \max(S(\theta)) \leq m_1\left(z_1+2\left(h_{a1}^*+x\right)\right)\pi. \tag{13}$$

For the condition 2, not having the interference between the gears, will be derived. When the variable stroke ratio is given, the geometric relationship used for the case of constant stroke ratio does not hold because the rack tooth profile curve is no longer a straight line. In this case, the interference not only near the root of the pinion but also near the root of the rack shown in Fig. 11(a) and Fig. 11(b), respectively, should be taken into account. Therefore, the condition that the tooth profile curve does not interfere is considered to be "the meshing point always moves in a fixed direction on the pinion tooth profile curve". Now, attention is paid to the solution u of the equation $f(u)=0$, which is used when deriving the meshing point. Since the solution u is also a parameter representing a point on the involute curve, the moving direction can be evaluated by whether u is increased or decreased with respect to the increment or decrement of the rotation angle. Therefore, the condition 2 is expressed by the following equation, where Δu is the increment of solution $\Delta u = u_{n,i} - u_{n,i-1}$ with respect to the increment of rotation angle $\Delta\theta = \theta_i - \theta_{i-1}$,

$$\Delta u \geq 0. \tag{14}$$

By evaluating Δu, the presence or absence of interference at $\theta=\theta_i$ can be examined. In order for the rack and pinion gears to satisfy the condition 2, it is necessary to satisfy the conditional Eq. (14) within the full rotation angle from the initial to the final ones.

The condition 3 can be expressed in the similar way to the above constant stroke case. However, in the following equation, pinion's pitch circle is used instead of the meshing pitch circle used in Eq. (8), since the radius of the latter one is not constant this time.

$$\begin{aligned} s_{a1} &= 2r_{a1}\left(\frac{s_1}{2r_1}+\mathrm{inv}\,\alpha_1-\mathrm{inv}\,\alpha\right)\geq 0 \\ \Leftrightarrow \frac{\pi+4x\tan\alpha_1}{2z_1} &+ \mathrm{inv}\,\alpha_1-\mathrm{inv}\left(\cos^{-1}\left(\frac{z_1\cos\alpha_1}{z_1+2(h_{a1}^*+x)}\right)\right)\geq 0. \end{aligned} \tag{15}$$

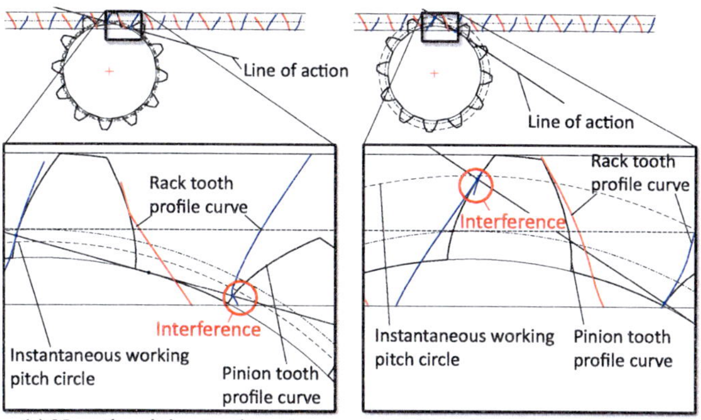

(a) Near the pinion tooth root. (b) Near the rack tooth root.

Fig. 11. Rack and pinion with interfering tooth curve.

For the condition 4, the contact ratio cannot be obtained from the geometric relationship when the variable stroke ratio is given. Therefore, the continuity of meshing motion is numerically evaluated. When a continuity of meshing motion is realized, there are one or more meshing points with respect to an arbitrary rotation angle. Hence, continuous meshing motion can be guaranteed if a solution u of the equation $f(u) = 0$ for at least one tooth exists within the full rotation angle. Condition 4 is expressed by

$$\forall i(1 \leq i \leq i_{end})\exists n(1 \leq n \leq z_1),\ \text{There exists}\ u_{n,i}\ \text{that satisfies}\ f_{n,i}\left(u_{n,i}\right) = 0. \quad (16)$$

In order for the rack and pinion gears to satisfy the condition 4, it is necessary to always satisfy the conditional Eq. (16).

For the condition 5, the pinion tooth root curve can be designed in the same way to the case of constant stroke ratio. On the other hand, the rack tooth fillet curve cannot be smoothly connected to the rack tooth profile curve by an arc of a constant radius R_R for all teeth. Therefore, an arc centered on the line passing through the end point of meshing is introduced as the tooth fillet curve.

3 Numerical Examples

An example of rack and pinion gears with a constant stroke ratio is shown. In this example, the module m_1, number of teeth z_1, pressure angle α_1, addendum coefficient h^*_{a1}, h^*_{a2}, working tooth depth h_{w2} and initial rotation angle θ_0 are given as constants as Table 1. In the calculation, a range of the shift coefficient x with respect to a target stroke ratio S should be taken into account. Figure 12 shows the obtained range of the set of x and S satisfying the conditional Eqs. (5), (7), (8), and (10). In the figure, $x_{min}, x_{max}, x_{max,\varepsilon}$ are the maximum/minimum shift coefficients that satisfy conditional Eqs. (7), (8), and (10), respectively. In the design conditions for the stroke ratio and shift coefficient that satisfy conditions 1 to 4 shown in Fig. 12, a rack and pinion gear that satisfy the geometric conditions can be derived by setting the fillet radius R_P, R_R that satisfy the conditional Eqs. (11) and (12).

Table 1. Given parameters of rack and pinion gear.

Parameter	Value
Module m_1	2 mm
Pressure angle α_1	20 degree
Number of teeth z_1	12
Addendum coefficient h_{a1}^*, h_{a2}^*	1, 0.6
Working tooth depth h_{w2}	3.2 mm
Initial rotation angle θ_0	0 degree

Table 2. Given parameters of variable rack and pinion gear.

Parameter	Value
Module m_1	2 mm
Number of teeth z_1	12
Addendum coefficient h_{a1}^*	1
Working tooth depth h_{w2}	3.2 mm
Profile shift coefficient x	0

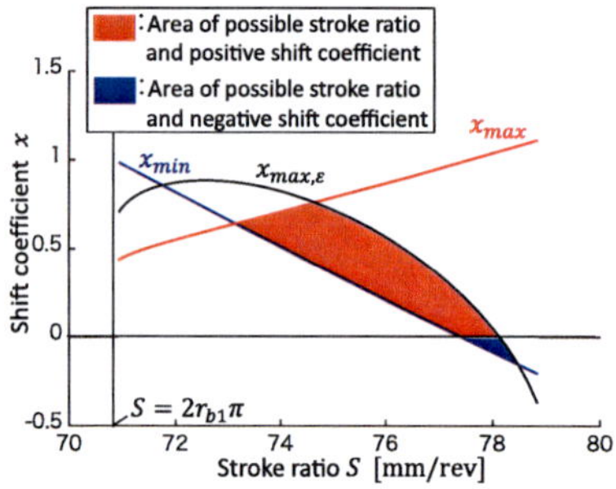

Fig. 12. Solution region formed by max/min shifted coefficient limits.

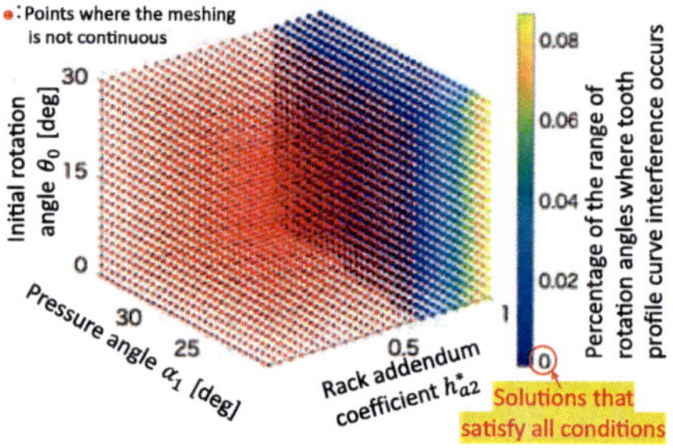

Fig. 13. Distribution of the tooth interference index value.

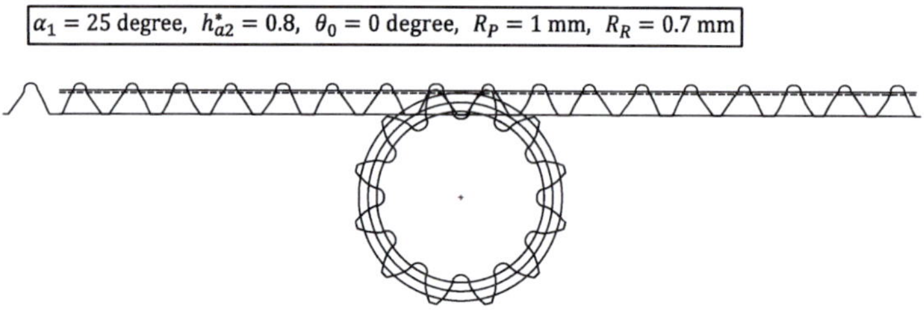

Fig. 14. Designed variable rack and pinion gears tooth profile curves.

An example of a variable rack and pinion gears is shown. In this example, the target stroke ratio is given by $S = 80 + 6\sin 5(\theta + \theta_0)$ mm/rev. Module m_1, number of teeth z_1, addendum coefficient h_{a1}^*, working tooth depth h_{w2} and the profile shift coefficient x are given as constants as Table 2. In the calculation, a possible range of the addendum coefficient h_{a2}^*, pressure angle α_1 and initial rotation angle θ_0 with respect to a given S has been taken into account. Evaluation points are discretely set for the design variables, and satisfaction of the conditional Eqs. (13), (14), (15) and (16) at each evaluation point has been judged. Figure 13 shows only the range of design variables that satisfy both conditional Eqs. (13) and (15). The blue evaluation points in the figure represent the solutions that satisfy the conditions. A pair of variable rack and pinion gears that satisfies all geometric conditions can be derived by setting the fillet radius to avoid interference at the tooth root with respect to the design conditions of the evaluation points that satisfy the conditions. An example of designed variable rack and pinion gears is shown in Fig. 14.

4 Conclusion

In this paper, a design methodology of variable rack and pinion gears tooth profile curves is shown. Module m_1, number of teeth z_1, pressure angle α_1, addendum coefficient h^*_{a1}, h^*_{a2}, shift coefficient x, working tooth depth h_{w2}, initial rotation angle θ_0 and fillet radius R_P, R_R are defined as design variables. The tooth profile curves of rack and pinion gears are derived to satisfy both a given target input/output relationship and essential geometric conditions. Especially, the interference between the tooth profile curves is avoided. In addition, especially when the variable stroke ratio is given, the essential geometric conditions cannot be formulated from the geometric relationship used for the case of constant stroke ratio, but it was clarified that the conditions can be expressed numerically and formulated. The issues to be solved are shown below.

1. Derivation of expression regarding tooth root curve to avoid interference.
2. Construction of optimal tooth profile curve calculation method.

References

1. Bishop, A.E.: Variable ratio power steering apparatus. US2865339A (1955)
2. Bishop, A.E.: Rack and pinion variable ratio steering gear. US3753378A (1971)
3. Morikawa, K., Higuchi, M., Yanai, T.: Method of manufacturing racks for variable ratio steering gears. US Patent #4516422A (1985)
4. Xu, M., Han, X., Hua, L., Zheng, F.: Modeling and methods for gear shaping process and cutting force prediction of variable transmission ratio rack. Int. J. Mech. Sci. **171**, 105364 (2020)

Tibial Motion Accuracy Using Circular Versus Noncircular Gears in Transfemoral Prosthetic Knees

Wen-Tzong Lee[1](✉) and Kevin Russell[2]

[1] National Pingtung University of Science and Technology, Neipu, Pingtung, Taiwan
wtlee@npust.edu.tw
[2] New Jersey Institute of Technology, Newark, USA

Abstract. This work presents the results of a study to determine if circular gears are comparable to noncircular gears (in terms of tibial position accuracy during gait) in a transfemoral prosthetic knee. In this study, the authors' design method for gear-based transfemoral prosthetic knees was utilized using clinical knee motion data during gait from several published sources. The data from these sources reflect knee motion distinctions due to gender, age, nationality, body mass, gait speed and knee condition. Unlike noncircular gears, which are custom-made, circular gears can offer advantages in manufacturing cost and design simplicity since it is a standard component. However, to justify the use of circular gears in a prosthetic knee design, it must be determined if the design can consistently exhibit tibial position accuracy that is at least comparable to noncircular gears when empirical knee motion data that reflect a range of differing human physical characteristics are considered.

Keywords: Prosthetic knee · RRSS linkage · Motion generation · Axode generation · Circle fitting · Gears

1 Introduction

1.1 Natural Human Knee Motion

The human knee joint connects the *tibia* and *femur* bones-the two largest bones in the human leg. The knee can be described as a spatial joint having 6 degrees of freedom. This mobility includes 3 principal rotations about and translations along a given spatial Cartesian frame within the knee [1]. In Fig. 1, δ_x, δ_y and δ_z represent the principal rotations (about the X, Y and Z-axes respectively) and Δ_x, Δ_y, and Δ_z represent the principal translations (also along the X, Y and Z-axes respectively). While specific clinical terms are often used to describe knee motion (e.g., δ_x is often described as *flexion-extension* motion and Δ_z as *distal-proximal* motion, etc.), the descriptions in Fig. 1 will be used throughout this work for brevity and simplicity.

© The Author(s), under exclusive license to Springer Nature Switzerland AG 2022
J. Beran et al. (Eds.): TMM 2020, MMS 85, pp. 44–51, 2022.
https://doi.org/10.1007/978-3-030-83594-1_5

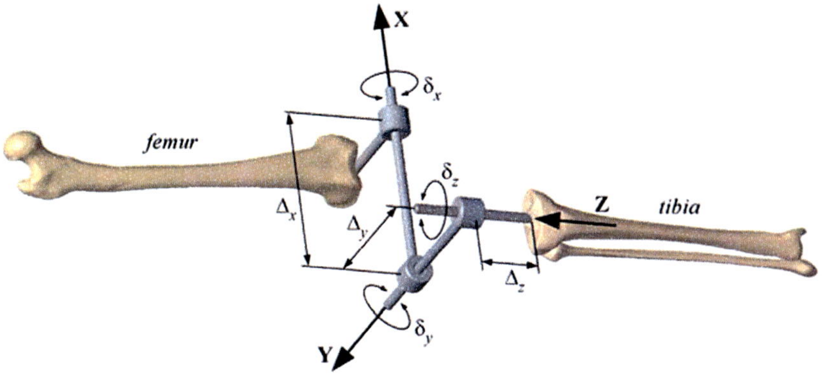

Fig. 1. Human knee displacements and rotations

1.2 Transfemoral Prosthetic Knee Design Method

A *transfemoral prosthetic knee* is the type of artificial knee used by above-the-knee amputees. There are two major categories of transfemoral prosthetic knee designs: *pin joint-based* and *polycentric* designs. While pin joint-based knee designs only allow pure rotation of the knee, polycentric knee designs allow a combination of rotational and translational knee motion. The latter is exhibited in natural knee motion.

The authors presented both a design method and a design for a gear-based transfemoral prosthetic knee [2–4]. The design method consists of the 4 principal stages as illustrated in Fig. 2. In stage 1, positions of the tibia bone over knee motion are produced. These positions are represented by the spatial Cartesian coordinates of 3 points attached to the tibia. The clinical approach to acquire this data is to attach bone-mounted or skin-mounted markers to the leg (in either a living subject or a cadaveric leg specimen) and to measure the point coordinates over knee motion. In this work however, tibial spatial Cartesian points over knee motion were calculated using plots for δ_x, δ_y, δ_z, Δ_x, Δ_y, and Δ_z which were obtained from published sources.

In stage 2, the tibial spatial Cartesian points are incorporated (as precision positions) in a motion generation model for the spatial four-bar *Revolute-Revolute-Spherical-Spherical* or RRSS linkage. The result of this stage are calculated dimensions for an RRSS linkage with a coupler link that approximates the tibial spatial Cartesian points from stage 1.

In stage 3, the dimensions of the synthesized RRSS linkage are incorporated in an axode generation model. The result of this stage is the calculated fixed and moving axodes for the RRSS linkage. The rolling motion of the moving axode over the fixed axode precisely replicates the coupler motion achieved by the synthesized RRSS linkage (and subsequently, the tibial spatial Cartesian points). The axodes can be used as pitch curves for a gear pair where rolling motion of the moving axode over the fixed axode become the rolling motion of the moving axode gear over the fixed axode gear.

Because the RRSS linkage axodes are not perfect circles, they would become pitch curves for noncircular gears if used as calculated in stage 3. The purpose of stage 4 is to calculate the centers and radii of circles that best fit the fixed and moving axodes. These circles are used as pitch circles for true circular gears in a prosthetic knee design. Figure 3 illustrates a prosthetic knee design resulting from the authors' design method [2, 4]. The rolling motion of the external gear over the internal gear approximates the tibial positions.

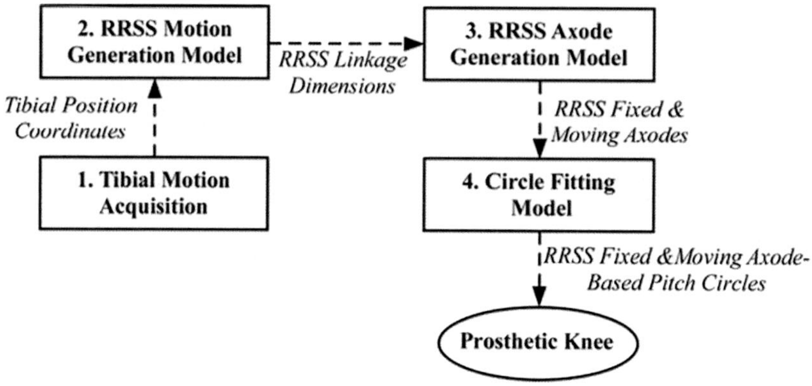

Fig. 2. Process stages for gear-based transfemoral prosthetic knee design

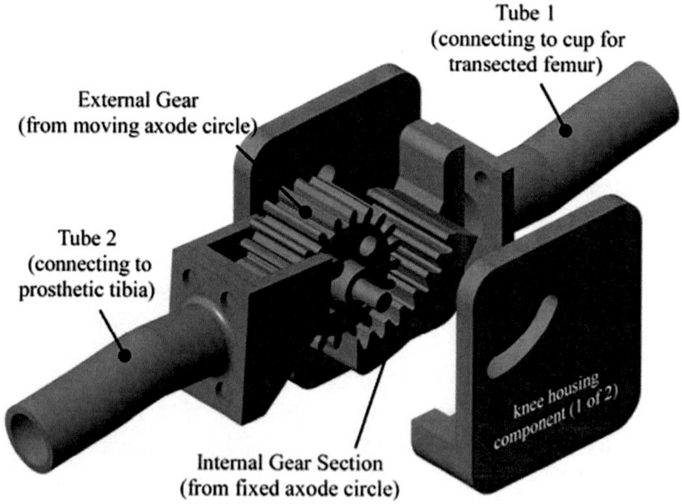

Fig. 3. Transfemoral prosthetic knee produced from design process

2 Empirical Knee Motion Data

Figure 4 includes plots of average principal knee rotations and translations from the work of Benoit et al. [5]. Six healthy subjects (all male) with an average age of 26 years and no history of knee injury or lower-limb surgery were recruited for this study. The subjects performed a series of normal walking trials along a 12-m walkway at a self-selected pace. Five points were selected from these knee rotation and translation plots and used (in stage 1) in the authors' design method to ultimately produce the transfemoral prosthetic knee illustrated in Fig. 3 [2, 4]. Because 5 groups of values for δ_x, δ_y, δ_z, Δ_x, Δ_y, and Δ_z were selected, the resulting transfemoral prosthetic knee approximates 5 tibial positions over the walking gait cycle. In this particular knee design, circular gears produced tibia positions that were consistently more accurate (in each tibial position) than the original noncircular gears [2].

Circular gears were demonstrated to be more accurate using the knee motion data from Benoit et al. This knee motion data however (representing only 6 young, healthy all-male subjects) do not reflect distinctions in human physical characteristics. Incorporating distinctions such as those due to gender, age, nationality, body mass, gait speed and knee condition would enable one to better evaluate the performance of circular versus noncircular gears in the knee design. To account for such distinctions, knee motion data like Fig. 4 were selected from published sources that do consider them.

This work includes knee motion data from the work of Kozánek et al. [6]. Eight healthy subjects (6 males and 2 females) aged 32 to 49 years, with an average body mass index of 23.5 kg/m^2 were recruited for this study. The subjects had no history of knee injury, surgery or systemic disease. The subjects performed treadmill gait for 1 min at a speed of 0.67 m/s.

This work also includes knee motion data from the work of Zhang et al. [7]. Twenty-eight healthy chinese subjects (14 males and 14 females) aged 20 to 30 years with an average body mass index of 20.8 kg/m^2 were recruited for this study. All of the subjects had no history of major trauma, surgery, knee-related symptoms, or obvious disorders in their lower extremities. The subjects performed treadmill gait for 15 s at a speed of 0.83 m/s.

This work also includes knee motion data from the work of Zeng et al. [8]. Twenty-six subjects (12 males and 14 females) with early medial knee osteoarthritis (KOA), an average age of 53.6 years and an average body mass index of 24.0 kg/m^2 were recruited for this study. Also, thirty-eight subjects (10 males and 28 females) with severe medial knee osteoarthritis, an average age of 63.5 years and an average body mass index of 23.1 kg/m^2 were recruited for this study. The subjects all performed treadmill gait for 15 s at a speed of 0.56 m/s.

This work also includes knee motion data from the work of Zheng et al. [9]. Nine subjects (3 males and 6 females) aged 18 to 53 years with an average body mass index of 25.0 kg/m^2 and each having a knee meniscectomy (KM), were recruited for this study. Four of the 9 subjects had each undergone a medial knee meniscectomy and the remining subjects had each undergone a lateral knee meniscectomy. The subjects all performed 15°-decline, dual belt treadmill gait (recorded over a 1-s duration) at a speed of 1 m/s.

Also included in this work are knee motion data from the work of Li et al. [10, 11]. Ten obese subjects (2 males and 8 females) with an average age of 42.8 years and an average body mass index of 39.6 kg/m^2 were recruited for this study. The subjects all reported experiencing knee pain but also reported being able to walk without assistance. The subjects performed treadmill gait (over durations limited by knee pain) at a speed of 0.67 m/s.

So as conveyed in this section, data were selected from knee motion studies whose participants reflect a much broader range of physical distinctions than those reflected in the work of Benoit et al. This data was utilized in stage 1 of the authors' design method to calculate tibial positions as spatial Cartesian coordinates.

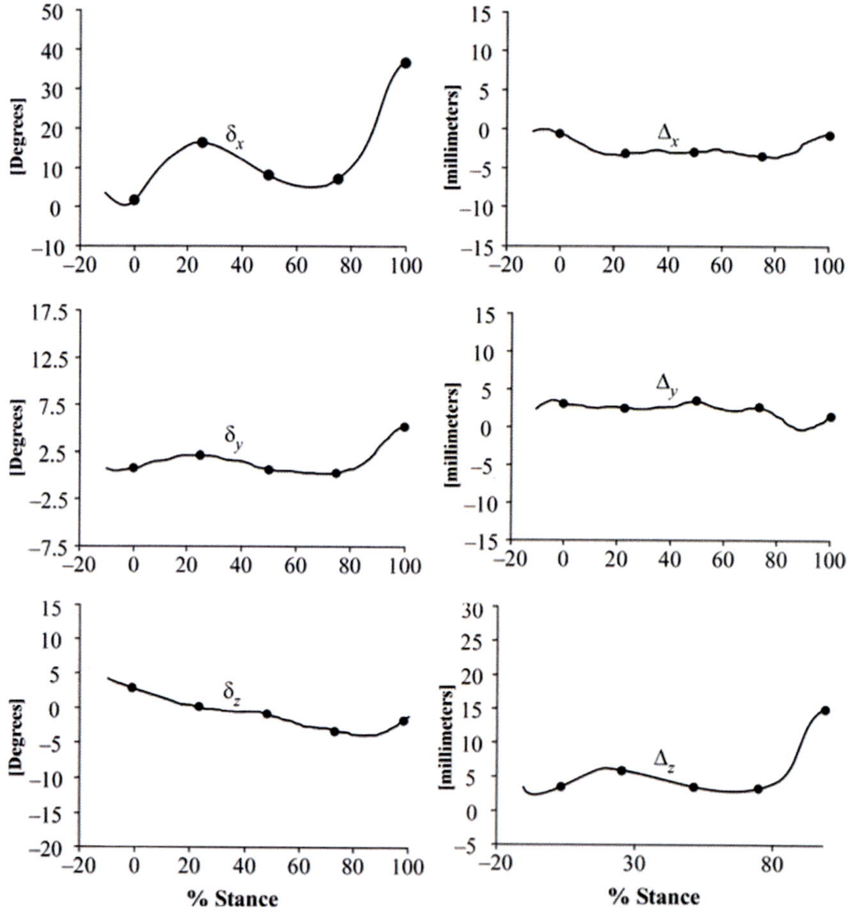

Fig. 4. Knee rotation and translation plots and selected values (from Benoit et al.)

3 Motion and Axode Generation and Pitch Circle Calculation

The RRSS linkage is a four-link spatial mechanism having 2 degrees of freedom. One degree of freedom however is the rotation of the follower link about its own length. This degree of freedom is known as a *passive degree of freedom* because it does not compromise the overall kinematics of the RRSS linkage (giving it an effective mobility of 1). The authors presented a constrained optimization model for the motion generation of defect-free RRSS linkages [12]. The precision positions in this model are the tibial position coordinates-the output from stage 1 of the authors' design method.

Axodes for the RRSS linkage are analogous to what *centrodes* are for the planar four-bar linkage. Therefore, like centrodes, the rolling motion of the moving axode over the fixed axode replicates the coupler motion of the linkage. The authors presented a method to calculate the fixed and moving axodes for the RRSS linkage [13].

A variety of published methods are available for fitting circles through data points. To determine the centers and radii of gear pitch circles to replace the noncircular axodes produced by the RRSS linkage, the method of least squares was employed in this work. With the method of least squares, "best fit" means Eq. (1) is minimized.

$$F(h, k, r) = \sum_{i=1}^{N} \left[(x_i - h)^2 + (y_i - k)^2 - r^2 \right]^2 \quad (1)$$

4 Tibial Position Accuracy Summary

Figure 5 include error plots between circular gears and noncircular gears at each tibia position. The error values are the squared norms of the differences between the tibial positions from stage 1 and those achieved by the prosthetic knee at each tibial position. Except for the plots based on data from Zang et al. and Zheng et al. (medial KM), tibial motion using circular gears outperformed noncircular gears in terms of the error sum. In the error plot based on data from Zang et al., noncircular gears are more accurate by only 0.34%. In the error plot based on data from Zheng et al. (medial KM), noncircular gears are more accurate by 7.95% – but this is largely due to the error at position 5. While circular gears are preferred over noncircular gears in terms of cost and design simplicity, Fig. 5 shows that circular gears also offer little to no consistent loss in tibial position accuracy compared to noncircular gears.

So, while tibial position accuracy using circular gears was shown to be at least comparable to noncircular gears using the empirical data from Benoit et al., the same conclusion can be reached from this study. And unlike the knee motion data from Benoit et al., the data used in this study reflect knee motion distinctions due to gender, age, nationality, body mass, gait speed and knee condition.

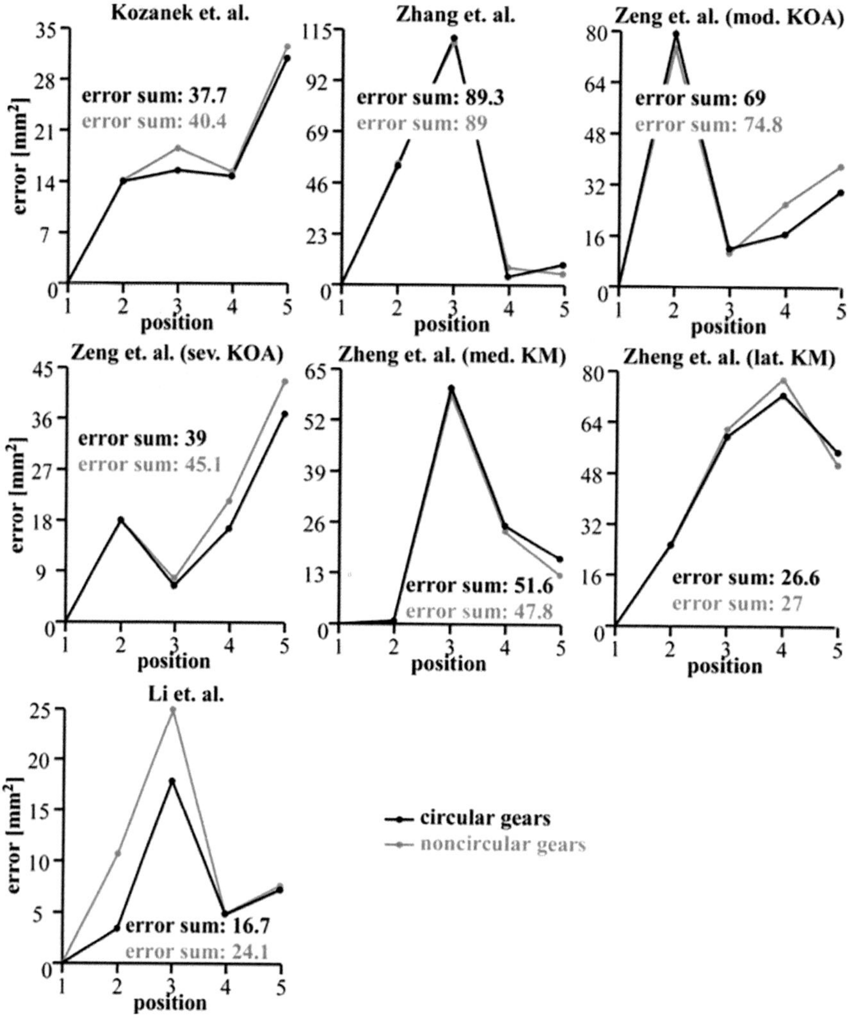

Fig. 5. Tibia position error plots using circular and noncircular gears

5 Conclusions

The objective of this work was to determine if circular gears were comparable to noncircular gears in terms of tibial position accuracy during gait in a transfemoral prosthetic knee. To determine this, the authors' design method for gear-based transfemoral prosthetic knees was utilized with empirical knee motion data from 7 clinical studies. These studies considered factors such as age, gender, nationality body mass, gait speed and knee condition. It was observed that the tibial position accuracy using circular gears was at least comparable to noncircular gears over each of the clinical studies. Subsequently, the authors' design method for gear-based transfemoral prosthetic knees is not a specialized method (effective only using knee motion data from Benoit et al.), but it was equally effective using data from 7 different clinical studies.

References

1. Grood, E.S., Suntay, W.J.: A joint coordinate system for the clinical description of three-dimensional motions: application to the knee. J. Biomech. Eng. **105**(2), 136–144 (1983)
2. Lee, W., Russell, K., Sodhi, R.S.: On transfemoral prosthetic knee design for natural human knee motion. Recent Pat. Mech. Eng. **13**(1), 49–59 (2020)
3. D'Alessio, J., Russell, K., Sodhi, R.S.: On Transfemoral Prosthetic Knee Design Using RRSS Motion and Axode Generation. In: Beran, J., Bílek, M., Žabka, P. (eds.) Advances in Mechanism Design II. MMS, vol. 44, pp. 29–35. Springer, Cham (2017). https://doi.org/10.1007/978-3-319-44087-3_4
4. Lee, W., Russell, K., D'Alessio, J., Sodhi, R.S.: Method of Designing and Manufacturing Artificial Joint, TW I642419 (2018)
5. Benoit, D.L., Ramsey, D.K., Lamontagne, M., Xu, L., Wretenberg, P., Renström, P.: In vivo knee kinematics during gait reveals new rotation profiles and smaller translations. Clin. Orthop. Relat. Res. **454**(1), 81–88 (2007)
6. Kozánek, M., et al.: Tibiofemoral kinematics and condylar motion during the stance phase of gait. J. Biomech. **42**(12), 1877–1884 (2009)
7. Zhang, Y., et al.: Motion analysis of Chinese normal knees during gait based on a novel portable system. Gait Posture **41**(3), 763–768 (2015)
8. Zeng, X., et al.: Relationship between Kellgren-Lawrence score and 3D kinematic gait analysis of patients with medial knee osteoarthritis using a new gait system. Sci. Rep. **7**, 4080 (2017)
9. Zheng, L., Carey, R., Thorhauer, E., Tashman, S., Harner, C., Zhang, X.: In vivo tibiofemoral skeletal kinematics and cartilage contact arthrokinematics during decline walking after isolated meniscectomy. Med. Eng. Phys. **51**, 41–48 (2017)
10. Li, J., Tsai, T., Felson, D.T., Li, G., Lewis, C.L.: Six degree-of-freedom knee joint kinematics in obese individuals with knee pain during gait. PLoS ONE **12**(3), e0174663 (2017)
11. Li, J., Tsai, T., Felson, D.T., Li, G., Lewis, C.L.: Correction: six degree-of-freedom knee joint kinematics in obese individuals with knee pain during gait. PLoS ONE **149**(2), e0213084 (2019)
12. Russell, K., Shen, Q.: Expanded spatial four-link motion and path generation with order and branch defect elimination. Inverse Probl. Sci. Eng. **21**(1), 129–140 (2013)
13. Shen, Q., Russell, K., Lee, W., Sodhi, R.S.: On cam system design to replicate spatial four-bar mechanism coupler motion. Inverse Prob. Sci. Eng. **19**(2), 251–265 (2011)

Curvature Interference Characteristic of Plane Worm Gear in Offsetting Cylindrical Worm Drive

Qingxiang Meng[1], Yaping Zhao[1,2(✉)], Jian Cui[1], and Gongfa Li[3]

[1] School of Mechanical Engineering and Automation, Northeastern University, Shenyang 110819, People's Republic of China
zhyp_neu@163.com
[2] Key Laboratory of Dynamics Reliability of Mechanical Equipment, Northeastern University, Shenyang 110819, Liaoning, People's Republic of China
[3] Key Laboratory of Metallurgical Equipment and Control of Ministry of Education, Wuhan University of Science and Technology, Wuhan 430081, People's Republic of China

Abstract. In this paper, the theory of the curvature interference characteristic is well established for the plane worm gear in the offsetting cylindrical worm drive. The offsetting cylindrical worm drive investigated in this paper is composed of an Archimedes cylindrical worm having the asymmetric tooth profile and a plane worm gear. The equations of the tooth surfaces, the meshing function, the meshing limit function, and the curvature interference limit function are all obtained for the worm drive. In accordance with the geometric construction and the elimination method, the first type of limit point on the curvature interference limit line is determined by iteratively solving the established nonlinear equation set. The result of the numerical case demonstrates that the curvature interference limit line on the worm gear tooth surface and its conjugate line on the worm helicoid are all located out the meshing zone of the worm drive. The meshing zones of the worm drive are located on the non-undercutting side. The part most likely to be undercut for the plane worm gear is located at the tooth root of the toe of its concave surface.

Keywords: Curvature interference · Cylindrical worm · Worm gear · Nonlinear equation · Meshing function

1 Introduction

The offsetting cylindrical worm drive is composed of a cylindrical worm having the asymmetric tooth profile and a plane worm gear as shown in Fig. 1, and can be used to transmit the movement and the torque between the two skew axes. This kind of worm drive is also called the Helicon gearing in the literature [1–3].

The offsetting cylindrical worm drive investigated in this paper was originally invited by Oliver E. Saari in the Illinois Tool Works in the late 1950s [4]. The helicoid of the cylindrical worm is an Archimedes helicoid which can be lathed by a lathe tool having the straight cutting edge. Thus, the tooth profile of the cylindrical worm is a straight line

© The Author(s), under exclusive license to Springer Nature Switzerland AG 2022
J. Beran et al. (Eds.): TMM 2020, MMS 85, pp. 52–61, 2022.
https://doi.org/10.1007/978-3-030-83594-1_6

in its axial section. The tooth surface of the plane worm gear is processed by a cylindrical hob, whose generating surface is the same as the Archimedes helicoid of the cylindrical worm.

The literature on the offsetting cylindrical worm drive, especially on the meshing theory, is very scarce overall. A scientific introduction to the offsetting cylindrical worm drive has been presented in the literature [2, 3], and they mentioned that although the load-bearing capacity is slightly inferior to the classical conical worm pair, it is also considerable.

During machining the offsetting cylindrical worm drive, the curvature interference should be firstly focused on since this determines its meshing performance. According to the meshing theory for gear drive [5], the curvature interference limit line is the envelope to the family of the instantaneous contact lines on the enveloped surface, i.e. the tooth surface of the plane worm gear. Therefore, the curvature interference limit line can divide the tooth surface of the plane worm gear into two parts: non-undercutting zone and undercutting zone. For this, the whole meshing zone of the offsetting cylindrical worm drive should be located in the non-undercutting zone to avoid the appearance of curvature interference. Thus it can be seen that computing the curvature interference limit line is of great significance to study the meshing performance of the worm drive.

The authors have studied the curvature interference characteristics for some kinds of worm gear by solving iteratively the established nonlinear equation sets [6, 7]. Based on the experience, the curvature interference characteristic of the plane worm gear is investigated and the numerical results are provided for the verification.

2 Generation of Plane Worm Gear

2.1 Equation of Cylindrical Worm Helicoid

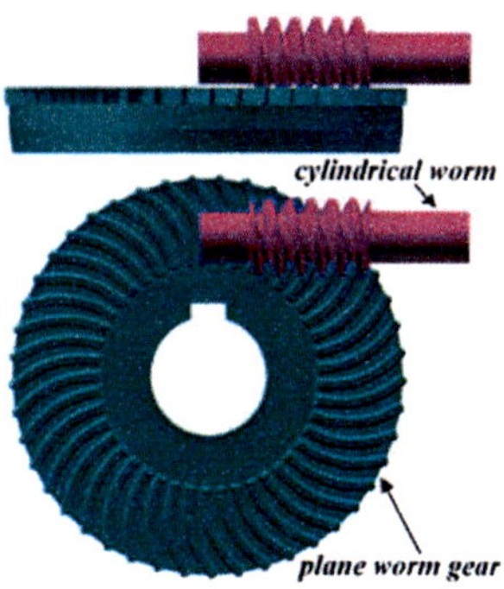

Fig. 1. Offsetting cylindrical worm drive.

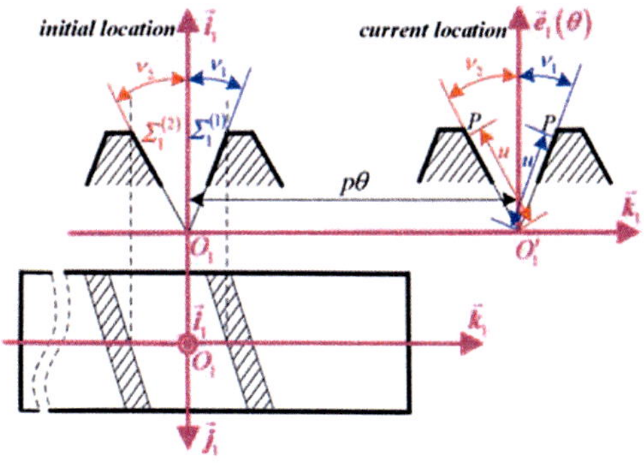

Fig. 2. Formation of Archimedes helicoid of cylindrical worm.

To acquire the vector equation of the Archimedes worm helicoid, a rotating coordinate system $\sigma_1\left\{O_1;\vec{i}_1,\vec{j}_1,\vec{k}_1\right\}$ is set to attach with the cylindrical worm as shown in Fig. 2.

The point O_1 is located at the middle point of the worm thread length, and the vector $\vec{k}_1$ is along the axis of the worm.

During the formation of the Archimedes helicoid, the cylindrical worm rough is fixed while the lathe tool makes a helical motion from the initial location to the current location as shown in Fig. 2. The moving distance of the lathe tool along the worm axis is equal to $p\theta$, in which the symbols p and θ are the helical parameter of the cylindrical worm and the rotating angle of the lathe tool, respectively. Due to the asymmetric tooth profile of the cylindrical worm, an auxiliary symbol S is adopted to distinguish the helicoids $\Sigma_1^{(1)}$ and $\Sigma_1^{(2)}$ on the two sides of one worm tooth. When $S = 1$, $\Sigma_1^{(1)}$ is the helicoid facing to the negative direction of the vector $\vec{k}_1$ and the tooth profile angle ν_1 of the cylindrical worm should be a positive value. When $S = 2$, $\Sigma_1^{(2)}$ is the helicoid facing to the positive direction of the vector $\vec{k}_1$ and the tooth profile angle ν_2 of the cylindrical worm should be a negative value.

Based on the setting above, the vector equation can be acquired for the worm helicoid $\Sigma_1^{(S)}$ and expressed in σ_1 as below

$$\left(\vec{r}_1\right)_1 = \overrightarrow{O_1P} = u\vec{m}_1\left(\theta, 90^\circ - \nu_S\right) + p\theta\vec{k}_1 = u\cos\nu_S\cos\theta\vec{i}_1 + u\cos\nu_S\sin\theta\vec{j}_1 + z_1\vec{k}_1, \tag{1}$$

in which $z_1 = u\sin\nu_S + p\theta$ and the two curvilinear coordinates of the worm helicoid $\Sigma_1^{(S)}$ are u and θ.

From Eq. (1), the first-order partial derivatives of $\vec{r}_1$ with respect to u and θ can be worked out in σ_1 as follows:

$$\left(\vec{r}_{1u}\right)_1 = \vec{m}_1\left(\theta, 90^\circ - \nu_S\right), \quad \left(\vec{r}_{1\theta}\right)_1 = u\cos\nu_S\vec{g}_1(\theta) + p\vec{k}_1. \tag{2}$$

In light of the theory of differential geometry [8], it is possible to work out the first kind of fundamental variables of the worm helicoid $\Sigma_1^{(S)}$ as follows

$$E = \left(\vec{r}_{1u}\right)_1 \cdot \left(\vec{r}_{1u}\right)_1 = 1, \quad F = \left(\vec{r}_{1u}\right)_1 \cdot \left(\vec{r}_{1\theta}\right)_1 = p\sin\nu_S, \quad G = \left(\vec{r}_{1\theta}\right)_1 \cdot \left(\vec{r}_{1\theta}\right)_1 = u^2\cos^2\nu_S + p^2. \tag{3}$$

By means of Eqs. (2) and (3), the unit normal vector of the worm helicoid $\Sigma_1^{(S)}$ can be achieved in σ_1 as

$$\left(\vec{n}_1\right)_1 = \frac{\left(\vec{r}_{1u}\right)_1 \times \left(\vec{r}_{1\theta}\right)_1}{\sqrt{EG - F^2}} = \frac{-1}{\sqrt{u^2 + p^2}}\left[u\sin\nu_S\vec{e}_1(\theta) + p\vec{g}_1(\theta) - u\cos\nu_S\vec{k}_1\right]. \tag{4}$$

2.2 Cutting Meshing of Plane Worm Gear

For the cylindrical hob used for processing the plane worm gear, its generating surface is the same as the helicoid of the cylindrical worm mentioned above, so that the cutting meshing between the cylindrical hob and the plane worm gear is the same as the working meshing of the offsetting cylindrical worm drive. For this reason, these two processes will be not distinguished in the following.

During the meshing of the offsetting cylindrical worm drive, the relative location and relative motion between the cylindrical worm and the plane worm gear are depicted in Fig. 3. The initial locations of the cylindrical worm and the plane worm gear are represented by the static coordinate systems $\sigma_{o1}\left\{O_1; \vec{i}_{o1}, \vec{j}_{o1}, \vec{k}_{o1}\right\}$ and $\sigma_{o2}\left\{O_2; \vec{i}_{o2}, \vec{j}_{o2}, \vec{k}_{o2}\right\}$, respectively. The two vectors $\vec{k}_{o1}$ and $\vec{k}_1$ are coincident. The rotating coordinate system $\sigma_2\left\{O_2; \vec{i}_2, \vec{j}_2, \vec{k}_2\right\}$ is rigidly attached to the plane worm gear and is utilized to denote its current position. The two unit vectors $\vec{k}_{o2}$ and $\vec{k}_2$ are all coincident with the axis of the plane worm gear. The unit vector $\vec{i}_{o2}$ is coincident with the common perpendicular of $\vec{k}_{o2}$ and $\vec{k}_{o1}$ is parallel to the unit vector $\vec{i}_{o1}$. The distance between the two unit vectors $\vec{i}_{o2}$ and $\vec{i}_{o1}$ along the axis of the cylindrical worm is equal to z_0 and $z_0 = z_A + L_w/2$, in which the symbol z_A and L_w are respectively the mounting distance and the face width of the cylindrical worm. Moreover, the distance between the two axes $\vec{k}_{o2}$ and $\vec{k}_{o1}$ is the center distance a of the worm drive, i.e. $\left|\overrightarrow{O_2O'}\right| = a$. The rotating angles between σ_{o1} and σ_1 and between σ_{o2} and σ_2 are φ and φ/i_{12}, respectively. Here, the symbol i_{12} is the drive ratio of the worm drive.

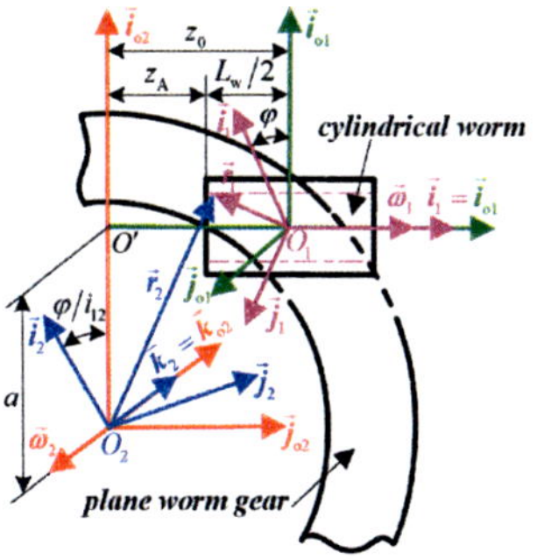

Fig. 3. Coordinate systems used in cutting meshing of plane worm gear.

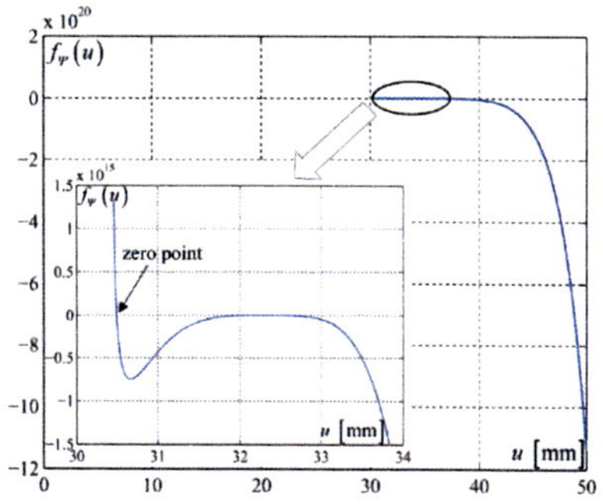

Fig. 4. Curve of function $f(u)$ when $L_\psi = 30$ mm and $S = 1$.

Based on the foregoing description, a family $\left\{\Sigma_1^{(S)}\right\}$ of worm helicoids can be formed in σ_{o1} when the worm rotates around its axis. Via transformation of coordinates [5], the equation of $\left\{\Sigma_1^{(S)}\right\}$ can be obtained as

$$\left(\vec{r}_1\right)_{o1} = R\left[\vec{k}_{o1}, \varphi\right]\left(\vec{r}_1\right)_1 = u\cos\nu_S\cos(\theta+\varphi)\vec{i}_1 + u\cos\nu_S\sin(\theta+\varphi)\vec{j}_1 + z_1\vec{k}_1. \quad (5)$$

During the meshing, it can be assumed that the angular velocity vector $\vec{\omega}_1$ of the worm is equal to $\vec{k}_{o1}$ and the corresponding angular velocity vector $\vec{\omega}_2$ of the worm gear is equal to $-\frac{1}{i_{12}}\vec{k}_{o2}$, respectively. Thus, the relative angular velocity vector during the meshing of the worm drive can be obtained in σ_1 as

$$\left(\vec{\omega}_{12}\right)_1 = \left(\vec{\omega}_1\right)_1 - \left(\vec{\omega}_2\right)_1 = \left(\vec{\omega}_1\right)_1 - R\left[\vec{k}_1, -\varphi\right]R\left[\vec{k}_{o1}, 90^\circ\right]\left(\vec{\omega}_2\right)_{o2} = -\frac{\sin\varphi}{i_{12}}\vec{i}_1 - \frac{\cos\varphi}{i_{12}}\vec{j}_1 + \vec{k}_1. \quad (6)$$

From Eqs. (1) and (6), the relative velocity vector during the meshing can be acquired in σ_1 as below

$$\left(\vec{V}_{12}\right)_1 = (\vec{\omega}_{12})_1 \times (\vec{r}_1)_1 - (\vec{\omega}_2)_1 \times \left(\overrightarrow{O_2O_1}\right)_1 = V_x\vec{i}_1 + V_y\vec{j}_1 + V_z\vec{k}_1, \quad (7)$$

in which the expression of the vector $\overrightarrow{O_2O_1}$ can be acquired in σ_1 as $\left(\overrightarrow{O_2O_1}\right)_1 = a\cos\varphi\vec{i}_1 - a\sin\varphi\vec{j}_1 + z_0\vec{k}_1$; the expressions of V_x, V_y, and V_z respectively are $V_x = -u\cos\nu_S\sin\theta - \frac{1}{i_{12}}(z_1+z_0)\cos\varphi$, $V_y = u\cos\nu_S\cos\theta + \frac{1}{i_{12}}(z_1+z_0)\sin\varphi$, and $V_z = \frac{1}{i_{12}}[u\cos\nu_S\cos(\theta+\varphi)+a]$.

According to the definition, the meshing function during the meshing of the worm drive can be figured out from Eqs. (4) and (7) as below

$$\Phi(u,\theta,\varphi) = (\vec{n}_1)_1 \cdot \left(\vec{V}_{12}\right)_1 = \frac{1}{i_{12}\sqrt{u^2+p^2}}[A\sin(\theta+\varphi) + B\cos(\theta+\varphi) + C], \quad (8)$$

where $A = -p(z_1+z_0)$, $B = u^2\cos^2\nu_S + u\sin\nu_S(z_1+z_0)$, and $C = u\cos\nu_S(a - i_{12}p)$.

With the aid of Eqs. (5) and (8), the vector equation of the tooth surface $\Sigma_2^{(S)}$ of the plane worm gear can be represented in σ_2 by right of some transformations of coordinates as below

$$(\vec{r}_2)_2 = R\left[\vec{k}_2, \frac{\varphi}{i_{12}}\right]R\left[\vec{i}_{o2}, -90^\circ\right]\left[(\vec{r}_1)_{o1} + \left(\overrightarrow{O_2O_1}\right)_{o1}\right] = x_2\vec{i}_2 + y_2\vec{j}_2 + z_2\vec{k}_2, \quad \Phi(u,\theta,\varphi) = 0, \quad (9)$$

in which $x_2 = [u\cos\nu_S\cos(\theta+\varphi)+a]\cos\frac{\varphi}{i_{12}} - (z_1+z_0)\sin\frac{\varphi}{i_{12}}$, $y_2 = [u\cos\nu_S\cos(\theta+\varphi)+a]\sin\frac{\varphi}{i_{12}} + (z_1+z_0)\cos\frac{\varphi}{i_{12}}$, and $z_2 = -u\cos\nu_S\sin(\theta+\varphi)$.

In Eq. (9), the tooth surfaces $\Sigma_2^{(1)}$ and $\Sigma_2^{(2)}$ of the plane worm gear are corresponding to the convex tooth surface and concave tooth surface, respectively.

2.3 Meshing Characteristic Parameters

The meshing limit function during the meshing of the worm drive can be yielded by taking the partial derivative of Eq. (8) with respect to the rotating angle φ as

$$\Phi_\varphi(u,\theta,\varphi) = \frac{1}{i_{12}\sqrt{u^2+p^2}}[A\cos(\theta+\varphi) - B\sin(\theta+\varphi)]. \quad (10)$$

Moreover, to determine the normal vector $\vec{N}$ of the contact line during the meshing of the worm drive in the natural frame $\sigma_P\{P; \vec{r}_{1u}, \vec{r}_{1\theta}, \vec{n}_1\}$ [9], the partial derivatives of Eq. (8) with respect to the variables u and φ should also be figured out. The obtained results are

$$\Phi_u(u,\theta,\varphi) = \frac{-1}{i_{12}\sqrt{u^2+p^2}}\left[p\sin\nu_S\sin(\theta+\varphi) - B_u\cos(\theta+\varphi) - \cos\nu_S(a - i_{12}p)\right], \quad (11)$$

$$\Phi_\theta(u,\theta,\varphi) = \frac{-p}{i_{12}\sqrt{u^2+p^2}}\left[p\sin(\theta+\varphi) - u\sin\nu_S\cos(\theta+\varphi)\right] + \Phi_\varphi, \quad (12)$$

in which $B_u = 2u\cos^2\nu_S + \sin\nu_S(z_1+z_0)$.

From Eqs. (2), (3), and (10)–(12), the normal vector $\vec{N}$ of the contact line can be expressed in σ_1 by means of the natural frame $\sigma_P\{P;\vec{r}_{1u},\vec{r}_{1\theta},\vec{n}_1\}$ as follows

$$\left(\vec{N}\right)_1 = \frac{1}{\left(u^2+p^2\right)\cos^2\nu_S}\left[(p\sin\nu_S\Phi_\theta - G\Phi_u)\left(\vec{r}_{1u}\right)_1 + (p\sin\nu_S\Phi_u - \Phi_\theta)\left(\vec{r}_{1\theta}\right)_1\right]. \quad (13)$$

Then the curvature interference limit function during the meshing of the worm drive can be easily calculated from Eqs. (7), (10) and (13) as

$$\begin{aligned}\Psi = \left(\vec{N}\right)_1\cdot\left(\vec{V}_{12}\right)_1 + \Phi_\varphi = \frac{1}{(u^2+p^2)\cos^2\nu_S}\Big[&(p\sin\nu_S\Phi_\theta - G\Phi_u)\left(\vec{r}_{1u}\right)_1\cdot\left(\vec{V}_{12}\right)_1 \\ &+(p\sin\nu_S\Phi_u - \Phi_\theta)\left(\vec{r}_{1\theta}\right)_1\cdot\left(\vec{V}_{12}\right)_1\Big] + \Phi_\varphi\end{aligned}, \quad (14)$$

in which $\left(\vec{r}_{1u}\right)_1\cdot\left(\vec{V}_{12}\right)_1 = -\frac{1}{i_{12}}\{(p\theta+z_0)\cos\nu_S\cos(\theta+\varphi) - a\sin\nu_S\}$ and $\left(\vec{r}_{1\theta}\right)_1\cdot\left(\vec{V}_{12}\right)_1 = \frac{1}{i_{12}}\{(z_1+z_0)u\cos\nu_S\sin(\theta+\varphi) + p[u\cos\nu_S\cos(\theta+\varphi)+a]\} + u^2\cos^2\nu_S$.

According to the direction of the normal vector $\vec{n}_1$ of the worm helicoid in Eq. (4), the value of Ψ in the non-undercutting zone of the worm drive should be positive when $S = 1$ and negative when $S = 2$, respectively.

3 Computing Principle of Curvature Interference Limit Line

The curvature interference limit line existing on the tooth surface of the plane worm gear can be acquired by connecting a set of the first type of limit points. To establish the nonlinear equation set for determining the first type of limit point, the coordinate of the first type of limit point along the axis of the plane worm gear in σ_2 is set as a constant L_Ψ, i.e. $L_\Psi = z_2 = -u\cos\nu_S\sin(\theta+\varphi)$. Then the nonlinear equation set for determining the first type of limit point can be built based on Eqs. (8) and (14) as

$$u\cos\nu_S\sin(\theta+\varphi) = -L_\Psi,\quad A\sin(\theta+\varphi) + B\cos(\theta+\varphi) + C = 0,\quad \Psi(u,\theta,\varphi) = 0. \quad (15)$$

System (15) has three unknowns u, θ and φ. For the sake of judging the existence of the solution of System (15) and determining its reasonable iterative initial value, the two variables θ and φ in System (15) can be reduced.

Firstly, from the first two expressions of in System (15), the two trigonometric functions $\sin(\theta+\varphi)$ and $\cos(\theta+\varphi)$ can be expressed by the two variables θ and φ as follows

$$\sin(\theta+\varphi) = -\frac{L_\Psi}{u\cos\nu_S},\quad \cos(\theta+\varphi) = \frac{AL_\Psi - Cu\cos\nu_S}{Bu\cos\nu_S}. \quad (16)$$

Based on the trigonometric function knowledge, the following quadratic equation with one unknown $z_1 + z_0$ can be acquired from Eq. (16) as

$$a_\theta(z_1 + z_0)^2 + 2b_\theta(z_1 + z_0) + c_\theta = 0, \tag{17}$$

where $a_\theta(u) = u^2 \sin^2 v_S\left(L_\Psi^2 - u^2\cos^2 v_S\right) + p^2L_\Psi^2$,
$b_\theta(u) = u^3\cos^2 v_S \sin v_S\left(L_\Psi^2 - u^2\cos^2 v_S\right) + upL_\Psi C\cos v_S$,
$c_\theta(u) = u^4\cos^4 v_S\left(L_\Psi^2 - u^2\cos^2 v_S\right) + u^2C^2\cos^2 v_S$.

The solutions of Eq. (17) can be easily figured out and then the variable θ can be denoted by the variable u. The acquired results are

$$z_1(u) = \frac{\hat{z}_1}{a_\theta}, \quad \theta(u) = \frac{\hat{z}_1 - ua_\theta\sin v_S}{pa_\theta}, \tag{18}$$

in which $\hat{z}_1 = -b_\theta \pm \sqrt{b_\theta^2 - a_\theta c_\theta} - a_\theta z_0$ and the plus-minus sign should be reasonably selected during the numerical simulation.

Secondly, the components A and B in Eq. (8) and the trigonometric function $\cos(\theta + \varphi)$ in Eq. (16) can be successively represented by the variable u as follows

$$A(u) = \frac{-p\left(\hat{z}_1 + a_\theta z_0\right)}{a_\theta}, \quad B(u) = \frac{u\hat{B}}{a_\theta}, \quad \cos(\theta + \varphi) = \frac{T_c}{\hat{B}}, \tag{19}$$

where $\hat{B} = \hat{z}_1\sin v_S + a_\theta\left(u\cos^2 v_S + z_0\sin v_S\right)$ and $T_c = -a_\theta\cos v_S(a - i_{12}p) - \frac{pL_\Psi(\hat{z}_1 + a_\theta z_0)}{u^2\cos v_S}$.

Similarly, the three partial derivatives in Eqs. (10)–(12) and the two dot products in Eq. (14) can also be successively expressed by the variable u as follows

$$\Phi_\varphi(u) = \frac{\hat{\Phi}_\varphi}{a_\theta\hat{B}}, \quad \Phi_u(u) = \frac{\hat{\Phi}_u}{a_\theta\hat{B}}, \quad \Phi_\theta(u) = \frac{\hat{\Phi}_\theta}{a_\theta\hat{B}}, \quad \left(\vec{r}_{1u}\right)_1\cdot\left(\vec{V}_{12}\right)_1 = \frac{V_{ru}}{a_\theta\hat{B}}, \quad \left(\vec{r}_{1\theta}\right)_1\cdot\left(\vec{V}_{12}\right)_1 = \frac{V_{r\theta}}{a_\theta\hat{B}}, \tag{20}$$

where $\hat{\Phi}_\varphi = \frac{\hat{B}^2L_\Psi - pT_c\cos v_S(\hat{z}_1 + a_\theta z_0)}{i_{12}\cos v_S\sqrt{u^2+p^2}}$, $\hat{\Phi}_u = \frac{u\hat{B}_uT_c + a_\theta\hat{B}(pL_\Psi\tan v_S + C)}{ui_{12}\sqrt{u^2+p^2}}$,
$\hat{B}_u = 2a_\theta u + \left[\hat{z}_1 + a_\theta(z_0 - u\sin v_S)\right]\sin v_S$, $\hat{\Phi}_\theta = \frac{pa_\theta}{i_{12}\sqrt{u^2+p^2}}\left(uT_c\sin v_S + \frac{p\hat{B}L_\Psi}{u\cos v_S}\right) + \hat{\Phi}_\varphi$,
$V_{ru} = \frac{1}{i_{12}}\left\{a_\theta\hat{B}a\sin v_S - T_c\cos v_S\left[\hat{z}_1 + a_\theta(z_0 - u\sin v_S)\right]\right\}$, and
$V_{r\theta} = \frac{1}{i_{12}}\left[pa_\theta\left(uT_c\cos v_S + a\hat{B}\right) - \hat{B}L_\Psi\left(\hat{z}_1 + a_\theta z_0\right)\right] + a_\theta\hat{B}u^2\cos^2 v_S$.

Finally, substituting Eq. (20) into the third expression of System (15) leads up to an equation with one unknown u which is equivalent to System (15) as below

$$\Psi = \frac{f(u)}{a_\theta^2\hat{B}^2} = 0, \tag{21}$$

in which $f(u) = \frac{1}{(u^2+p^2)\cos^2 v_S}\left[V_{ru}\left(p\sin v_S\widehat{\Phi}_\theta - G\widehat{\Phi}_u\right) + V_{r\theta}\left(p\sin v_S\widehat{\Phi}_u - \widehat{\Phi}_\theta\right)\right] + a_\theta\hat{B}\widehat{\Phi}_\varphi$.

The existence of the solution of Eq. (21) can be explored by judging whether there is the zero point or not in the given solution domain by means of the geometric construction [10]. If the zero point exists, the reasonable iterative initial value can be acquired. Then Eq. (21) can be solved iteratively and the corresponding first type of limit point can be determined. Via changing the value of L_{Ψ}, other first type of limit points can be determined as well, and then the first type of limit line can be obtained.

4 Numerical Example

The main parameters of the offsetting cylindrical worm drive discussed in this paper are: the center distance $a = 100$ mm, the drive ratio $i_{12} = 51$, the modules $m = 4$ mm, the number of cylindrical worm threads $Z_1 = 1$, the radius of the cylindrical worm reference circle $r_1 = 26$ mm, the tooth profile angle of the cylindrical worm $\nu_1 = 10°$ and $\nu_2 = -35°$, the face width of the cylindrical worm $L_w = 0.73a = 73$ mm, and the mounting distance of cylindrical worm $z_A = 0.65a = 65$ mm.

According to the preceding parameters, the boundary of the plane worm gear can be determined by means of the computation of the boundary of the conical worm gear in Ref. [10] and based on this, five first type of limit points are selected to determine the curvature interference limit line of the offsetting cylindrical worm drive. The values of L_{Ψ} at these first type of limit points are equal to 15 mm, 20 mm, 25 mm, 30 mm, and 35 mm, respectively. Due to the same determining method for these first type of limit points, the point with $L_{\Psi} = 30$ mm when $S = 1$ is used to explain the computing process. At this time, the curves of the function $f(u)$ in Eq. (21) is plotted in Fig. 4, in which the abscissa and ordinate are the independent variable u and the functional value of $f(u)$, respectively. The data range of the independent variable u is set as (0, 50 mm] reasonably according to the computation of the meshing zone.

Figure 4 displays that the function $f(u)$ has only one zero point within the given data range, and thus the equation $f(u) = 0$ has only one solution within the preceding data range. Since the zero point in Fig. 4 is approximately equal to 30.5 mm, the iterative initial value for solving the unique solution of the equation $f(u) = 0$ within the given data range can be reasonably set as $u = 30.5$ mm. After the solution of the equation $f(u) = 0$ is solved, the values of the parameters θ and φ can be figured out from Eqs. (18) and (16) in sequence. Then the first type of limit point with $L_{\Psi} = 30$ mm when $S = 1$ can be determined. Based on this, other first type of limit points can also be determined, and the numerical results of these points are provided in Table 1.

The curvature interference limit line on the tooth surface of the plane worm gear and the conjugate line of the curvature interference limit line on the cylindrical worm helicoid are acquired via connecting the first type of limit points determined above, and are drawn in the axial sections of the plane worm gear and the cylindrical worm as shown in Figs. 5 and 6, respectively.

The points ①–⑤ in Figs. 5 and 6 are the obtained first type of limit points. The area $M_{1i}M_{2i}M_{3i}M_{4i}M_{1i}$ in Fig. 5 and area $M_{1e}M_{2e}M_{3e}M_{4e}M_{1e}$ in Fig. 6 are the meshing zones of the offsetting cylindrical worm drive which can be acquired by connecting some key meshing points [10]. Obviously, the curvature interference limit line on the worm gear tooth surface and its conjugate line on the worm helicoid are all located out

Table 1. Numerical results of the first type of limit points.

Parameter		Limit point				
		①	②	③	④	⑤
L_Ψ [mm]		15	20	25	30	35
$S = 1$	u [mm]	15.2353	20.3298	25.4173	30.4984	35.5763
	θ [°]	−2636.6977	−2721.1415	−2791.0767	−2850.0484	−2901.3141
	φ [°]	2907.9899	2993.7654	3063.9348	3122.8184	3173.9051
$S = 2$	u [mm]	19.2649	25.3425	31.4408	37.5508	43.6678
	θ [°]	−1734.5624	−1414.8888	−1099.1624	−785.5821	−473.3207
	φ [°]	2022.6625	1700.4337	1383.0680	1068.3436	755.2350

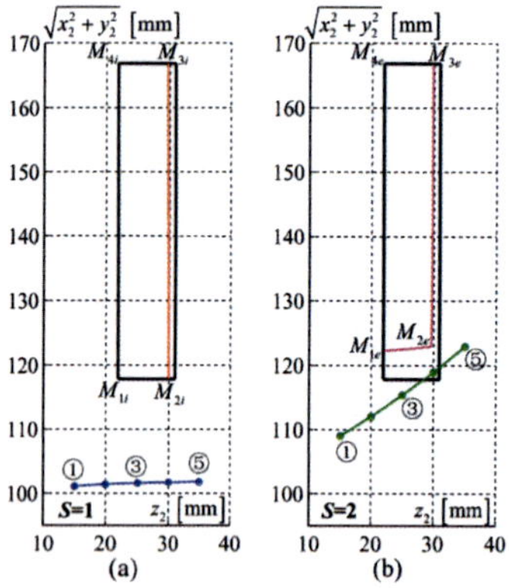

Fig. 5. Curvature interference limit line in axial section of plane worm gear.

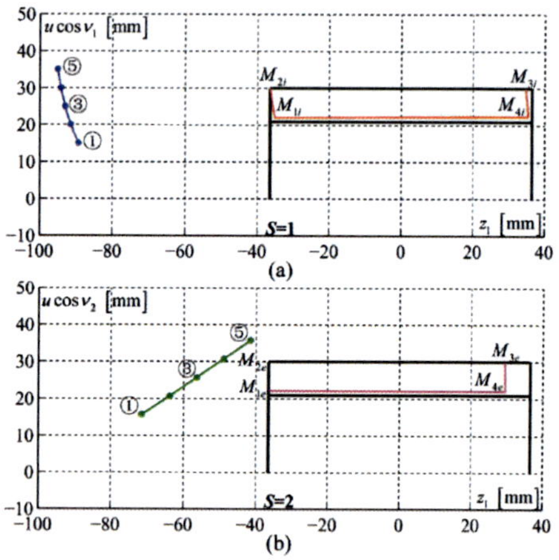

Fig. 6. Conjugate line of curvature interference limit line in axial section of cylindrical worm.

the meshing zone of the worm drive. Moreover, the values of Ψ at the meshing point M_{1i} and M_{1e} can be worked out as 0.0317 and −0.0211 from Eq. (14), respectively, and therefore the meshing zones of the offsetting cylindrical worm drive are all located on the non-undercutting side. When $S = 2$, although the curvature interference limit line comes into the entity boundary of the worm gear at its tooth root of toe, the effect of this phenomenon on the occurrence of the curvature interference is little since there is a long distance between the curvature interference limit line and the meshing zone of the worm drive as shown in Fig. 5b. Moreover, overall speaking, the part most likely to be undercut on the tooth surface of the plane worm gear is located at the tooth root of the toe when $S = 2$.

5 Conclusions

The theory of the curvature interference characteristic of the plane worm gear in the offsetting Archimedes cylindrical worm drive is fully established and the method to compute the curvature interference limit line is proposed.

Some basic and important formulas are obtained, such as the equations of the tooth surfaces of the cylindrical worm and the plane worm gear, the meshing function, the meshing limit function, the curvature interference limit function of the worm drive. The nonlinear equation set to determine the first type of limit points on the curvature interference limit line is established and is properly solved iteratively by means of the elimination method and the geometric construction.

The numerical example is implemented and analyzed, and the result indicates that the curvature interference limit line on the worm gear tooth surface and its conjugate line on the worm helicoid do not come into the meshing zone of the worm drive. The meshing zones of the worm drive are located on the non-undercutting side. The dedendum of the toe of the concave tooth surface of the plane gear is the part most likely to be undercut.

Acknowledgments. This study was funded by the National Natural Science Foundation of China (52075083) and the Open Fund of the Key Laboratory for Metallurgical Equipment and Control of Ministry of Education in Wuhan University of Science and Technology (MECOF2020B03 and 2018B05).

References

1. Dudas, I.: The Theory and Practice of Worm Gear Drives. Penton Press, London (2000)
2. Dudley, D.W. (ed.): Gear Handbook. McGraw-Hill Book Company Inc, New York (1962)
3. Radzevich, S.: Dudley's Handbook of Practical Gear Design and Manufacture. CRC Press (2012). https://doi.org/10.1201/b11842
4. Saari, O.E.: Skew Axis Gearing. United States Patent Office (1960)
5. Dong, X.: Foundation of Meshing Theory for Gear Drives. China Machine Press, Beijing (1989). (in Chinese)
6. Meng, Q., Zhao, Y., Yang, Z.: Curvature interference characteristic of conical worm gear. Forschung Ingenieurwesen **83**(3), 759–773 (2019). https://doi.org/10.1007/s10010-019-00372-3
7. Meng, Q., Zhao, Y.: Curvature Interference Characteristic of ZC1 Worm Gear. In: Goldfarb, V., Trubachev, E., Barmina, N. (eds.) New Approaches to Gear Design and Production. MMS, vol. 81, pp. 513–529. Springer, Cham (2020). https://doi.org/10.1007/978-3-030-34945-5_24
8. Kristopher, T.: Differential Geometry of Curves and Surfaces. Springer International Publishing, Switzerland (2016)
9. Zhao, Y., Zhang, Y.: Computing method for induced curvature parameters based on normal vector of instantaneous contact line and its application to Hindley worm pair. Adv. Mech. Eng. **9**, 1–15 (2017)
10. Zhao, Y., Kong, X.: Meshing principle of conical surface enveloping spiroid drive. Mech. Mach. Theory **123**, 1–26 (2018)

Equivalent Contact Length of Load Disks and Specimen

Jiří Ondrášek(✉)

1VÚTS, a.s., Liberec, Czech Republic
jiri.ondrasek@vuts.cz

Abstract. The paper deals with a method of performing experimental tests to test the service life of a general kinematic pair of cam mechanisms with a roller follower. In the case of the test rig, the general kinematic pair is formed by the contact surface of a cylindrical specimen in contact with three load disks. The specimen represents the loaded cam and the disks a roller follower. The disk profile shape has a significant effect on the stress distribution in the surface layers of specimens due to their load. The disk profile is expressed by an equivalent width, which is determined by the finite element method.

Keywords: Cam mechanism · General kinematic pair · Test rig · Contact stress · Disk shape · Equivalent contact length

1 Introduction

Conventional cam mechanisms are characterized by the ability to transfer high powers at high speed and positional accuracy; their applicability is mainly connected with the so-called hard automation that is characterized by unchangeable or difficult to change actions of the given technical equipment. A cam mechanism is a three-link mechanical system with one degree of freedom that contains at least one cam connected to the other links by at least one general kinematic pair (GKP). Through the working surface of the cam (driving link), the motion of the driven link via GKP is derived. The driven link is called a follower. The follower performs a translating, oscillating or general motion. The general kinematic pair is formed by the contact of the cam and the follower.

The force ratios in the general kinematic pair formed by the contact of the cam and the follower cause the contact stresses. The main stresses that have the character of pulses with a period of 2π depending on the cam angular displacement ψ characterize the state of the stress on the cam working surface and thereunder. These are transient compressive stresses. In the surface layers of the material stressed in this way, variable elastic or elastoplastic strains occur. Exceeding a certain limit of these stresses can cause fatigue damage of the working surfaces of the cam and the follower after a certain number of cycles of the cam mechanism operation. This damage (pitting) results in the form of pits that develop from the cracks on the working surface.

In general, the fatigue is caused by progressive and localized structural damage which occurs in the material under cyclic loading. The maximum load stress values σ_{red} do not

© The Author(s), under exclusive license to Springer Nature Switzerland AG 2022
J. Beran et al. (Eds.): TMM 2020, MMS 85, pp. 62–69, 2022.
https://doi.org/10.1007/978-3-030-83594-1_7

reach the ultimate tensile strength σ_U and may be below the tensile yield strength σ_Y of the material. The maximum load stress values have to be limited by the tension-loaded material fatigue strength σ_C. The fact that during the operation of the cam mechanisms, no destructive action of elastic deformation occurs in the general kinematic pair, it is described by the conditional inequality [1]

$$\max(\sigma_{red}(\psi, z)) < \sigma_Y, \psi \in \langle 0, 2\pi), z \geq 0 \tag{1}$$

where variable z expresses the depth under the loaded surface. The conditional inequality Eq. (1) is valid for steel materials. Thus, the mechanical properties of the material of a cam or follower working surface have a significant effect on the occurrence of fatigue damage. Equally important is the information describing the heat or chemical-heat treatment and its effect on the properties of the internal structure of the material. The material of bodies in contact must be of high quality, any imperfection or inclusion acts as a crack initiation site. The surface of the material should be as smooth as possible, without unevenness, to prevent the propagation of cracks from the body surface. Lubrication has a significant effect on the contact fatigue as well.

2 Experimental Method

The aim of the experiments is to achieve the fatigue damage of the specimen contact surface after a certain number of cycles depending on the magnitude of the loading force, specimen material, lubrication of contact surfaces, type of lubricant used, type of coatings being applied to the specimen contact surface, disk crown profile etc. Experiments can take place: when the contact surfaces are mutually rolling or at a certain value of the slippage of the contact surfaces. In this way, we can obtain complete lifetime curves for a certain material with specific properties and then we can use them in designing other cam mechanisms.

2.1 Test Rig

Testing can be carried out on a device which allows a simulation of the state on a radial cam or an axial cam with a roller follower. As shown in Fig. 1, the cylindrical test specimen 5 of a defined width is placed on the shaft between three disks, one of which is a pressure disk 4 and the other two are fixed (2 and 3). The disks are arranged on a common plane at the vertices of an equilateral triangle with the length of its sides s. The axes of rotation of the disks are parallel. Pressure disk 4 is rotationally mounted on one of the arms of the pivotally mounted two-arm lever 6 with the arm length l. The two-arm lever is coupled through the other arm to the application place of the pressing force N of a defined magnitude. In terms of the geometric arrangement of the loading disks and the loaded specimen, the specimen is loaded by three equal reactions N, individually from each disk. Due to the rotation of the tested body 5 between three disks rotating with the same angular velocity Ω_2, the pressing force N has the character of pulses with the period of $2\pi/3$. It is therefore a transient loading. Due to the contact at three locations, the testing time is reduced to 1/3, which greatly simplifies testing as each specimen is tested for 10^7 to 10^8 cycles. In analogy with the cam mechanism with a cam and a roller follower, loading bodies 2, 3 and 4 represent a cylindrical roller and the loaded body 5 corresponds to the cam.

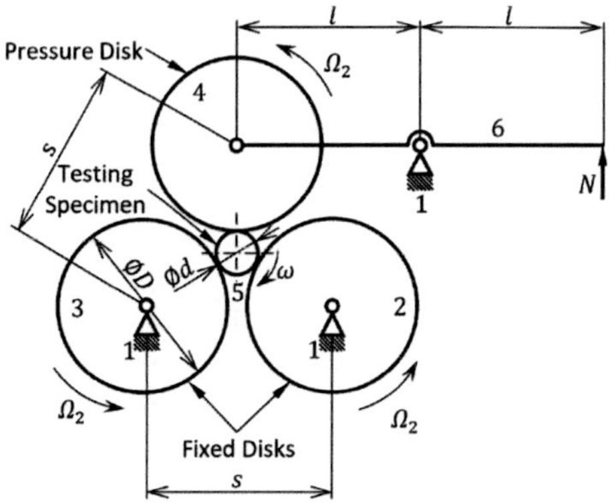

Fig. 1. Kinematic scheme of the conceptual design of a test rig.

2.2 Shape of Surfaces in Contact

When examining the factors that influence the contact fatigue, it was published [2] that the shape of the surfaces in contact has a significant effect. Due to the nature of the course of tests, it is advisable that the contact stress is distributed as evenly as possible for various load levels.

The disk profile has a significant effect on the stress distribution due to the load in the surface layers of specimens. In the case of a conventional cylindrical profile of the disks, there are discontinuities at the intersections of their cylindrical profile with the cylindrical profile of the specimen, i.e. if one contact part is axially shorter than the other, and also when the edges of the disks are chamfered, see Fig. 2. The mentioned discontinuities cause a very sharp increase in pressure distribution in the respective contact area of the bodies. These local increases in the pressure distribution can exceed the ultimate strength of the given material and thus cause plastic deformations, residual stresses in the material or change in the steel structure. Furthermore, the region concerned will be more prone to fatigue damage of the contact surfaces, i.e. to pitting or spalling. In the case of the cylindrical disk, the contact between the disk and the specimen cannot be considered as simply straight, but as a more complex three-dimensional type of contact.

The excessive edge stress in the case of the cylindrical disks can be reduced via such a shape of the axial cross-section of the disk crown that includes a straight line and one circular arc or a combination of several circular arcs, see Fig. 2. Such a shape of the disk crown one segment of which is cylindrical and the adjoining one is convex, causes a certain stress concentration during the transfer from the cylindrical region to the convex one. The radius of curvature ρ_y characterizes the convex segments in the plane Oyz. The size of the radius of curvature ρ_y is the subject of optimization of the disk profile in order that the distribution of the contact stress generated by the effects of the force N was as evenly as possible. This disk type was chosen for the test rig because of the acceptable technological complexity of its manufacturing in desired quality and accuracy.

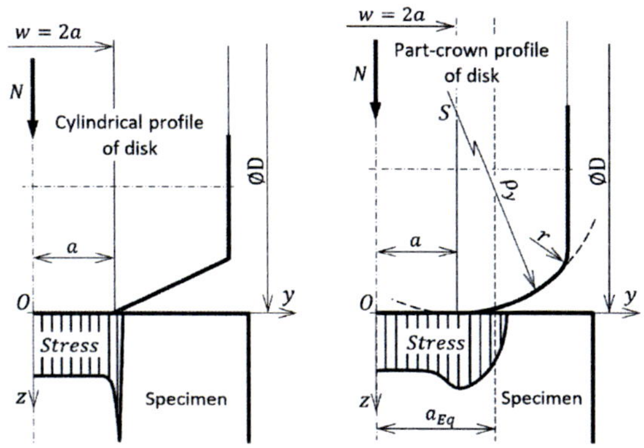

Fig. 2. Schematic drawing of the disk profiles.

2.3 Contact of Two Cylindrical Bodies

In order to calculate the contact stresses distribution in the contact regions of the load disks and the specimen, it is possible to use both the results of the contact mechanics [3] for the respective case of the contact of two elastic bodies and the finite element method [4, 5]. Hertzian contact stress theory [3] is applied to the contact of cylindrical bodies with parallel axes, see Fig. 3. Using this theory, it is possible to calculate the strain and stress components in both bodies in the contact surface and its vicinity.

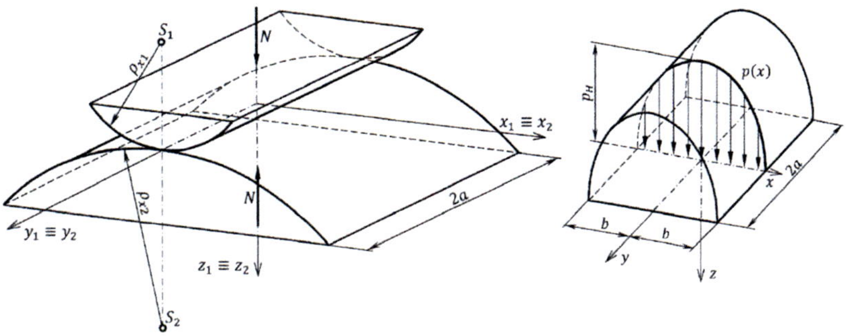

Fig. 3. Contact of two cylindrical bodies.

When two cylindrical bodies are brought into contact, they touch initially along a line. Under the action of a slightest load N, they will deform and contact is made over a finite area of length $2a$ and width $2b$ which is relatively small, compared with the dimensions of both bodies. The coordinate system $Oxyz$ is located in its center. The contact pressure $p(x)$ is distributed according to the elliptical cylinder and its maximum value is called Hertzian pressure, which is derived based on Hertzian contact theory [3]

$$p_H = \sqrt{\frac{NE^*}{\pi 2aR_e}}, \frac{1}{E^*} = \frac{1-\nu_1^2}{E_1} + \frac{1-\nu_2^2}{E_2}, \frac{1}{R_e} = \frac{\text{sign}(\rho_{x2})}{\rho_{x1}} + \frac{1}{|\rho_{x2}|} = \frac{2(D+d)}{Dd} \quad (2)$$

E^* is the effective modulus of elasticity, E_i and ν_i are the respective Young's modulus of elasticity and Poisson's ratio of the individual solids. The geometry of the two contacting cylinders is described with the equivalent radius of curvature R_e where $\rho_{x1} = D/2$ denotes the radius of the loading disks and $\rho_{x2} = d/2$ is the radius of the specimen. The half-width b of the contact area is determined as [3]

$$b^2 = 4f\,R_e/(\pi\,E^*),\ f = N/(2a) \tag{3}$$

The contact stress state existing on the symmetry plane yz is determined by the actual main stress components $\sigma_x, \sigma_y, \sigma_z$. These quantities are assumed as compressive stresses and their absolute value decreases in proportion to a distance z from the contact area. The stress components are given by the expressions [3]

$$\sigma_x = -p_H\left(\frac{1+2\xi}{\sqrt{1+\xi^2}} - 2\xi\right), \sigma_y = -2\nu p_H\left(\sqrt{1+\xi^2} - \xi\right), \sigma_z = -p_H\frac{1}{\sqrt{1+\xi^2}} \tag{4}$$

where the proportional independent variable $\xi = z/b$ is introduced. The stress distribution in the planar interface is given by the relations [3]

$$\sigma_x = \sigma_z = -p_H\sqrt{1-(x/b)^2},\ \sigma_y = -2\nu p_H\sqrt{1-(x/b)^2} \tag{5}$$

The Tresca yield criterion can be used to find the reduced stress σ_{red}, according to which this theory can be considered in terms of the maximum sliding stress, proportional to the difference of the main stresses [6]

$$\sigma_{red} = \max\{\,|\sigma_x - \sigma_y|,\ |\sigma_z - \sigma_x|,\ |\sigma_z - \sigma_y|\}\,,\ \max(\sigma_{red}) = 0.6p_H \tag{6}$$

When the cylindrical bodies come into contact, the largest value of the reduced stress $\max(\sigma_{red})$ is at a given depth z_e below the surface and reaches the size according to Eq. (6) [1]. Based on Eqs. (1), (2) and (6), we can formulate a condition for determining the limit value of the amplitude of the loading force N so that plastic deformations will not occur under the contact surface during the operation

$$N \le \frac{\pi 2aR_e}{E^*}\left(\frac{K_R\sigma_U}{0.6}\right)^2,\ K_R = \frac{\sigma_Y}{\sigma_U} \in \langle 0.55,\ 0.8\rangle \tag{7}$$

Constant K_R expresses the ratio of the yield tensile strength $\sigma_Y R_{p0.2}$ just to the ultimate strength σ_U.

2.4 Equivalent Contact Length of Load Disks and Specimen

In order to determine the shape of the contact area of the load disks and the specimen, the components of strains and stresses in both bodies around the contact area, it is necessary due to the above given shape of the disks profile to determine equivalent contact length $2a_{Eq}$ in contact of the load disks with the specimen. This equivalent length $2a_{Eq}$ will be greater than the width w of the cylindrical part of the disk $2a_{Eq} > w$ and will depend on the size of the loading force N. In accordance with [4, 5], FEM models of the disk and specimen contact have been created for a set of loading forces $N = \{0.25,0.50,1,2,3,4,6,8,10\}\cdot 10^3 N$ and $w = 3$ mm. By analyzing the results of the distribution of the contact stress and the contact pressure listed in Table 1, we determined the computational relation for the estimation of the equivalent length $2a_{Eq}$ of the contact surface in the contact of the load disks and the specimen

$$N(w = 3\,\mathrm{mm}) = 3532.7\left(2a_{Eq}\right)^3 - 36235\left(2a_{Eq}\right)^2 + 124108\left(2a_{Eq}\right) - 141594 \quad (8)$$

assuming that both parts are made of steel. In Table 1, the variables $\bar{\sigma}_{red}$ and $\bar{p}_H$ represent respectively the mean value from the course of the largest values of the reduced contact stress $\max(\sigma_{red})$ in the plane of symmetry Oxz and from the course of the values of the Hertz pressure p_{H} on the contact surface Oxy. From Eq. (8), e.g. by Newton's numerical method, we can determine the appropriate length $2a_{Eq}$ for a given load N and a set of load disks. Examples of the distribution of the reduced stress in the plane of symmetry Oxz and the contact pressure in the contact plane Oxy are shown in Fig. 4 to Fig. 7.

Table 1. Contact stress for the width of the cylindrical part of the disk $w = 3$ mm.

N [N]	$\bar{\sigma}_{red}$[MPa]	$\bar{p}_H$[MPa]	$2a_{Eq}$ [mm]
250	320.00	570.00	3.22
500	430.00	780.00	3.73
1000	600.00	1050.00	3.95
2000	800.00	1400.00	4.17
3000	950.00	1700.00	4.30
4000	1100.00	1950.00	4.41
6000	1300.00	2300.00	4.57
8000	1500.00	2600.00	4.70
10000	1650.00	2900.00	4.80

We can then use the values of the variable a_{Eq} in the relations for the calculation of the main stresses, the geometry of the two contacting cylinders, the contact pressure and the limit value of the amplitude of the loading force N according to the Eqs. (2), (3), (4) and (7).

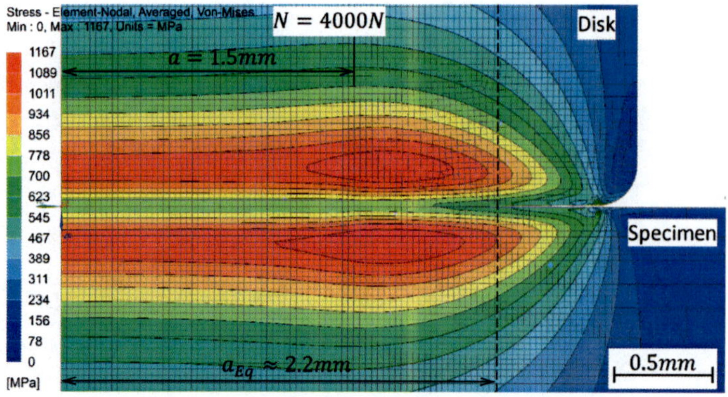

Fig. 4. Reduced stress.

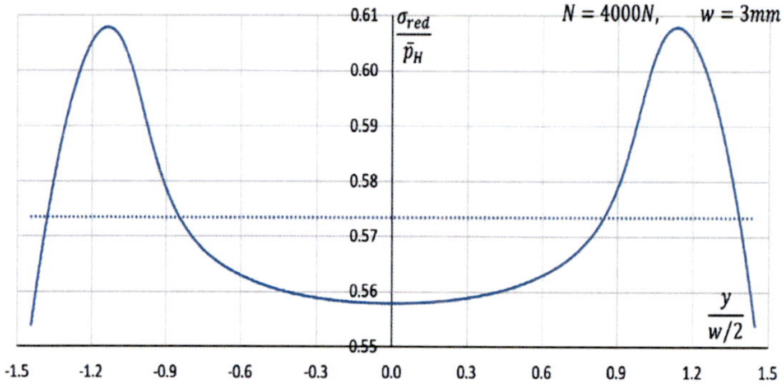

Fig. 5. Relative expression of reduced stress.

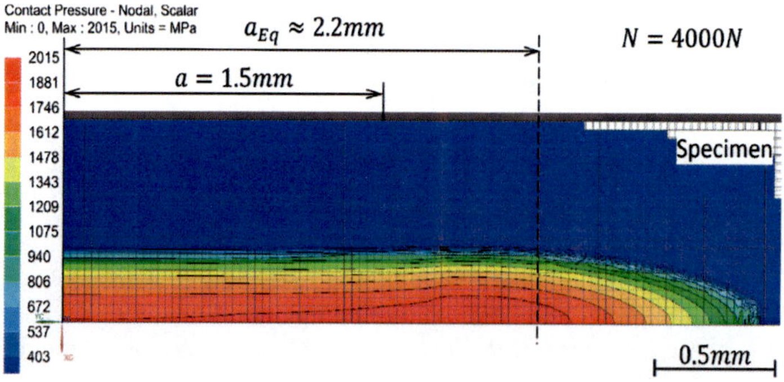

Fig. 6. Contact pressure.

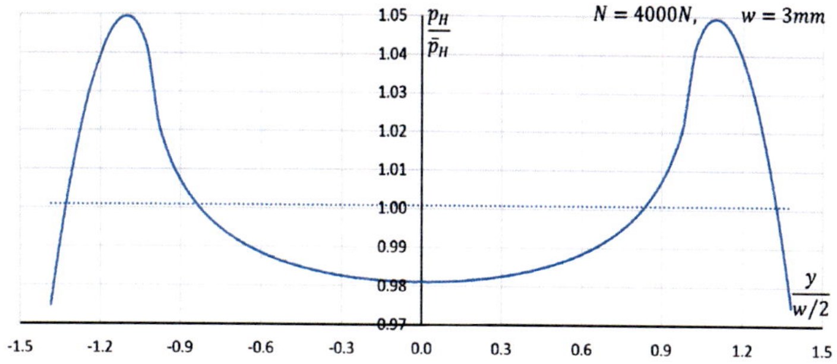

Fig. 7. Relative expression of contact pressure.

3 Conclusion

The paper presents a method for evaluating the contact stress of a cylindrical sample, which is loaded in a test device with three rotating disks. In order to distribute the contact stress as evenly as possible, the loading disks are provided with a part-crown shape of the profile. In analogy with a cam mechanism with a roller follower, the loaded specimen corresponds to the cam and the loading disks represent a cylindrical roller. In this case, the disk profile is expressed by an equivalent width, which is determined based on the finite element method. We can then use the value of the equivalent width in the formulas to evaluate the components of the contact stresses and the strains of cylindrical bodies with the parallel axes according to the methods of contact mechanics.

References

1. Koloc, Z., Václavík, M.: Cam mechanisms, 1st edn. Elsevier, Amsterdam (1993)
2. Norton, R.L.: Cam design and manufacturing handbook. 2nd ed. Industrial Press, Inc., New York (2009). ISBN 978–0–8311–3367–2.
3. Johnson, K., L.: Contact mechanics. 1st edn. Cambridge University Press, Cambridge (1985). ISBN 0 521 34796 3.
4. Ondrášek, J.: Effect of the roller crown shape on the cam stress. In: Corves, B., Wenger, P., Hüsing, M. (eds.) EuCoMeS 2018. MMS, vol. 59, pp. 223–230. Springer, Cham (2019). https://doi.org/10.1007/978-3-319-98020-1_26
5. Jiří, O. The general kinematic pair of a cam mechanism. In: Joseph, M. (ed.) Kinematics - analysis and applications, IntechOpen, (2019). https://doi.org/10.5772/intechopen.86682. https://www.intechopen.com/books/kinematics-analysis-and-applications/the-general-kinematic-pair-of-a-cam-mechanism.
6. Vable, M.: Mechanics of materials, 2nd ed., Section 10.3, Failure theories, pp. 486–492, MTU. http://www.me.mtu.edu/~mavable/MoM2nd.html.

Dynamics of Machines and Mechanisms, Computational Mechanics, Rotor Dynamics, Biomechanics, Vibration and Noise in Machines

Vibration Reduction of a Steam Turbine Wheel by Means of Intentional Mistuning

Bernd Beirow[1](✉), Mark Golze[1], and Frederik Popig[2]

[1] Brandenburg University of Technology, Cottbus, Germany
beirow@b-tu.de
[2] Siemens Energy Global GmbH & Co. KG, Görlitz, Germany

Abstract. A last stage steam turbine wheel is analyzed with the objective to alleviate the flutter susceptibility by employing intentional mistuning (IM). In particular, the operation at nominal speed under part-load conditions may cause unfavorable flow conditions facilitating flow separation. In consequence, negative aerodynamic damping ratios occur for the first bending mode family in some circumstances. Employing intended alternate mistuning of adequate magnitude has proved to be a promising measure to stabilize rotors in terms of avoiding self-excited vibration phenomena. From the manufacturing point of view, this two-blade design is advantageous as well and hence, chosen here as a first measure to attenuate flutter susceptibility. Two prototypes of bladed disks series have been made, which are exhibiting small but unavoidable deviations from the design intention due to manufacturing. The real blade alone frequencies have been identified within foregoing experimental investigations. Numerical modal analyses carried out for the prototypes as manufactured finally reveal that there is an additional positive contribution of random mistuning in terms of further enhancing the least aerodynamic damping ratio. Another promising and robust IM pattern is found by using generic algorithms to optimize the least aerodynamic damping ratio yielding stable conditions at any time as well. Moreover, it shows that IM combined with random mistuning also mitigates the maximum forced response at part-speed conditions.

Keywords: Intentional mistuning · Blade vibration · Flutter · Forced response · Optimization · Aeroelastics

1 Introduction

Commonly, bladed disks and blade integrated disks of turbine and compressor applications are designed as cyclic symmetric structures featuring identical blades. However, actually the wheels are exhibiting small variations from blade-to-blade, and consequently different mechanical properties of each blade. This characteristic is known as mistuning, which typically results from manufacturing, wear, damage, repair measures or even strain gauge instrumentation. On the one hand, mistuning has a beneficial effect on flutter suppression [1]. On the other hand, unfortunately, unavoidable random mistuning may cause a severe magnification of the maximum forced response compared to the tuned counterpart with identical blades. Already more than 50 years ago, Whitehead

© The Author(s), under exclusive license to Springer Nature Switzerland AG 2022
J. Beran et al. (Eds.): TMM 2020, MMS 85, pp. 73–82, 2022.
https://doi.org/10.1007/978-3-030-83594-1_8

[2] or Ewins [3] contributed trendsetting publications addressing this issue. Until today hundreds of papers dealing with the mistuning phenomenon have been published, which indicates the significance of the problem.

Over the years, it was found that the application of intentional mistuning (IM) is well suited to mitigate the susceptibility of the forced response towards random mistuning. Castanier and Pierre [4] conducted comprehensive structural analyses with a focus on harmonic IM patterns without considering aerodynamic coupling for both academic blisks and an industrial blisk with 29 blades. They proved a clearly enhanced robustness towards large forced response amplitudes for both cases. Lim et al. [5] continued this work for the industrial blisk by considering other set-ups of IM patterns, each effecting a mitigation of the forced response compared to the blisk without IM. Han et al. [6] proved that optimized IM patterns composed by only two different blade types A and B are also suitable to attenuate the susceptibility of the maximum forced response to random mistuning.

Novel strategies target on increasing the aerodynamic damping by employing IM in order to decrease the maximum forced response even below that of the reference design with identical blades. According to the publications of Petrov [7] and Schoenenborn et al. [8] favorable conditions are required with respect to the dependence of aerodynamic damping on the inter blade phase angle and the engine order (EO) to be considered. Martel and Sánchez-Álvarez [9] have presented a successful forced response reduction by means of IM superimposed with random mistuning for different academic blisks. Furthermore, they introduced a theoretical limit of the achievable forced response reduction depending on the exciting EO, the dedicated aerodynamic damping and the mean value of aerodynamic damping. The practical implementation of an optimized IM pattern is addressed in [10] for a turbine blisk by means of varying fillet radii, which affects a 70% reduction of the maximum forced response caused by low engine order excitation of the fundamental bending mode. Figaschewsky et al. [11] demonstrated another practical application of IM for a blisk fan. Both forced response caused by low engine order excitation and flutter susceptibility of the first blade mode could be reduced crucially.

Among others, Miyakozawa [1] addressed the increasing effect of IM for stabilizing aerodynamically unstable rotors. The same applies to the work of Kaza and Kielb [12] who carried out aeroelastic analyses for a high-aspect ratio turbofan, which demonstrated that alternate mistuning has the potential to alleviate flutter problems. Equally, Srinivasan [13] exemplarily showed that an originally tuned but unstable rotor in terms of torsional flutter is stabilized due to the presence of mistuning. The case study of Kielb et al. [14] also demonstrates the positive effect of mistuning on flutter suppression, which becomes more effective with increasing mistuning magnitude. Martel et al. [15] employed their well-established asymptotic mistuning model to enhance the stability of two aerodynamically unstable low-pressure turbine rotors by means of optimized intentional mistuning patterns. Considering a tip-shrouded bladed disk, Corral et al. [16] presented the experimental evidence of how alternate mistuning is applicable to mitigate or even completely remove flutter instabilities.

The present paper focuses on stability analyses of the last stage low-pressure steam turbine wheel highlighted in Fig. 1 with respect to flutter suppression aided by IM. The academic case study makes use of a modally reduced structural model introduced by Yang and Griffin [17], which is known as subset of nominal system modes (SNM). The SNM easily allows for considering both frequency mistuning and aeroelastic interaction. Alternate IM is chosen first with the objective to stabilize the originally unstable wheel for the first bending mode (1B) at nominal speed and part-load conditions. The stability is assessed based on occurring negative aerodynamic damping ratios and indeed, it becomes apparent that the flutter susceptibility could be enhanced in terms of a satisfying and robust solution from the engineering point of view, i.e. even the least aerodynamic damping ratio takes a positive value. Basically, further improvements are achievable by employing an IM pattern optimized by means of genetic algorithms even if still only two different blade designs are admitted in order to keep the manufactural efforts small. However, the solution found turns out to be less robust compared to alternate IM. Further analyses are addressing the impact of mistuning magnitude. Beyond that, complementary analyses prove that IM superimposed with random mistuning also has a positive impact on the maximum forced response compared to the reference design with identical blades.

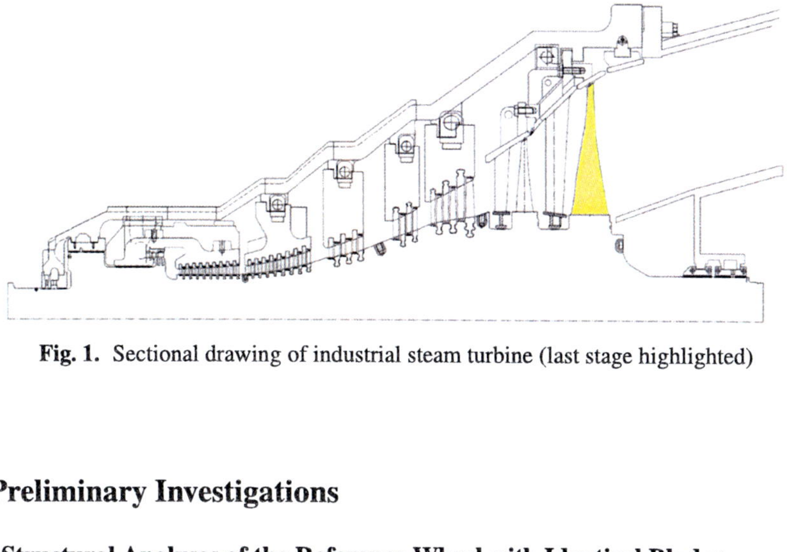

Fig. 1. Sectional drawing of industrial steam turbine (last stage highlighted)

2 Preliminary Investigations

2.1 Structural Analyses of the Reference Wheel with Identical Blades

Both identifying modal characteristics of the turbine wheel with 51 blades and preparing input data for the following reduced order modelling require to set up a finite element (FE) sector model, which features 43.000 tetrahedral elements. As long as mode shapes are dominated by blade motion and the strain energy is concentrated in the blades, the modes are grouped in blade mode families. The first three of them (Fig. 2, right) are named conformable to their appearance with first bending mode (1B), second bending mode (2B) and first torsional mode (1T). Focusing on the first bending mode, the sector mode shapes of the 1B-family are used to set up SNM-models with only 51 degrees of freedom, which are consequently only valid for the frequency range of the 1B. Three different models are created due to the speed dependence of natural frequencies revealed in the Campbell plot of Fig. 2, namely for 100% speed, where self-excited vibration may occur at part-load conditions, and for 30% and 50% speed, where resonance crossings may cause large forced response.

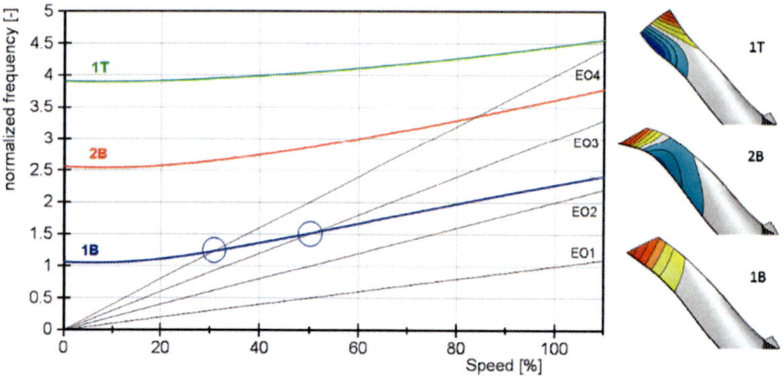

Fig. 2. Campbell diagram involving the first three blade modes

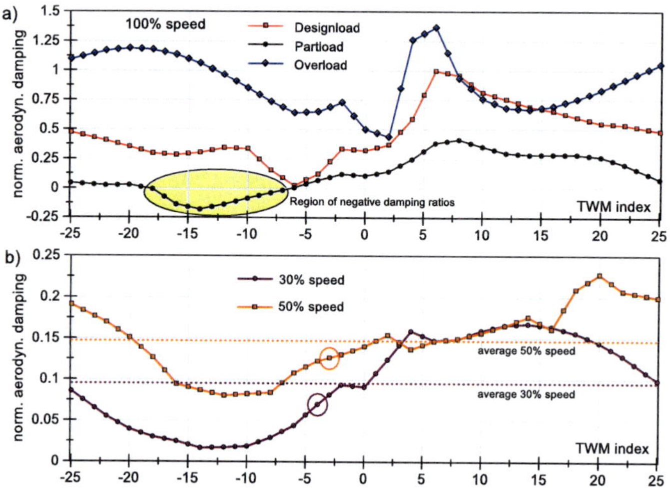

Fig. 3. Travelling wave mode dependent aerodynamic damping a) 100% speed (nominal), and b) part-speed (30% and 50%)

2.2 Aeroelastic Simulation

Starting with 2D computations of the whole steam turbine, steady-state CFD analyses are carried out to identify the flow boundary conditions for the second step, 3D computations of the last two turbine stages only. Circumferentially periodic boundary conditions and the two-equation k-ω-SST turbulence model [18] are chosen for last-named analyses. Fillets and tip gaps are neglected to keep the simulation as simple as possible. Results are calculated at nominal speed (100%) for three different pairs of condenser pressure and exhaust velocity: (i) part load, (ii) design load, and (iii) over load. In contrast, only one relevant load condition each is considered for 30% and 50% part-speed conditions.

Aeroelastic eigenvalues of the last stage are determined by means of single passage computations using the results of steady state computations to define both initial conditions and boundary conditions. Structural natural frequencies and mode shapes are carried over from FE-analyses. A time-linearized flow solver is used to provide aerodynamic damping values and aeroelastic natural frequencies. Since natural frequencies of the 1B blade mode family hardly change due to the impact of aeroelastic interaction, the following discussions are focused on travelling wave mode (TWM) dependent aerodynamic damping ratios. According to the yellow highlighted region in Fig. 3a, negative damping ratios are occurring at part load conditions and 100% speed for TWM between −18 and −7, so that unstable conditions may take place for this operating point. Increasing condenser pressure and exhaust velocity take a positive effect, nevertheless the least damping ratio at design load is approaching the stability margin whereas overload conditions prove to be always stable. As previously announced IM is employed to ensure safe operation at any load condition and time.

Since IM will affect the forced response at part-speed conditions as well, additional simulations are necessary for speeds close to resonance crossings highlighted in Figs. 2 and 3b (1B excitation in EO 3 and 4). Comparatively low aerodynamic damping ratios occur for negative or backwards travelling waves, respectively. Consequently, the damping ratios dedicated to regularly excited TWMs −3 and −4 are located beneath the average damping lines so that IM offers the opportunity to increase the resulting aerodynamic damping by coupling in higher damped TWMs [9]. Hence, the commonly response magnifying impact of mistuning can be stemmed or even inverted [10].

2.3 Speed Dependence of Structural Mistuning

The manufacturer has chosen alternate IM first since it has proved to be effective for flutter suppression in the past. That is why two different sets of blades have been produced, which are arranged in IM patterns denoted as A and B. Foregoing bonk tests have shown, that these patterns feature small deviations from the design intention (Fig. 4). However, the blade-to-blade frequency differences change due to the speed dependence of natural frequencies. In consequence, a speed-dependent correction factor $c(\Omega)$ is multiplied to relative mistuning patterns $\Delta f_i(0)$ determined at rest:

$$\Delta f_i(\Omega) \quad = c(\Omega) \cdot \Delta f_i(0) \tag{1}$$

The determination of c(Ω) requires FE modal analyses of the wheel with alternate IM taking into account speed and stiffening effects. Next, SNM models are employed to adopt the mistuning magnitude as long natural frequencies of FEM and SNM are matching. This procedure is repeated for every speed of interest. Figure 5 reveals the tremendous impact of speed on alternate IM yielding a reduction of mistuning magnitude up to 63% at 100% speed compared to the resting wheel. As a side note, the popular correction suggested by Feiner [19] and by Nipkau [20] later on overrates the speed impact on mistuning in this particular case.

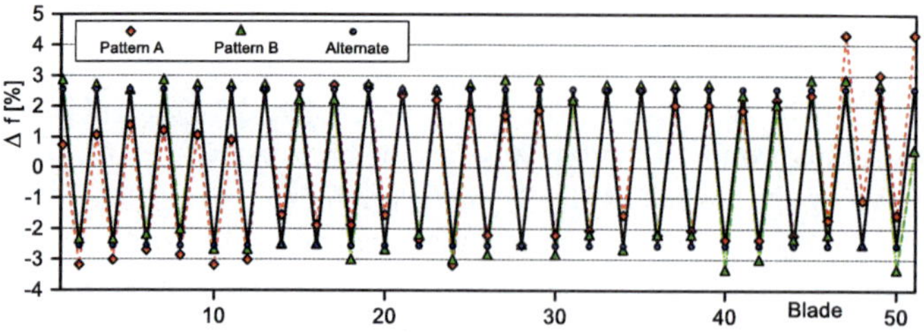

Fig. 4. Alternate frequency mistuning at rest: As manufactured (A and B) and original

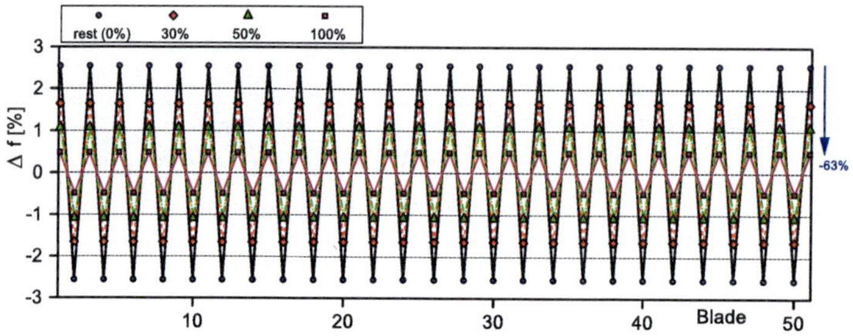

Fig. 5. Alternate frequency mistuning: Impact of speed

3 Stability Analyses

3.1 Alternate Intentional Mistuning

Previously shown aerodynamic damping curves (Fig. 3a) reveal that negative damping ratios become apparent at 100% speed under part load conditions. First, alternate IM is employed with the objective to stabilize the rotor at any load condition. The success of this measure is evaluated based on the least aerodynamic damping ratios, which are computed together with aeroelastic natural frequencies by solving the aeroelastic SNM eigenvalue problem. Starting with design load conditions (Fig. 6), a number of particularities become apparent:

(a) Pure alternate mistuning causes a splitting of the tuned frequency group (squared symbols) into two groups, one located at greater frequencies and one located at lower frequencies (triangles).
(b) Additional random mistuning as in case of patterns A and B affects first, a widening of the frequency range of both groups, and second, that the co-domain of aerodynamic damping ratios is significantly narrowed.
(c) In general, mistuning causes an increase of the least aerodynamic damping ratio, which is highlighted by circle marks.
(d) Alternate IM superimposed with random mistuning due to manufacturing (patterns A and B) yields an insignificantly greater increase of the least aerodynamic damping ratio compared to pure alternate mistuning.

Moving on to part load conditions (Fig. 7), the observations made before prove to be true. Moreover, alternating mistuning increases the least aerodynamic damping ratio such that completely stable conditions are achieved i.e. the system does not feature negative damping ratios beneath the stability margin anymore. Additional random mistuning affects further stabilization as indicated by patterns A and B.

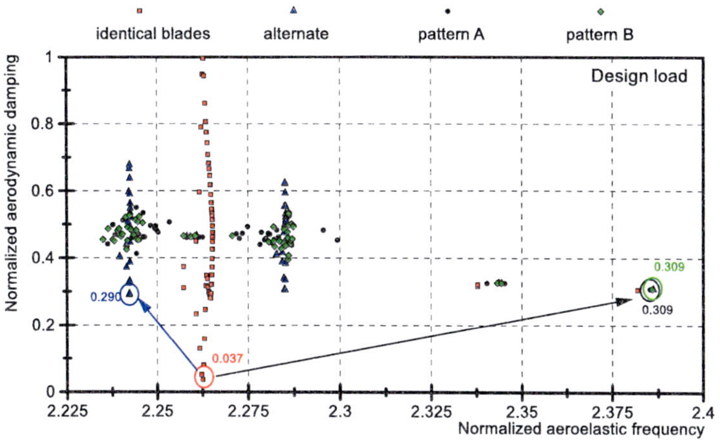

Fig. 6. Aeroelastic eigenvalues at design load conditions and 100% speed (Alternate IM)

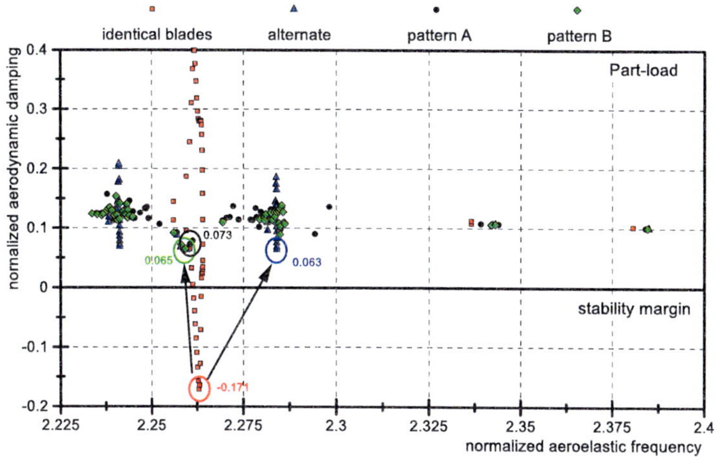

Fig. 7. Aeroelastic eigenvalues at part load conditions and 100% speed (Alternate IM)

3.2 Mistuning Magnitude, Optimized Mistuning and Robustness

The impact of both, IM magnitude and additional random mistuning on flutter suppression is analyzed focusing on part-load conditions. Starting with alternate IM, the solid red line in Fig. 8a reveals that positive damping ratios or stable conditions, respectively, are obtained for frequency mistuning standard deviations ranging from $\Delta f = 0.18 \ldots 5\%$. If additional random mistuning of $\Delta f = \pm 1\%$ is superimposed in terms of Monte Carlo analyses (40000 samples per standard deviation) the solutions remain always stable with respect to the 1% percentile between 0.42 and 5% standard deviations. If the standard deviation is greater than 1% a positive impact of additional random mistuning becomes apparent. The choice of block by block IM with two, three or four blades per package do not improve flutter susceptibility compared to alternate IM. The three (AAABBB…) or four blades per package solutions even yield unstable conditions in terms of negative least aerodynamic damping ratios at any time. Figure 8b shows, that a pure optimized IM pattern (solid red line) could further increase the least aerodynamic damping ratio compared to pure alternate IM. However, again superimposing 1% random mistuning reveals that the 1% percentiles are clearly less than those of alternate IM. Nevertheless, the least aerodynamic damping ratio never falls below the stability margin if the basic standard deviation of optimized IM is greater than 0.43%.

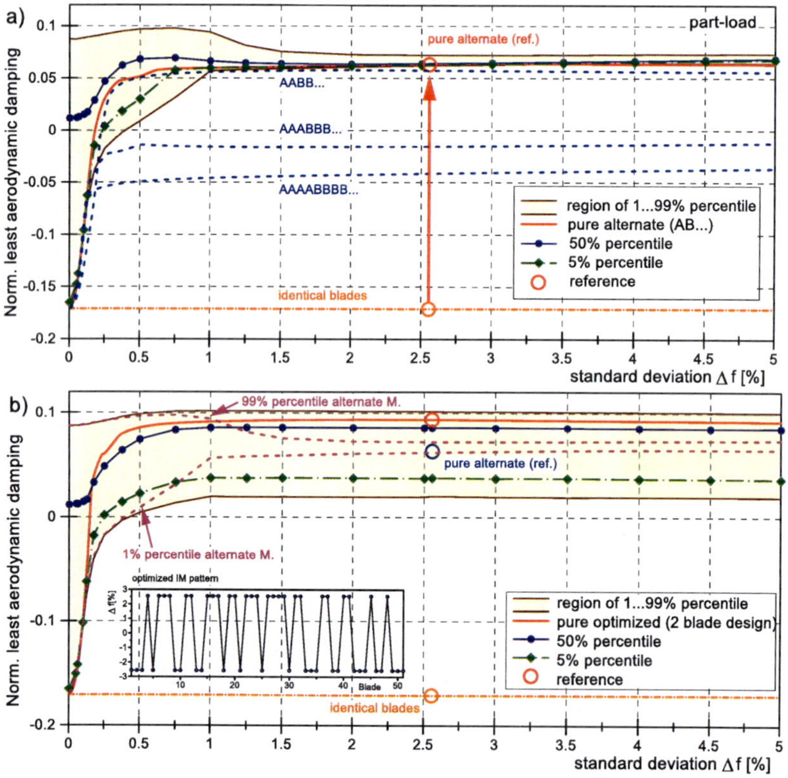

Fig. 8. Impact of mistuning magnitude superimposed by $\Delta f = \pm 1\%$ random mistuning on least aerodynamic damping, a) alternate IM, and b) optimized IM.

4 Forced Response

Since two resonance crossings occur at part speed conditions (Fig. 2, engine orders 3 and 4) forced response analyses are carried out considering the various IM patterns. Again, the impact of random mistuning is taken into account in terms of probabilistic analyses. The results given in Table 1 illustrate first that the tuned reference with nominally identical blades suffers a maximum displacement amplification (99% percentile) of up to 115.9% (EO 3) due to the impact of small random mistuning. IM patterns superimposed with random mistuning exhibit smaller magnification factors compared to the tuned design intention. In particular, alternate IM give favorable results with tremendously lesser magnification factors (47.6% at EO3). Hence, the IM approach has positive effect on both, flutter susceptibility and maximum forced response.

Table 1. 50% and 99% percentiles of maximum displacement magnification superimposed with $\Delta f = 0.5\%$ random mistuning (10000 samples)

Case	EO 3 ($\gamma_{50\%}$)	EO 3 ($\gamma_{99\%}$)	EO 4 ($\gamma_{50\%}$)	EO 4 ($\gamma_{99\%}$)
Identical blades	1.631	2.159	1.276	1.869
Alternate IM	1.212	1.476	1.001	1.210
Optimized IM	1.332	1.644	1.061	1.322

5 Conclusions

The case study could show that intentional mistuning alleviates both flutter susceptibility and maximum forced response of a last stage turbine wheel. The SNM-based reduced order models used for this purpose allow for considering aeroelastic interaction and speed dependence of frequency mistuning in a simple manner. Focusing on the fundamental bending mode and IM patterns composed of only two different blade designs, the impact of superimposed and unpreventable random mistuning has been taken into account, whereby the robustness of the approach has been proved. Both, an alternate IM pattern and an IM pattern designed by means of optimization based on genetic algorithms effectuate the most satisfying suppression of flutter from the engineering point of view. Stable conditions are defined by completely positive aerodynamic damping ratios within the eigenvalue solutions. In this regard, always stable conditions are computed by applying these IM patterns. Simultaneously, the impact of random mistuning on the maximum forced response at part speed proves to be less negative compared to the tuned counterpart with nominally identical blades. In consequence, the danger of getting severe magnifications of the maximum forced response is reduced as well, in particular alternate IM yields the most promising results in this regard.

6 Permission of Use

The content of this paper is copyrighted by Siemens Energy, Inc., and is licensed to Springer for publication and distribution only. Any inquiries regarding permission to use the content of this paper, in whole or in part, for any purpose must be addressed to Siemens Energy, Inc., directly.

References

1. Miyakozawa, T.: Flutter and Forced Response of Turbomachinery with Frequency Mistuning and Aerodynamic Asymmetry. PhD-Thesis, Duke University, Durham (2008)
2. Whitehead, D.S.: Effect of mistuning on the vibration of turbomachine blades induced by Wakes. J. Mech. Eng. Sci. **8**, 15–21 (1966)
3. Ewins, D.J.: The effects of detuning upon the forced vibrations of bladed disks. J. Sound Vib. **9**, 65–79 (1969)
4. Castanier, M.P., Pierre, P.: Using intentional mistuning in the design of turbomachinery rotors. AIAA J. **40**(10), 2077–2086 (2002)
5. Lim, S.-H., Castanier, M.P., Pierre, C.: Intentional mistuning design space reduction based on vibration energy flow in bladed disks. ASME-Paper GT2004-53873 (2004)
6. Han, Y., Murthy, R., Mignolet, M.P., Lentz, J.: Optimization of intentional mistuning patterns for the mitigation of effects of random mistuning. J. Eng. Gas Turb. Power **136**(6), 1–9 (2014)
7. Petrov, E.P.: A Method for Forced Response Analysis of Mistuned Bladed Disk with Aerodynamic Effects Included. ASME-Paper GT2009-59634, Orlando/FL, June 8–12, 2009
8. Schoenenborn, H., Junge, M., Retze, U.: Contribution to Free and Forced Vibration Analysis of an Intentionally Mistuned Blisk. ASME-Paper GT2012-68683 (2012)
9. Martel, C., Sánchez-Álvarez, J.J.: Intentional mistuning effect in the forced response of rotors with aerodynamic damping. J. Sound Vibr. **433**, 212–229 (2018)
10. Beirow, B., Figaschewsky, F., Kühhorn, A., Bornhorn, A.: Vibration analysis of an axial turbine blisk with optimized intentional mistuning pattern. J. Sound Vibr. **442**, 11–27 (2019)
11. Figaschewsky, F., Kühhorn, A., Beirow, B., Nipkau, J., Giersch, T., Power, B.: Design and Analysis of an Intentional Mistuning Experiment Reducing Flutter Susceptibility and Minimizing Forced Response of a Jet Engine Fan. ASME Paper GT2017-64621 (2017)
12. Kaza, K.R.V., Kielb, R.E.: Flutter of turbofan rotors with mistuned blades. AIAA J. **22**, 1618–1625 (1984)
13. Srinivasan, A.V.: Influence of Mistuning on Blade Torsional Flutter. Paper No. NASA CR-165137 (1980)
14. Kielb, R.E., Hall, K.C., Hong, E., Pai, S.S.: Probabilistic Flutter Analysis of a Mistuned Bladed Disk. ASME-Paper GT2006-90847, Barcelona, Spain, May 8–11, 2006
15. Martel, C., Corral, R., Llorens, J.M.: Stability increase of aerodynamically unstable rotors using intentional mistuning. J. Turbomach. **130**(1), 011006 (2008)
16. Corral R., Beloki J., Calza P., Elliott R.: Flutter Generation and Control Using Mistuning in a Turbine Rotating Rig. ASME-Paper GT2016-57949, Seoul, Korea, June 13–17, 2016
17. Yang, M.T., Griffin, J.H.: A reduced-order model of mistuning using a subset of nominal system modes. J. Eng. Gas Turb. Power **123**, 893–900 (2001)
18. Menter, F.R.: Two-equation eddy-viscosity turbulence models for engineering applications. AIAA-J. **32**(8), 269–289 (1994)
19. Feiner, D.: A fundamental model of mistuning for system identification and forced response prediction. PhD-Thesis, Pittsburgh (2003)
20. Nipkau, J.: Analysis of mistuned blisk vibrations using a surrogate lumped mass model with aerodynamic influences. PhD-Thesis at Brandenburg University of Technology Cottbus (2010)

Unsteady Resonant Oscillations of a Gyroscopic Rigid Rotor with Non-linear Damping and Non-linear Rigidity of the Elastic Support

Zharilkassin Iskakov[1](✉), Nutulla Jamalov[2], and Kuatbay Bissembayev[3]

[1] Institute of Mechanics and Engineering, Almaty University of Power Engineering and Telecommunications, Almaty, Kazakhstan

[2] Institute of Mechanics and Engineering, Al-Farabi Kazakh National University, Almaty, Kazakhstan

[3] Institute of Mechanics and Engineering, Abay Kazakh National Pedagogical University, Almaty, Kazakhstan

Abstract. The article is concerned with the effect of linear and cubic non-linear damping of an elastic bearing on forced resonant vibrations of a gyroscopic vertical rigid rotor taking into account non-linear stiffness of the cubic nature of the bearing material. It is confirmed that non-linear cubic damping of the support can suppress not only the maximum amplitude, but also the amplitudes of forced unsteady oscillations behind the rotation speed corresponding to the maximum amplitude and the variation of its values in time along the main curve, around its mean values. It shifts the speed of rotation of the amplitude maximum, with rigid and soft non-linear elastic characteristics of the support material downwards and upwards, respectively. It is shown that with a "slow" increase in the shaft rotation speed, an increase in the absolute value of the angular acceleration is accompanied by a shift of the amplitude peak towards high speeds, with a "slow" decrease in the shaft rotation speed – towards low speeds with a decrease in the amplitude of oscillations. It is shown that during the rotor takeoff run, the maximum amplitude for the case with a rigid non-linear elasticity characteristic of the support material is greater than the same value for the case with a soft non-linear elasticity characteristic of the support material, and conversely, during the rotor run-down for similar cases.

Keywords: Gyroscopic rotor · Non-linear rigidity · Non-linear damping · Unsteady oscillation

1 Introduction

The operating speeds of rotary machines can be above or between critical speeds. In the practice of rotary machines operation there were cases when the machines had unacceptably high vibrations during the transition through the critical speed(s). It is known that one of the main causes of shaft vibration is the inertial forces of unbalanced masses.

© The Author(s), under exclusive license to Springer Nature Switzerland AG 2022
J. Beran et al. (Eds.): TMM 2020, MMS 85, pp. 83–93, 2022.
https://doi.org/10.1007/978-3-030-83594-1_9

A simplified model with lumped parameters of the rotor system, as a rule, is used to study the dynamics of the shaft of one rotor on the bearing supports. It is very important to use properties and characteristics of the material of the supports for attenuation and damping of vibration in order to stabilize movement of an unbalanced rotor and vibration systems. Supports are the means of connecting the device between the rotor and the supporting structure, which have various shapes and designs, depending on specific assumptions.

A convenient way to introduce attenuation to support bearings in a rotor system on viscoelastic flexible rubber supports [1]. In parallel with the development of viscoelastic material modeling, which helps to describe the complexity of material properties, the use of viscoelastic components in the dynamics of the rotor and vibration systems also, increased as a whole, in particular with non-linear elastic characteristics and damping. So, for example, in works [2, 3] the influence of quadratic non-linear damping on resonant oscillations and stability of a gyroscopic rotor with quadratic or cubic non-linear stiffness of an elastic support was considered. Studies [4–8] show that linear and non-linear cubic damping can significantly suppress the resonant peak of the fundamental harmonic, eliminate the jump-like phenomena of the non-linear system. In non-resonance regions, where the vibration frequency is higher than its resonance value, non-linear cubic damping, unlike linear damping, can reduce the amplitude of the rotor vibration. Therefore, in all regions of oscillation frequency (rotation speed), only non-linear cubic damping can support the performance characteristics of the vibration isolator. The work [4] provides an excellent overview of research on linear and non-linear vibration-isolating systems.

Non-linear damping suspension can affect the stability of the flexible rotor in short journal bearings. In the work [9] a numerical method is used to solve the equations of motion, and bifurcation diagrams, orbits, Poincaré maps, maps and amplitude spectra are used to display motions. The results of works [2–8] are confirmed.

Under unsteady oscillations, the amplitude and frequency of disturbances change, they differ significantly from the oscillations observed under constant frequency and amplitude of disturbances.

Recently, unsteady oscillations have begun to be studied as transient processes in systems in connection with the spread of methods of direct and analytical-numerical modeling of the equations of the oscillatory process. Therefore, there are ample opportunities for new research in this direction.

This article examines the unsteady vibrations of a gyroscopic rotor with a vertical rigid shaft mounted on the lower hinge and upper elastic bearings. An ideal system is modeled, non-linear differential equations of the rotor motion are solved analytically by the method of varying amplitude, which allows obtaining a system of shortened equations and equations of unsteady oscillations of the rotor. Assuming that the speed of rotation of the shaft is a function of "slow" time, these equations are solved and their results are compared. The influence of non-linear cubic damping of elastic support, non-linear characteristics of support elasticity, rate of "slow" change in shaft rotation speed on the amplitude-frequency dependence of rotor oscillations is investigated. The numerical results of solving the rotor motion equations are compared with the analytical results.

2 Equations of Motion

The rotor is considered, the structural diagram of which is shown in Fig. 1. Rotor consists of a shaft with a length L, mounted vertically by means of a lower hinge and an upper elastic support spaced from it at a distance l_0 and a disk fixed at the free end of the shaft, having a mass m, a polar moment of inertia I_P and a transverse moment of inertia I_T the same for any direction. The elastic support has linear stiffness k_1, non-linear stiffness k_3, linear damping μ_{d1}, non-linear cubic damping μ_{d3}. The speed of the shaft rotation $\dot{\varphi} = \omega$ is such that the rotor can be viewed as a gyroscope, the fixed point of which is the lower shaft support. The position of the geometric center of the disk S is determined by coordinates x, y in a fixed coordinate system $Oxyz$, and the position of the shaft and the rotor as a whole in space by the Euler angles α, β and the angle of rotation φ. The angles α, β are small, the movement of the rotor in the direction of the coordinate axis z is neglected. Next, denote the coordinates of the center of mass m of the disk through x_m and y_m. Assume also that the linear eccentricity e lies in the direction of the N axis of the $ONKZ$ coordinate system rotating with the rotor. Restrict to small deviations of the rotor axis.

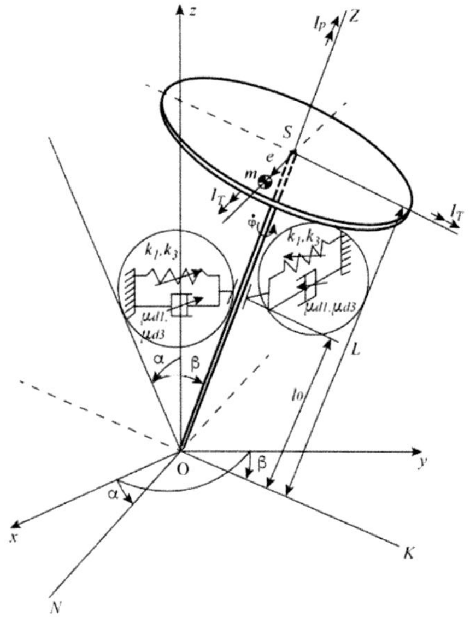

Fig. 1. Rotor geometry

Expressing the projections of the angular velocity of the rotor in the coordinate axes of the $ONKZ$ system, the coordinates of the center of mass of the disk and the coordinates of the upper support through the angular coordinates α, β and φ, finding expressions for the kinetic energy, potential energy of the rotor, the Rayleigh function and the projections of the moments of forces acting on the system, substituting them into the Lagrange equations of the second kind using the following dimensionless parameters

$$
\begin{gathered}
l = l_0/L; \bar{t} = t\omega_0; \overline{\tau} = \tau\omega_0; \Omega(\overline{\tau}) = \omega(\tau)/\omega_0; \overline{I}_T = I_T/(mL^2); \\
\overline{I}_P = I_P/(mL^2); \overline{K}_1 = k_1/(m\omega_0^2); e_r = e/\left[L(1+\overline{I}_T)\right]; \\
I_{P1} = \overline{I}_P/(1+\overline{I}_T); \overline{G} = g/(L\omega_0^2); K_3 = k_3 l_0^4/\left[mL^2\omega_0^2(1+\overline{I}_T)\right]; \\
\mu_1 = \mu_{d1}/\left[mL^2\omega_0(1+\overline{I}_T)\right]; \mu_3 = \mu_{d3}\omega_0/\left[mL^2(1+\overline{I}_T)\right],
\end{gathered} \tag{1}
$$

where is $\omega_0 = \sqrt{(k_1 l_0^2 - mgL)/\left[mL^2 - (I_p - I_T)\right]}$ the natural frequency of the damped rotor system, obtain the equations of motion of the rotor in the form

$$
\begin{aligned}
\alpha'' + I_{P1}\Omega(\bar{\tau})\beta' + \mu_1\alpha' + \mu_3\alpha'^3 + \omega_n^2\alpha + K_3\alpha^3 = e_r\left(\Omega^2(\bar{\tau}) + \bar{G}\right)\cos\varphi, \\
\beta'' - I_{P1}\Omega(\bar{\tau})\alpha' + \mu_1\beta' + \mu_3\beta'^3 + \omega_n^2\beta + K_3\beta^3 = e_r\left(\Omega^2(\bar{\tau}) + \bar{G}\right)\sin\varphi,
\end{aligned} \tag{2}
$$

where $\Omega(\overline{\tau})$ is the dimensionless rate of the shaft rotation, depending on the "slow" dimensionless time $\overline{\tau} = \varepsilon\, \bar{t}$, $\varepsilon \ll 1$, is a small parameter [10].

On the right-hand part of the system of Eqs. (8) perturbations containing φ'', were discarded, since in the region close to the resonance velocity $\varphi'' \ll \Omega^2$, and perturbations having a parameter $\overline{I}_P$ (in what follows, assuming that $\overline{I}_P \ll \overline{I}_T$) and values of the second and higher orders of smallness with respect to α, β, their derivatives, and their combinations. The indicated disturbances are small in comparison with disturbances, the amplitudes of which are proportional to the angular velocity squared.

Consider a rotor system close to a linear system. Therefore, choose one of the asymptotic methods, for example, the method of slowly varying amplitudes [11]. For the direct use of this method, the following restrictions are taken to solve Eqs. (2). The projections of the moments of the damping forces $\mu_1\alpha'$, $\mu_1\beta'$ and $\mu_3\alpha'^3$, $\mu_3\beta'^3$, as well as the moment of the cubic component of the restoring force $K_3\alpha^3$, $K_3\beta^3$, the moments of the centrifugal force of the imbalance of mass and gravity $e_r\left(\Omega^2(\overline{\tau}) + \overline{G}\right)\cos\varphi$, $e_r\left(\Omega^2(\overline{\tau}) + \overline{G}\right)\sin\varphi$ are considered small in comparison with the projections of the moments of the vibration inertia force and the linear restoring force acting in the system. Assuming that $\overline{I_P} \ll \overline{I_T}$ the projections of the moment of the passive gyroscopic force can also be considered small, $I_{P1}\Omega(\overline{\tau})\alpha'$ $I_{P1}\Omega(\overline{\tau})\beta'$. we will also limit ourselves to considering a spinning rotor: $\Omega^2(\overline{\tau}) \gg \overline{G}$ and motion in the resonance range, where the frequency of free oscillations ω_n is close to the frequency of forced oscillations Ω, i.e.$\xi = \varepsilon\xi_1 = \Omega(\overline{\tau}) - \omega_n \ll \omega_n$.

Equations (2), at small values of the quantity ξ and restrictions accepted above will take the following form:

$$\begin{gathered}\alpha'' + \Omega^2(\overline{\tau})\alpha = \\ e_r\Omega^2(\overline{\tau})\cos\varphi - I_{P1}\Omega(\overline{\tau})\beta' - \mu_1\alpha' - \mu_3\alpha'^3 - \omega_n^2\alpha - K_3\alpha^3 + 2\xi\alpha, \\ \beta'' + \Omega^2(\overline{\tau})\beta = \\ e_r\Omega^2(\overline{\tau})\sin\varphi + I_{P1}\Omega(\overline{\tau})\alpha' - \mu_1\beta' - \mu_3\beta'^3 - \omega_n^2\beta - K_3\beta^3 + 2\xi\beta,\end{gathered} \tag{3}$$

where is $\omega_n = \sqrt{\left(\overline{K}_1 l^2 - \overline{G}\right)/\left(1 + \overline{I}_T\right)}$ the dimensionless natural frequency of the linear rotor system (3) at $\overline{I_T} \gg \overline{I_P}$.

Equations (3) are a system of second order nonlinear ordinary differential equations with respect to α, β.

3 Solutions of Motion Equations

In an oscillatory system, under the influence of damping forces, which cause attenuation of higher harmonics, single-frequency oscillations of the fundamental tone are established with a frequency close to the frequency of the disturbing force. The single-frequency method allows us to consider both stationary oscillations and the process of the rotor transition through critical speeds under very general conditions - causing the variability of the coefficients of the differential equations, in the presence of elastic supports with a non-linear characteristic of elasticity and non-linear damping. Although the law of variation of angular speed of the rotor can be obtained only on the basis of processing the results of experimental studies of acceleration and running down of the machine, but to determine the general nature of the transient process, the single-frequency method allows solving the problem with the arbitrary law of variation of angular speed of the rotor. The only limitation that determines the applicability of this method is the requirement for a slow change in the angular velocity with respect to the value of the natural frequency of the system under study.

Therefore, search for solutions (3) in the form:

$$\alpha = A(\overline{t})\cos\left[\varphi + \theta(\overline{t})\right], \beta = A(\overline{t})\sin\left[\varphi + \theta(\overline{t})\right]. \tag{4}$$

Here is $A(\overline{t})$ the slowly varying amplitude, $\theta(\overline{t})$ is the phase shift of the oscillations relative to the forced harmonic moment.

Further, using the method of varying amplitudes [11], obtain the equations of the transient process in the form

$$A^{'} = [e_r\Omega^2(\overline{\tau})\cos\varphi + \Big(2\xi\Omega(\overline{\tau}) - I_{P1}\Omega^2(\overline{\tau})\Big)Acos(\varphi+\theta) + \mu_1\Omega(\overline{\tau})Asin(\varphi+\theta)+$$

$$\mu_3\Omega^3(\overline{\tau})A^3sin^3\mu_3\Omega^3(\overline{\tau})A^3sin^3(\varphi+\theta) - K_3A^3cos^3(\varphi+\theta)]sin(\varphi+\theta), \quad (5)$$

$$A\theta^{'} = -[e_r\Omega^2(\overline{\tau})\cos\varphi + \Big(2\xi\Omega(\overline{\tau}) - I_{P1}\Omega^2(\overline{\tau})\Big)Acos(\varphi+\theta) + \mu_1\Omega(\overline{\tau})Asin(\varphi+\theta)$$
$$+ \mu_3\Omega^3(\overline{\tau})A^3sin^3(\varphi+\theta) - K_3A^3cos^3(\varphi+\theta)]cos(\varphi+\theta). \quad (6)$$

After performing averaging of Eqs. (5) and (6), the system of equations for the transient process of the rotor is obtained in the following form

$$A^{'} = -e_r\Omega^2(\overline{\tau})sin\theta/2 - \mu_1\Omega(\overline{\tau})A/2 - 3\mu_3\Omega^3(\overline{\tau})A^3/8 \quad (7)$$

$$\theta^{'} = -e_r\Omega^2(\overline{\tau})cos\theta/(2A) - \left[\Omega(\overline{\tau}) - \omega_n - \frac{1}{2}I_{P1}\Omega(\overline{\tau})\right]\Omega(\overline{\tau}) + 3K_3A^2/8 \quad (8)$$

4 Unsteady Oscillations

To illustrate the influence of the value of non-linear cubic damping of the support on the development of the oscillatory process when passing through the resonant region, consider calculation of the unsteady mode of motion of the rotor system under the assumption that the speed of the shaft rotation Ω is also a "slowly" changing parameter according to the law $\Omega = \Omega_0 + \nu\overline{t}$. The equations of the unsteady process (7) and (8), (5) and (6) were modeled in the Mathlab-Simulink package. The angular speed of the shaft rotation Ω increased "slowly" uniformly ($\nu > 0$) or decreased uniformly over ($\nu < 0$) time.

The system parameters have the following values: $e_r = 0.0346$, $\omega_n \approx 1$, $I_{P1} = 0.021$, $\mu_1 = 0.01$.

For K_3=0.1, choose the initial conditions for the case with $\nu > 0 : \overline{t} = 0 : 1)$ $\Omega_0 = 0.81$, $A_0 = 0.067$, $\theta_0 =$ -0.02521 with $\mu_3 = 0.01$; 2) $\Omega_0 = 0.79$, $A_0 = 0.06254$, $\theta_0 =$ -0.02297 with $\mu_3 = 0.02$; 3) $\Omega_0 = 0.79$, $A_0 = 0.0625381$, $\theta_0 =$ -0.0230614 with $\mu_3 = 0.043$, for the case with $\nu < 0 : \overline{t} = 0 : 1)$ $\Omega_0 = 1.39$, $A_0 = 0.0640693$, $\theta_0 = 0.0134013$ with $\mu_3 = 0.01$; 2) $\Omega_0 = 1.39$, $A_0 = 0.0640692$, $\theta_0 = 0.0134806$ with $\mu_3 = 0.02$; 3) $\Omega_0 = 1.39$, $A_0 = 0.0640691$, $\theta_0 = 0.0136628$ with $\mu_3 = 0.043$.

For $K_3 =$ -0.1, accept the initial conditions for the case with $\nu > 0 : \overline{t} = 0 : 1)$ $\Omega_0 = 0.80$, $A_0 = 0.0664575$, $\theta_0 =$ -0.0240624 with $\mu_3 = 0.01$; 2) $\Omega_0 = 0.79$, $A_0 = 0.0626448$, $\theta_0 =$ -0.0230045 with $\mu_3 = 0.02$; 3) $\Omega_0 = 0.79$, $A_0 = 0.0626447$, $\theta_0 =$ -0.0231013 with $\mu_3 = 0.043$, for the case with $\nu < 0 : \overline{t} = 0 : 1)$ $\Omega_0 = 1.39$, $A_0 = 0.0640315$, $\theta_0 = 0.0133933$ with $\mu_3 = 0.01$; 2) $\Omega_0 = 1.39$, $A_0 = 0.0640315$, $\theta_0 = 0.0134724$ with $\mu_3 = 0.02$; 3) $\Omega_0 = 1.39$, $A_0 = 0.0640313$, $\theta_0 = 0.0136543$ with $\mu_3 = 0.043$.

The abscissa axis has two scales: the scale Ω and the corresponding time scale $\bar{t}$ Resonance curves of non-stationary oscillations of the rotor, constructed on the results of modeling Eqs. (7) and (8), (5) and (6), are shown in Figs. 2 - 6. All the plots clearly show that increase in the value of the non-linear cubic damping of the elastic support μ_3 from 0.01 to 0.043 suppresses not only the maximum amplitude and its variation around the mean value, but also the oscillation amplitude and its variation below the resonance and over the resonance rotation speed. It shifts the shaft rotation speed corresponding to the maximum amplitude with a rigid non-linear elastic characteristic ($K_3 > 0$) of the support material downward, and with a soft non-linear elastic characteristic ($K_3 < 0$) of the support material toward an increase. Comparison of the plots in Fig. 2a and Fig. 2b shows that with an increase in the value of ν from 0.00025 to 0.0005 with a rigid non-linear characteristic ($K_3 > 0$) of the support elasticity, the resonance peak of the amplitude shifts towards high speeds of rotation and its value decreases [10].

Comparison of Fig. 3 and Fig. 2, shows the identity of the results of solving the equations of the non-stationary process before averaging (5) and (6) with the results of solving the equations of the non-stationary process after averaging (7) and (8) over time, although in Fig. 3, there is a variation in the values of the amplitude of oscillations in time along the main curve, around its mean values.

From Figs. 2 and 4, it is clearly seen that at $K_3 > 0$ and $\nu > 0$ (run-up) the resonance amplitude is greater than at $\nu < 0$ (run-down).

Changes in the nonlinear stiffness characteristics of an elastic support significantly affect the description of the resonance curves. The amplitude-frequency characteristics of the rotor during the transient process and the soft characteristic of the non-linear elasticity of the support ($K_3 < 0$) are shown for the take-off run ($\nu > 0$) of the machine in Fig. 5, for run-down ($\nu < 0$) of the machine – in Fig. 6. From these graphs, it is noticed that ($\nu > 0$) the resonance peak of the amplitude is less during the run-up than during the run-down ($\nu < 0$), i.e. on the contrary, than in the case with a rigid non-linear characteristic of the support elasticity ($K_3 > 0$) [12]. When the absolute value of the angular acceleration ν changes from 0.00025 to 0.0005 in the case of starting ($\nu > 0$) the machine, the resonance peak of the amplitude shifts towards an increase in the rotation speed [10], then in the case of braking ($\nu < 0$) towards a decrease in the shaft rotation speed.

Comparison of the amplitude-frequency characteristics of the rotor for the case with a rigid non-linear elastic characteristic (Fig. 2, 3, 4) and for the case with a soft non-linear elastic characteristic (Fig. 5, 6) of the support material shows that during the run-up, ($\nu > 0$) the maximum amplitude of the resonance curves for $K_3 > 0$ is greater than the analogous value for $K_3 < 0$; at the run-down ($\nu < 0$), the maximum amplitude for $K_3 > 0$ is less than the similar parameter for $K_3 < 0$.

In order to check whether the considered transient process is really "resonant", the equation of the reference line of the resonance curve is derived from the equations of motion (3)

$$\Omega = \omega_n/(2 - I_{P1}) + \sqrt{[\omega_n/(2 - I_{P1})]^2 + 3K_3A^2/[4(2 - I_{P1})]}. \tag{9}$$

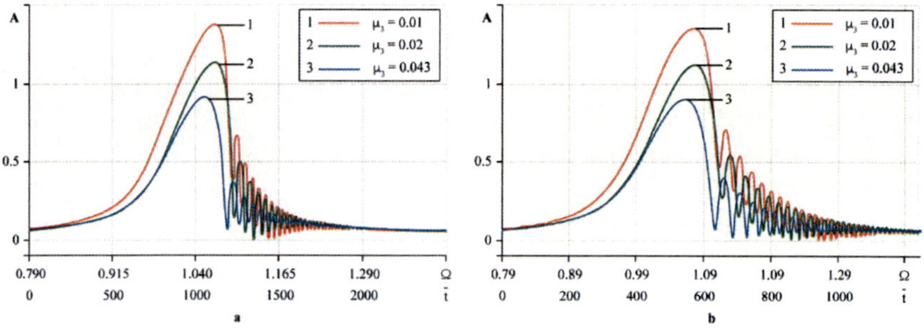

Fig. 2. Transition through resonance at $K_3 > 0$, according to the results of modeling Eqs. (7) and (8) with a - ν = 0.00025, b - ν = 0.0005

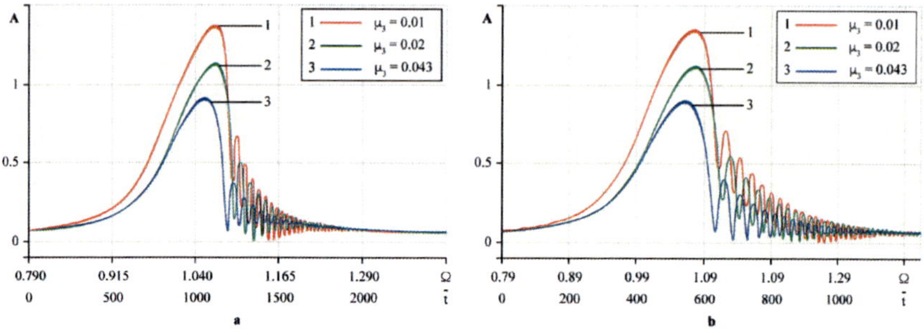

Fig. 3. Transition through resonance at $K_3 > 0$, according to the results of modeling Eqs. (5) and (6) with a - ν = 0.00025, b - ν = 0.0005.

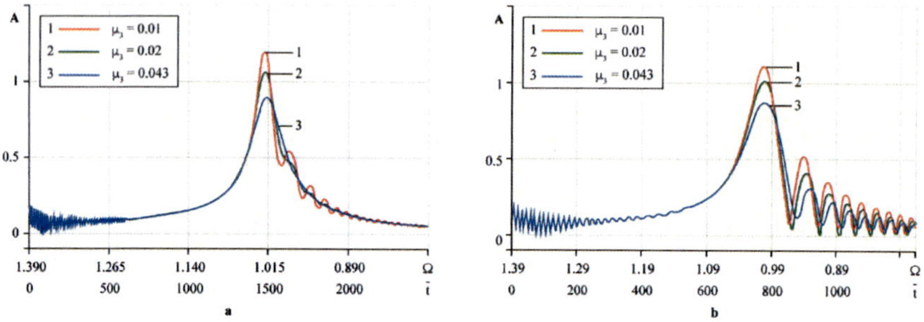

Fig. 4. Transition through resonance at $K_3 > 0$, according to the results of modeling Eqs. (7) and (8) with a - ν = -0.00025, b - ν = -0.0005

Assuming that, $\nu \ll \Omega^2$ the peak amplitudes and the corresponding rotational speeds of the resonance curves approximately satisfy Eq. (9). So, for example, for $K_3 = 0.1$, $\mu_1 = 0.01$, $\mu_3 = 0.01$, ν = 0.00025 the maximum amplitude $A = 1.360$ corresponds to the rotation speed Ω = 1.075 (1 resonance curve in Fig. 2), for K_3 = -0.1, $\mu_1 = 0.01$, $\mu_3 = 0.01$, ν = 0.00025 the maximum amplitude $A = 1.163$ – the rotation speed Ω = 0.9570 (1 resonance curve in Fig. 5).

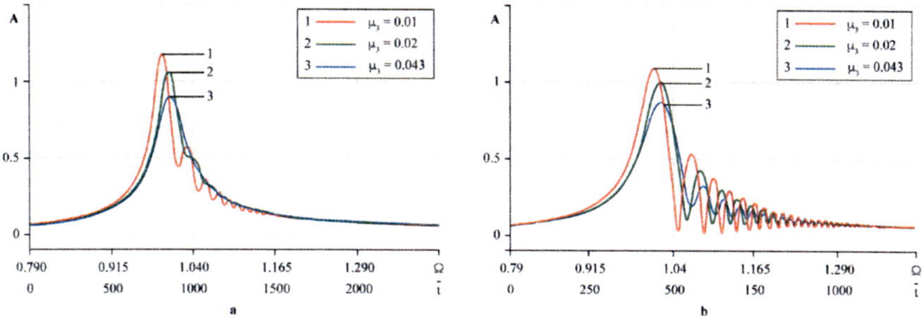

Fig. 5. Transition through resonance at $K_3 < 0$, according to the results of modeling Eqs. (7) and (8) with *a* - ν = 0.00025, *b* - ν = 0.0005

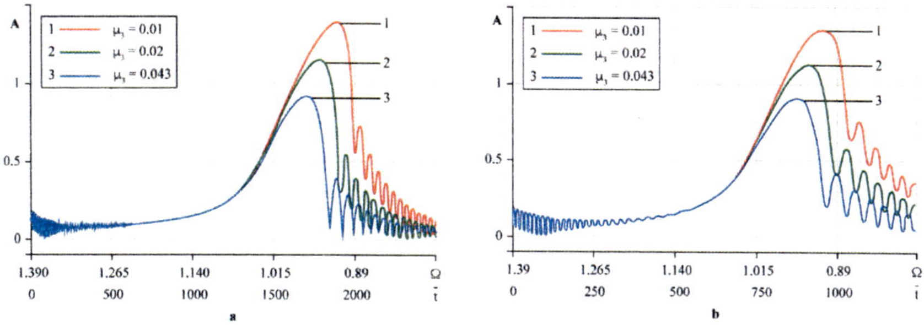

Fig. 6. Transition through resonance at $K_3 < 0$, according to the results of modeling Eqs. (7) and (8) with *a* - ν = -0.00025, *b* - ν = -0.0005

To confirm the analytical study, Eqs. (3) were solved directly numerically. Figure 7 shows the numerical results for passing through the resonance with a rigid non-linear elastic characteristic of the support of the support material and a "slowly" varying value of the angular velocity of rotation Ω. In this figure, the effects of damping of the oscillations amplitude having the value of μ_3 and beating of similar oscillations are observed. These results are consistent with previous analytical results shown in Fig. 2 and Fig. 4. The differences lie in the width of the region of clearly visible vibrations, the magnitude

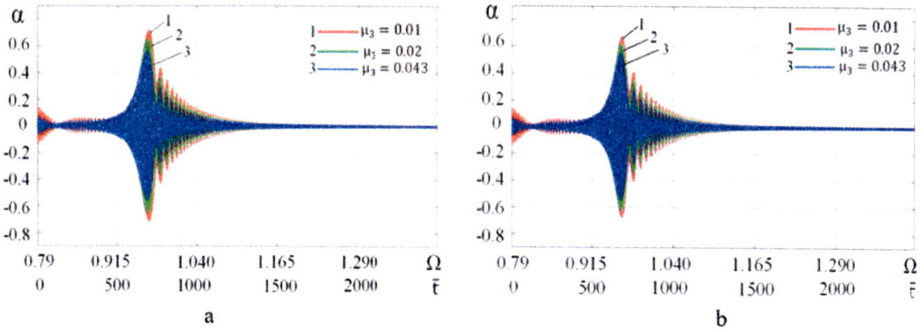

Fig. 7. Transition via resonance with ν = 0.00025 according to the results of the numerical solution of Eqs. (3) for *a* - $K_3 > 0$, *b* - $K_3 < 0$

of the maximum vibration amplitude and the displacement of the corresponding shaft rotation speed. Despite this, the basic behavior of the transient process persists. Jumping effects are not detected.

5 Conclusions

Differential equations of motion of a gyroscopic rigid unbalanced rotor with non-linear cubic stiffness and non-linear cubic damping are constructed and solved by the method of varying amplitude. Differential equations of unsteady oscillations of the rotor are obtained, which were solved numerically for the transient process through the resonance region.

It is shown that non-linear cubic damping significantly suppresses not only the maximum amplitude and its variation, but also the oscillation amplitude and its variation below the resonance and over the resonance rotation speed. It shifts the resonant rotation speed of the shaft downwards with the rigid non-linear elastic characteristic ($K_3 > 0$) of the support material and upwards with the soft non-linear elastic characteristic ($K_3 < 0$) of the support material.

It was confirmed that with a "slow" increase in the shaft rotation speed ($\nu > 0$), an increase in the absolute value of the angular acceleration is accompanied by a shift of the amplitude maximum towards high rotation speeds, with a "slow" decrease in the shaft rotation speed ($\nu < 0$) - towards low rotation speeds with a decrease in the amplitude of oscillations.

The results from the analysis of studies of influence of non-linear characteristics of elasticity ($K_3 > 0$ and $K_3 < 0$) of the support material during the run-up ($\nu > 0$) and run-down ($\nu < 0$) of the rotor on the peak amplitude of oscillations are presented.

There is an agreement between the results of analytical solutions and numerical solutions of the equations of rotor motion.

The research results can be used in the manufacture of a vibration isolator, which significantly suppresses the peak amplitude, and the amplitude of oscillations below the resonance and over the resonance rotation speed, for a vibrating system, incl. rotary one.

Acknowledgments. This research is funded by the Science Committee of the Ministry of Education and Science of the Republic of Kazakhstan (Grant No. AP08856763).

References

1. Zakaria, AA, Rustighi, E., Ferguson, NS: A numerical investigation into the effect of the supports on the vibration of rotating shafts. In: Proceedings of the 11-th International Conference on Engineering Vibration, Ljubljana, Slovenia, pp. 539–552 (2015)
2. Iskakov, Zh.: Resonant Oscillations of a Vertical Unbalanced Gyroscopic Rotor with Non-linear Characteristics. In: Proceedings of the 14th IFToMM World Congress, Taipei, Taiwan, 3, 505–513 (2015)
3. Iskakov, Zh.: Dynamics of a vertical unbalanced gyroscopic rotor with non-linear characteristics. Mech. Mach. Sci. **46**, 107–114 (2017)

4. Peng, C., Meng, L.Z., Zhang, W.M.: Study of the effects of cubic non-linear damping on vibration isolations using harmonic balance method. Int. J. Nonlin. Mech., 47, 1065–1166 (2012)
5. Ho, C., Lang, Z., Billings, S.A.: The benefits of non-linear cubic viscous damping on the force transmissibility of a Duffing-type vibration isolator. In: Proceedings of UKACC International Conference on Control, UK, Cardiff, UK, pp. 479–484 (2012)
6. Xiao, Z., Jing, X., Cheng, L.: The transmissibility of vibration isolators with cubic nonlinear damping under both force and base excitations. J. Sound. Vib, 332 (5), 1335-1354 (2013)
7. Iskakov, Zh.: Resonant oscillations of a vertical hard gyroscopic rotor with linear and nonlinear damping. Mech. Mach. Sci., 73, 3353–3362 (2019)
8. Iskakov, Z., Bissembayev, K.: The nonlinear vibrations of a vertical hard gyroscopic rotor with nonlinear characteristics. Mech. Sci., 10, 529–544 (2019)
9. Yan, S., Dowell, E.H., Lin, B.: Effects of nonlinear dampingsuspension on nonperiodic motions of a flexible rotor in journal bearings. Nonlin. Dyn, 78 (2), 1435–1450 (2014)
10. Grobov, V.A.: Asymptotic methods for calculating bending vibrations of turbomachine shafts. Moscow: Publishing house of the Academy of Science, USSR (1961)
11. Parshakov, A. N. Physics of vibratory motion. Publishing House of Perm State University, Perm (2010)
12. Banakh, L., Nikiforov, A.: Non stationary oscillations of high-speed rotor systems at start-up and braking. In Proceedings of the 10th Biennial International Conference on Vibration Problems (ICOVP), Prague, Czech Republic, pp. 328–333 (2011)

Check for updates

Modeling of Nonlinear Dynamics of Planar Mechanisms with Elastic and Flexible Pre-stressed Elements

L. A. Khajiyeva[1(✉)], A. K. Kudaibergenov[1], G. A. Abdraimova[2], and R. F. Sabirova[1]

[1] Al-Farabi Kazakh National University, Almaty, Kazakhstan
[2] Satbayev University, Almaty, Kazakhstan

Abstract. The motion of planar hinge-lever mechanisms with flexible and elastic links in a closed pre-stressed contour is considered. Modeling of the mechanism motion is carried out on the basis of their kinetic-elastodynamic analysis, which takes into account the inertial relationship between the large-scale motion of mechanisms as a rigid body and nonlinear vibrations of the links as a result of their elastic deformation. This work pays attention to both longitudinal and lateral vibrations of elastic links. The equations of motion of the mechanisms are obtained by the use of Novozhilov's nonlinear theory of elasticity, according to which the link deformations are assumed to be finite. Based on Biot's theory of incremental deformations, the field of initial stresses in flexible elements is taken into account due to their preliminary tension, which determines the geometric nonlinearity of dynamic models. As an example, the dynamics of a planar five-link hinge-lever mechanism with closed pre-stressed contour is studied.

Keywords: Dynamics · Mechanisms · Links · Elasticity · Nonlinearity · Pre-stress

1 Introduction

The idealization of the links of mechanisms and machines as rigid ones significantly narrows the range of problems under consideration for most of the dynamic problems of modern mechanical engineering. Basically, they are limited to quasi-static and pre-critical operating modes [1, 2]. Deformations of flexible and elastic elements of mechanisms and machines cause, when unaccounted for, complex dynamic processes in them. They affect the strength and operating characteristics of the system as a whole, arousing scientific and practical interest amongst researchers.

In contrast to rigid elements when the main problems of the machine dynamics are

1) studying the forces acting on the linkages under conditions of a given law of motion of the machine;

2) studying the true law of motion of a machine under the influence of given forces; which are solved separately and quite easily, in case of elastic elements, a separate

© The Author(s), under exclusive license to Springer Nature Switzerland AG 2022
J. Beran et al. (Eds.): TMM 2020, MMS 85, pp. 94–103, 2022.
https://doi.org/10.1007/978-3-030-83594-1_10

consideration of the problems cannot take place since the kinematics of the mechanisms is complicated by deformations of the linkages [1].

These questions are reflected in the method of kinetic-elastodynamic analysis, widely used nowadays for the dynamic analysis of mechanisms and machines, considering the deformability of links and elements. It takes into account the nonlinear inertial relationship between the motion of a mechanism as a rigid body and the vibrational process resulting from the elastic deformation of its linkages. Amongst the first works in this area, one can note the works of Sadler, Sandor [3, 4], Chu, Pan [5] and others, where the elastic motion of planar mechanisms is modeled with limiting the number of elastic elements and their representation by the Euler-Bernoulli beam. In case of mechanisms of spatial topology, the models are presented discretely with limited degrees of freedom of elastic elements by Dubowsky et al. [6, 7], Winfrey [8, 9], Shabana [10]. It is worth mentioning the work of Erdman, Sandor, Oakberg [11], where the method of kinetic-elastodynamic analysis and synthesis of mechanisms using a flexibility matrix was developed. Beams with different loading options depending on the boundary conditions are taken as a model for elastic analysis of the mechanism links. One of the first works on the dynamic analysis of planar mechanisms accounting for finite deformations of elastic links is the work of Viscomi, Ayre [12]. Despite the fact that only one link was assumed elastic, it was the most complete, meaningful work and underlie modern research on modeling the nonlinear dynamics of elastic elements of structures, mechanisms and machines when removing restrictions on the magnitude of their deformations.

Modern works aim at studying the spatial motions of rotational-vibrational mechanical elastic systems, which are widely used in mechanical and instrument engineering, power engineering, transport, and many other fields of modern technology as driving elements, elements of percussion mechanisms, drilling equipment, etc. These systems are based on finite segments of elastic rectilinear rod elements. Modern works on the nonlinear dynamics of rod elements with no restrictions on the magnitude of their deformations are of particular interest. Amongst them are the works of Erofeev [13, 14], Asghari [15], Gulyaev [16] and others. Flexible elements by their nature are also nonlinear; they are presented in sufficient detail in the works of Svetlitskii. As a rule, flexible elements are tensioned to eliminate their sagging. There is little research on the influence of initial stress caused by the constructive necessity of mechanisms on their dynamics. At the same time, as shown by the research results of Guz' and others [17, 18], the wave speed depends significantly on the initial stress tensor. This fact was noted in Ogden's works studying the influence of initial stresses of pre-stressed media subject to finite deformations. Ogden's works regarding the study of incremental motion superimposed on an initially stressed configuration subject to finite deformation are known. They investigated the effect of initial stress on infinitesimal wave propagation [19, 20]. Chadwick and Ogden obtained formulas for a pre-stressed material in the absence of residual stress.

Therefore, study of the mutual influence nature of the static fields of initial stress and the disturbed state of the elastic links of mechanisms during their operation is of practical interest.

The authors of this paper investigate the planar hinge-lever mechanisms with elastic links and pre-tensioned flexible elements in a closed elastic contour. The considered approach is based on the kinetic-elastodynamic analysis of mechanisms with flexible pre-stressed elements and elastic links taking into account their finite deformations. The nonlinear dynamics of such mechanisms is modeled using Hamilton's variation principle, as well as the widely known Biot's theory of incremental deformations [21] and Novozhilov's theory of finite deformations [22].

2 Kineto-Elastodynamic Analysis of Planar Hinge-Lever Mechanisms

In most works on kineto-elastodynamic analysis of planar hinge-lever mechanisms, the number of elastic elements is restricted to one link. Moreover, its elastic displacements are assumed to be small. The nature of the link deformation is defined by the elastic deflection, and axial forces are assumed to be constant along the length of the elastic elements. It restricts the model since the variability of the axial forces along the link length significantly affects the vibrational process, causing the effect of time-varying stiffness or "frequency modulation". This phenomenon is of particular importance for mechanisms with a sufficiently large ratio of the lengths of connecting rods to the length of a crank.

This paper proposes an extension to the model by assuming all links to be elastic, increasing the degrees of freedom of the link deformation and considering the finiteness of their magnitudes.

Based on the generalized model of spatial deformation of the rod element [23] and the transition to its particular topologies, a dynamic model of elastic motion of the entire mechanism is constructed. The field of longitudinal and lateral displacements $u_i(x, t)$ and $v_i(x, t)$ of the links is given by

$$\begin{aligned} U(x, y, z, t) &= u(x, t) - \frac{\partial v(x, t)}{\partial x} y \,, \\ V(x, y, z, t) &= v(x, t) \,, \\ W(x, y, z, t) &= 0 \,, \end{aligned} \tag{1}$$

where $u_i(x, t)$ is the translational displacement of the section of the i-th link along the x-axis, $v_i(x, t)$ the displacement of the flexural center of the i-th link cross-section along the y-axis owing to bending.

The generalized model of elastic motion of links is shown in Fig. 1. Figure 1a corresponds to the simply supported driven links, whereas Fig. 1b shows elastic deformation of the driving link as a cantilever beam.

The position of the points of the i-th link of the deformed mechanism in the inertial coordinate system OXY is determined by the radius vector R_i;$X_i(t)$, $Y_i(t)$, $\theta_i(t)$ determine the plane motion of the undeformed i-th element relative to the inertial (global) coordinate system OXY; $O_i x_i y_i$ is a moving coordinate system associated with the i-th link and describing its deformed configuration (local coordinate system); u_i, v_i longitudinal and lateral displacements of the i-th link in the section x as a result of the link deformation, respectively. The kinematics of the mechanisms is defined by methods known in the literature.

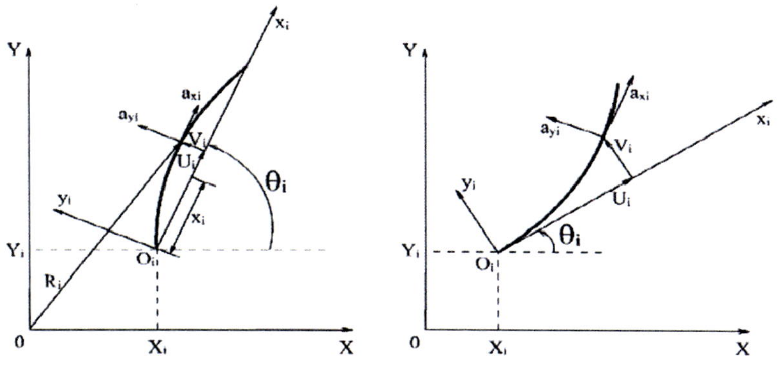

Fig. 1. Generalized model of deformation of elastic links for hinge-lever mechanisms.

When deriving the equations of motion of the mechanism as a system of rod elements, it is necessary to ensure the connection between the elements through kinematic pairs in view of its multi-link structure. For that the force analysis of the mechanisms is carried out. Figure 2a shows the diagram of the loading forces of the linkages; the force diagram in a kinematic pair connecting two adjacent links is given in Fig. 2b.

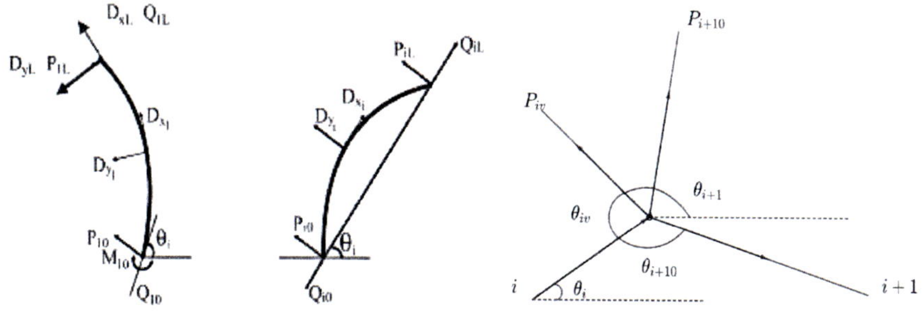

Fig. 2. Diagrams of the loading forces of the driving link and i-th driven links (a) and ones in a kinematic pair connecting two adjacent links (b).

The unknown components of reactions in the i-th link hinges are determined from the equilibrium condition of forces and moments acting on this link:

$$\begin{aligned} &Q_{i0} + Q_{iL} - \int_0^{l_i} m_i a_{xi} dx = 0, \\ &P_{i0} + P_{iL} - \int_0^{l_i} m_i a_{yi} dx = 0, \\ &P_{i0} l_i + \int_0^{l_i} m_i a_{xi} v dx - \int_0^{l_i} m_i a_{yi} x dx = 0. \end{aligned} \tag{2}$$

A system of $3N$ equations was obtained to govern the reactions in the hinges (N is the number of moving links of the mechanism). The missing conditions for the solvability of the system of Eqs. (2) are obtained from the conditions of equilibrium of forces in the hinges connecting adjacent links. According to Fig. 2, the equations of equilibrium of forces in the hinges in case of two adjacent driven links (simply supported beams) have the form:

$$\begin{aligned} &Q_{iL}\cos\theta_i + Q_{i+10}\cos\theta_{i+1} - P_{iL}\sin\theta_i + P_{i+10}\sin\theta_{i+1} = 0, \\ &Q_{iL}\sin\theta_i + Q_{i+10}\sin\theta_{i+1} + P_{iL}\cos\theta_i + P_{i+10}\cos\theta_{i+1} = 0. \end{aligned} \tag{3}$$

If a kinematic pair connects more than two links, then the number of terms in (3) increases.

Figure 3 shows a diagram of the force loading of a flexible element.

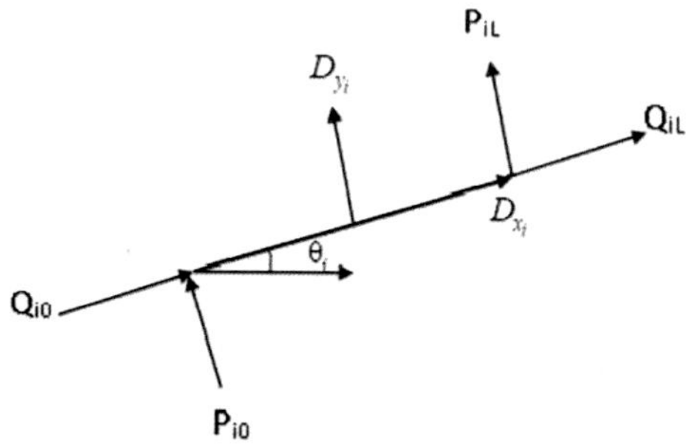

Fig.3. Diagram of loading forces for flexible elements.

Flexible elements have a number of advantages and benefits: reduced metal capacity, the absence of threat of longitudinal bending of the link, unpretentiousness in assembly and accuracy of manufacturing kinematic pairs, etc. Therefore, they are widely used in transmission mechanisms of machines. However, flexible elements in the form of a cable, belt, chain and other transmissions, carrying out a one-way connection, sag during operation. There are two main approaches to address this deficiency:

1) tension of flexible elements, creating an initial tension;

2) power closure of the kinematic chain for the implementation of alternating work of two flexible elements. Both these approaches are applied in mechanisms with closed elastic pre-stressed contour.

The mutual influence of elastic motion of the linkages is taken into account by transferring forces through their hinged joints. The force analysis of flexible elements is carried out similarly to the force analysis of elastic elements, i.e. reactions in hinges are determined from the condition of equilibrium of forces and moments of each link, and the missing equations for closing the system are taken from the condition of equality of forces in the hinges connecting flexible and elastic linkages.

3 Equations of Motion of Flexible and Elastic Linkages

The motion of a rotating rod element in cases of its plane and spatial topology of deformation is modeled in [24, 25]. The equations of motion are derived on the basis of Ostrogradsky-Hamilton's variation principle. Also, the finiteness of the link deformations is assumed according to Novozhilov's theory of finite deformations. Here, in contrast to an unlinked element, when its motion can be specified locally, elastic linkages between the system of elements lead to the complication of the model.

In this case, the motion of the entire system must be determined in the global coordinate system, specifying the nominal motion of elements and elastic displacements resulting from their deformation.

In accordance with the deformation scheme (1), the kinetic and potential deformation energies of any i-th link are defined as

$$\begin{aligned}\mathrm{T}_i &= \frac{m_i}{2}\int_0^{l_i}\left(\dot{X}_i\cos\theta_i + \dot{Y}_i\sin\theta_i + \dot{u}_i - v_i\dot{\theta}_i\right)^2 dx \\ &+ \frac{m_i}{2}\int_0^{l_i}\left(\dot{Y}_i\cos\theta_i - \dot{X}_i\sin\theta_i + \dot{v}_i + x_i\dot{\theta}_i + u_i\dot{\theta}_i\right)^2 dx.\end{aligned} \tag{4}$$

$$\mathrm{U_i} = \frac{E_iJ_{iy}}{2}\int_0^{l_i}\left(\frac{\partial^2 v}{\partial x^2}\right)^2 dx + \frac{E_iF_i}{2}\int_0^{l_i}\left[\left(\frac{\partial u}{\partial x} + \frac{1}{2}\left(\frac{\partial v}{\partial x}\right)^2\right)^2 + \frac{1}{4}\left(\frac{\partial v}{\partial x}\right)^4\right] dx. \tag{5}$$

As a result, a system of coupled nonlinear equations of motion of the i-th type is obtained:

$$E_iJ_iv_{xxxx} - E_iF_i\left(v_{xx}u_x + v_xu_{xx} + 3v_x^2v_{xx}\right) + \rho_iF_iv_{tt}$$

$$= \rho_iF_i\left[\sum_j\left[l_j\dot{\theta}_j\sin(\theta_i - \theta_j)(2\dot{\theta}_i - \dot{\theta}_j) - l_j\ddot{\theta}_j\cos(\theta_i - \theta_j)\right] + (x+u)\ddot{\theta}_i + v\dot{\theta}_i^2 + 2u_t\dot{\theta}_i\right],$$

$$E_iF_i(u_{xx} + v_xv_{xx}) - \rho_iF_iu_{tt} = \rho_iF_i\left[\sum_j\left[l\ddot{\theta}\sin(\theta_i - \theta_4) + l_j\dot{\theta}_j\cos(\theta_i - \theta_j)(\dot{\theta}_i - \dot{\theta}_j)\right]\right.$$

$$\left. + \sum_j l_j\dot{\theta}_j\dot{\theta}_i\cos(\theta_i - \theta_j) + \ddot{\theta}_iv - (x+u)\dot{\theta}_i^2 + 2\dot{\theta}_i\dot{v}\right], \qquad j = \overline{1 \div i - 1}. \tag{6}$$

It specifies the axial and transverse displacements of the links of the mechanism relative to their undeformed position.

The summation sign by j on the right-hand side of Eqs. (6) is associated with the determination of the position functions of the i-th deformable link through the previous $i - 1$ links of the mechanism. An "overdots" on the right-hand side of Eqs. (6) indicate derivatives with respect to time. Depending on the type of kinematic pairs of the mechanism links, the corresponding boundary conditions that include both external and internal forces and moments are chosen. For driven links (simply supported beams), they are specified as

$$v_i(0,t) = u_i(0,t) = \frac{\partial^2 v_i}{\partial x^2}(0,t) = 0, \quad v_i(l,t) = u_i(l,t) = \frac{\partial^2 v_i}{\partial x^2}(l,t) = 0. \tag{7}$$

The boundary conditions for driving links (cantilever beam) are given by

$$\begin{gathered} v_1(0,t) = u_1(0,t) = \frac{\partial^2 v_1}{\partial x^2}(0,t) = 0, \quad E_1F_1\left.\frac{\partial^3 v}{\partial x^3}\right|_{x=0} = P_{10}, \\ E_1F_1\left.\frac{\partial u_i}{\partial x}\right|_{x=0} = Q_{10}, \quad E_1I_1\left.\frac{\partial^2 v}{\partial x^2}\right|_{x=0} = M_{10}. \end{gathered} \tag{8}$$

In the case of the action of external forces and moments on the i-th element of the mechanism, the influence of the latter on the mechanism dynamics can be accounted for in the boundary conditions.

The equations of motion of flexible elements are based on the fundamental relations of Biot's theory of initial stresses [21]. The latter are obtained from the equations of equilibrium of a volume element of a deformed medium during the transition to the initial state of the medium:

$$\frac{\partial A_{xx}}{\partial x} + \frac{\partial A_{xy}}{\partial y} + \frac{\partial A_{xz}}{\partial z} + X(\xi, \eta, \varsigma) = 0, \tag{9}$$

where

$$A_{xx} = \bar{\sigma}_{\xi\xi}\frac{d(\eta,\varsigma)}{d(y,z)} + \bar{\sigma}_{\xi\eta}\frac{d(\varsigma,\xi)}{d(y,z)} + \bar{\sigma}_{\xi\varsigma}\frac{d(\xi,\eta)}{d(y,z)} \quad (10)$$

$\frac{d(\eta,\varsigma)}{d(y,z)}$ are Jacobians of transformation of pairs of $x,\ y,\ z$ variables into pairs ξ,η,ς;

$$\bar{\sigma}_{\xi\xi} = \sigma_{11}^{0} + \bar{s}_{\xi\xi}, \qquad \bar{\sigma}_{\xi\eta} = \sigma_{12}^{0} + \bar{s}_{\xi\eta}, \qquad \bar{\sigma}_{\xi\varsigma} = \sigma_{13}^{0} + \bar{s}_{\xi\varsigma} \quad (11)$$

σ_{11}^{0} initial stresses in a Cartesian coordinate system;
$\bar{s}_{\xi\xi}$ increment of stresses as a result of deformation.
The rest equations are obtained by cyclic substitution.
By introducing simplifications in (10), we can obtain various equations of motion with nonlinear effects and different topology of deformation.

4 Modeling of Elastic Motion of a Planar Five-Link Mechanism with Closed Pre-stressed Contour

As an example, the motion of a planar five-link hinge-lever mechanism with closed elastic pre-stressed contour is considered (Fig. 4). Such mechanisms are used as a drive for machines performing reciprocating motion. The motion of mechanisms is modeled taking into account the elastic properties of the links and the initial stresses in flexible elements.

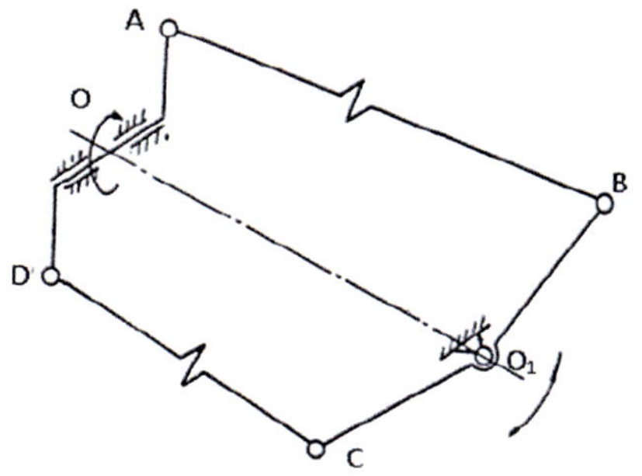

Fig. 4. Planar five-link hinge-lever mechanism with closed pre-stressed contour.

When the mechanism operates, the field of initial stresses of flexible elements transits from a static state to the dynamic one. Moreover, it interacts with the disturbed state of other elastic links of the mechanism. The equations of motion of elastic elements, taking into account the finiteness of deformations, will be determined by a nonlinear system of the form (6).

Passing to the one-dimensional case in Eqs. (9), (10) and generalizing it for the case of finite deformations, the equation of motion of flexible elements is obtained:

$$\frac{\partial^2 u}{\partial x^2}+\frac{\partial \sigma^{\circ}_{xx}/\partial x}{E+\sigma^{\circ}_{xx}(1+e_{xx})}\frac{\partial u}{\partial x}\left(1+\frac{1}{2}\frac{\partial u}{\partial x}\right)+\frac{\partial Q/\partial x}{F(E+\sigma^{\circ}_{xx}(1+e_{xx}))}=\frac{\rho}{E+\sigma^{\circ}_{xx}(1+e_{xx})}\frac{\partial^2 u}{\partial t^2}. \tag{12}$$

Depending on the type of supports and mountings of flexible elements with adjacent links, the appropriate boundary conditions are specified for the equations of motion (12). For hinged joints, they are written as

$$E_iF_i\left(\frac{\partial u}{\partial x}\right)\bigg|_{x=0}=-Q_{i0},\quad E_iI_i\left(\frac{\partial u}{\partial x}\right)\bigg|_{x=L}=-Q_{iL}. \tag{13}$$

Thus, the mathematical model of elastic motion of the considered mechanism is given by the following system of equations:

$$\begin{aligned}
&E_iF_i\frac{\partial^2 u_i}{\partial x^2}-m_i\frac{\partial^2 u_i}{\partial t^2}=_i(\varphi_i,\varphi_j,\dot{\varphi}_j,\ddot{\varphi}_j,u_i,v_i,\dot{v}_i),\\
&E_i\,J_i\frac{\partial^4 v_i}{\partial x^4}+m_i\frac{\partial^2 v_i}{\partial t^2}=G_i(\varphi_i,\varphi_j,\dot{\varphi}_j,\ddot{\varphi}_j,u_i,\dot{u}_i,v_i)\ ,\ j=1,\ \vec{i}-1,\\
&\frac{\partial^2 u}{\partial x^2}+\alpha(E,\sigma^{\circ}_{xx},e_{xx})\frac{\partial u}{\partial x}\left(1+\frac{1}{2}\frac{\partial u}{\partial x}\right)+q=\frac{1}{a^2}\frac{\partial^2 u}{\partial t^2}
\end{aligned} \tag{14}$$

with boundary conditions (7), (8), and (13).

5 Conclusion

Based on the kineto-elastodynamic analysis, the generalization of the model of motion of the planar hinge-lever mechanisms with pre-stressed flexible and elastic elements was carried out. The nonlinear inertial relationship between the nominal motion of the links as rigid bodies and the vibrational process as a result of deformation of all the mechanism links was taken into account. The geometric nonlinearity of the model due to the finiteness of deformations of elastic links and accounting for the field of initial stresses in flexible elements was observed.

Acknowledgements. This research is funded by the Science Committee of the Ministry of Education and Science of the Republic of Kazakhstan (Grant No. AP08857255).

References

1. Artobolevsky, I.I.: The main problems of modern machine dynamics. In: Proceedings of the All-Union Congress on Theoretical Physics and Applied Mechanics, pp. 5–24 (1962). (In Russ.)

2. Vulfson, I.I.: Introduction to the dynamics of mechanisms taking into account the elasticity of links. Leningrad Political Institute, Leningrad (1977).(In Russ.)
3. Sadler, J.P., Sandor, G.N.: A lumped parameter approach to vibration and stress analysis of elastic linkages. ASME J. Eng. Ind. **95**(2), 549–557 (1973)
4. Sadler, J.P., Sandor, G.N.: Nonlinear vibration analysis of elastic four-bar linkages. ASME J. Eng. Ind. **96**(2), 411–419 (1974)
5. Chu, S.C., Pan, K.C.: Dynamic response of a high speed slider – crank mechanism with an elastic connecting rod. ASME J. Eng. Ind. **97**(2), 542–550 (1975)
6. Dubowsky, S., Gardner, T.N.: Dynamic interactions of link elasticity and clearance connections in planar mechanical systems. ASME J. Eng. Ind. **97**(2), 652–661 (1975)
7. Sunada, W., Dubowsky, S.: The application of finite element methods to the dynamic analysis of flexible spatial and co-planar linkage systems. ASME J. Mech. Des. **103**(3), 643–651 (1981)
8. Winfrey, R.C.: Elastic link mechanism dynamics. ASME J. Eng. Ind. **93**(1), 268–272 (1971)
9. Winfrey, R.C.: Dynamic analysis of elastic link mechanisms by reduction of coordinates. ASME J. Eng. Ind. **94**(2), 577–581 (1972)
10. Shabana, A.A.: Automated analysis of constrained systems of rigid and flexible bodies. J. Vib., Acoust., Stress, and Reliab. 107(4), 431–439 (1985).
11. Erdman, A.G., Sandor, G.N., Oakberg, R.G.: A general method for kineto-elastodynamic analysis and synthesis of mechanisms. ASME J. Eng. Ind. **94**(4), 1193–1205 (1973)
12. Viscomi, B.V., Ayre, R.S.: Nonlinear dynamic response of elastic slider-crank mechanism. ASME J. Eng. Ind. **93**(1), 251–262 (1971)
13. Erofeev, V.I.: Nonlinear bending and torsional waves in rods and rod systems. Bulletin of scientific and technical development **20**(4), 46–50 (2009). (In Russ.)
14. Erofeev, V.I., Kazhaev, V.V., Semerikova, N.P.: Nonlinear bending stationary waves in the Timoshenko beam. Nonlinear World **6**(5), 348–358 (2008). (In Russ.)
15. Asghari, M., Kahrobaiyan, M.H., Ahmadian, M.T.: A nonlinear Timoshenko beam formulation based on the modified couple stress theory. Int. J. Eng. Sci. **48**, 1749–1761 (2010)
16. Gulyaev, V.I., Gaidaichuk, V.V., Koshkin, V.L.: Elastic deformation, stability and vibrations of flexible curved rods. Naukova Dumka, Kyiv (1992).(In Russ.)
17. Guz', A.N., Zhuk, A.P., Makhort, F.G.: Waves in a layer with initial stresses. Naukova Dumka, Kyiv (1976).(In Russ.)
18. Guz', A.N.: The mechanics of brittle fracture of materials with initial stress. Naukova Dumka, Kyiv (1983).(In Russ.)
19. Ogden, R.W.: Incremental statics and dynamics of pre-stressed elastic materials, in waves in nonlinear pre-stressed materials. In: Destrade, M., Saccomandi, G. (eds.) CISM Courses and Lectures, vol. 495, pp. 1–26. Springer, Vienna (2007)
20. Shams, M., Destrade, M., Ogden, R.W.: Initial stresses in elastic solids: constitutive laws and acoustoelasticity. Wave Motion **48**(7), 552–567 (2011)
21. Biot, M.: Mechanics of incremental deformation. JWS, New York (1965)
22. Novozhilov, V.V.: Foundations of the nonlinear theory of elasticity. Dover Publications, New York (1999)
23. Filippov, A.P.: Oscillations of deformable systems. Mechanical engineering, Moscow (1970).(In Russ.)
24. Sergaliyev, A.S., Khajiyeva, L.A.: Flat Flexural Vibration of Drill-String with an Initial Curvature. In: Beran, J., Bílek, M., Žabka, P. (eds.) Advances in Mechanism Design II. MMS, vol. 44, pp. 231–237. Springer, Cham (2017). https://doi.org/10.1007/978-3-319-44087-3_30
25. Khajiyeva, L.: Kudaibergenov, Askar, Kudaibergenov, Askat: The effect of gas and fluid flows on nonlinear lateral vibrations of rotating drill strings. Commun. Nonlinear Sci. **59**, 565–579 (2018)

Modeling the Separation Process in Vertical Rotor Systems

A. B. Kydyrbekuly and G. E. Ibrayev(✉)

Al-Farabi Kazakh National University, Almaty 050040, Kazakhstan

Abstract. In this paper, we study and analyze the features of the separation process in a centrifugal force field, i.e. centrifugation process in vertical rotor systems. The main parameters that determine the time of separation of particles are revealed, and the optimal modes are indicated both from a constructive and from an economic point of view. Special cases of a fixed rotor are considered. Nonlinear differential equations of motion of a suspension particle are obtained, which do not have an exact solution. The study is carried out by analytical and numerical methods. The dependences of the slope angles of the tubes on the angular velocity of rotation of the rotor, sedimentation curves that allow one to estimate the time of deposition of particles, as well as the effect of the dispersed composition on the separation process as a whole, are obtained. The results of the study of this work allow us to determine with sufficient accuracy all the necessary characteristics working process of separation and sedimentation, and also allow in certain cases to exclude experimental work.

Keywords: Centrifugation · Particles · Rotor system · Emulsion separation · Sedimentation

1 Introduction

The centrifugation process has a wide range of applications in scientific research, medical and industrial sectors. The first scientific studies conducted by Knight in 1806 [1], where primary forecasts were made of the effect of the angle of inclination of the glass on the separation process. Also, the foundations of theoretical and experimental studies of centrifuges are laid in the works of de Laval, Svedberg, Pickels, Brakke, Anderson, G.I. Bremer, V.I. Sokolov, P.G. Romankov, N.N. Lipatov, E.M. Goldin. Research Yu.N. Bochkov, S.A. Plyushkin, E.V. Semenov, A. V. Schlau significantly deepened the theory of centrifugation processes and contributed to the creation of new effective centrifuges [2, 3].

Today, centrifuges are commonly used in a wide variety of fields, from large-scale commercial applications to laboratory research. The number of designs and configurations of centrifuges used in the mineral, petrochemical, chemical, medical, pharmaceutical, industrial, dairy, food, polymer, energy and agricultural industries is numerous [4, 5]. Emerging new trends in centrifugal technology have led to the emergence of diverse designs of centrifuges, some of which are difficult to evaluate using a known

© The Author(s), under exclusive license to Springer Nature Switzerland AG 2022
J. Beran et al. (Eds.): TMM 2020, MMS 85, pp. 104–113, 2022.
https://doi.org/10.1007/978-3-030-83594-1_11

calculation method [6, 7]. Numerous scientific works are devoted to this topic, patents, copyright certificates are obtained, methods for mechanically separating solid particles from liquids using centrifugation, and devices for their implementation are proposed [8–11]. At the present time there are two ways of separation the liquid heterogeneous systems: centrifugal filtration and centrifugal sedimentation. The centrifugal filtration method is widely applied in the sugar industry for separating raw sugar under the action of the centrifugal forces field. As a consequence of the filtration centrifugation the solid particles accumulate on the inner wall of the rotor. The centrifugal sedimentation that is as well widely applied in the modern industry is divided into two methods of separation heterogeneous systems: centrifugal clarification and precipitation centrifugation. As compared with the centrifugal clarification by using which occurs removing impurities that are in the liquid in small quantities, the precipitation centrifugation implies the separation of the suspensions that contain the most amount of the solid particles. It should be noted that the separation of the suspensions, at applying the centrifugal deposition method, is carried out in the following way: particles of the solid phase, which have a weight greater than the particles of the liquid phase, are deposited on the inner wall of the rotor under the action of the centrifugal force in the form of a ring layer, whereas particles of the liquid phase further in the form of the ring layer are located closer to the axis of rotation of the rotor [3, 5, 12–18]. Currently, great attention is paid to micro and nanoobjects, nanostructures and nanoparticles, and, consequently, there is a question of their experimental separation. In this regard, the problem of finding ways to differentiate nanoparticles by size is important. One of the possible and promising ways to solve this problem is application of the centrifugation method for description the process of the sedimentation of the nanoparticles in the centrifugal forces field [19–22].

In this paper, we study the spatial motion of a particle and the time of its deposition in a vertical centrifuge, where, for a complete assessment of the process of separation of solid particles from a liquid, it is taken into account that the angular velocity of rotation of the rotor, tubes and their angle of inclination are time-dependent variables, which also complicates the search for a general solution to the differential equations of motion of the particle and the mechanical system as a whole. Particular cases that take place in industrial production are given and analyzed. An analytical research technique used in the industry of the separation method for the technical rotary unit used for processing the suspension was developed and presented in [17].

2 Statement of the Problem

The rotor system consists of a flexible, symmetrical relative to the shaft supports and a disk mounted on the shaft, on which tubes with emulsion are mounted symmetrically. The tubes can rotate around their own horizontal axis with an angular velocity Ω (Fig. 1). The angular velocity of rotation of the shaft ω. The angle of rotation of the tubes from the vertical α, then $\Omega = \dot{\alpha}$ – the angular speed of rotation of the containers, L_s – distance from the rotor axis to the axis of rotation of containers, L_m – distance to the particles of the suspension with a mass Δm, r – containers radius, g – acceleration of gravity, L – containers length.

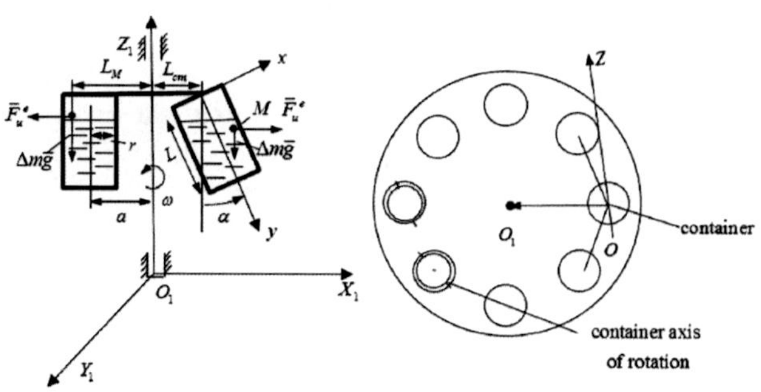

Fig. 1. The rotor system: shaft and disc with containers (test tube)

The rotor position is determined relative to a fixed coordinate system $Ox_1y_1z_1$ (see Fig. 1). The position of the studied particles *M(x,y,z)* is determined by the relative moving coordinate system *Oxyz*. *Oy* axis is directed along the axis of symmetry of the container bottom, an *Oz* axis directed along the rotational axis of the container. In the calculation the following assumptions were used: the angular speed of the rotor is large enough so that we can ignore the force of gravity suspension ($g \ll \omega^2 r$), spherical shape of particles assumed, particle mean free path is much larger its size, no turbulence suspension (laminar flow), the interaction forces between the particles and the walls of the container (test tube) absent, container (test tube) is rather narrow radius is much smaller than its length ($r_0 \ll L$), after deposition of the particles on the wall of the container stop moving, the friction force between the containers and their axes of rotation, as well as a change in gravity from the suspensions disregarded containers, the force of friction between the solid particles and the container wall disregarded.

3 The Equations of Motion

To compile the equation of motion of a particle, it is necessary to determine transfer $\bar{F}_e$ and Coriolis inertia $\bar{F}_c$ forces of the suspension particle. The transport acceleration of a particle M is determined by the formula.

$$\bar{a}_e = (\bar{\omega} + \bar{\Omega})((\bar{\omega} + \bar{\Omega}) \cdot \bar{r}) - \bar{r}\left(\bar{\omega} + \bar{\Omega}^2\right) + \bar{\varepsilon} \times \bar{r}, \tag{1}$$

Projecting the vector of transfer acceleration defined by Eq. (1) onto the moving coordinate axes *Ox*, *Oy*, *Oz*, we obtain:

$$\begin{aligned} a_{ex} &= -(L_s \cos\alpha + x)\left(\omega^2 + \Omega^2\right) + \omega^2 \sin\alpha (x \sin\alpha - y\cos\alpha) \\ &\quad + \dot{\Omega}(L_s \sin\alpha + y) - \dot{z}\omega\cos\alpha, \\ a_{ey} &= -(L_s \sin\alpha + y)\left(\omega^2 + \Omega^2\right) - \omega^2 \cos\alpha (x\sin\alpha - y\cos\alpha) \\ &\quad - \dot{\Omega}(L_s\cos\alpha + x) - \dot{z}\omega sin\alpha, \\ a_{ez} &= \dot{\omega}(L_s + y\sin\alpha + x\cos\alpha)\left(\omega^2 + \Omega^2\right) - 2\omega\Omega(x\sin\alpha - y\cos\alpha) - z\omega^2. \end{aligned} \tag{2}$$

The transfer and Coriolis inertia forces are equal respectively.

$$\bar{F}_e = -\Delta m\bar{a}_e,\ \bar{F}_c = -\Delta m\bar{a}_c.$$

Coriolis acceleration of the particle M of the suspension, taking into account the fact that $\Omega_{ox} = \omega \sin \alpha,\ \Omega_{oy} = -\omega \cos \alpha,\ \Omega_{oz} = -\Omega$, has the form

$$a_{cx} = -2(\dot{z}\omega \cos \alpha - \dot{y}\Omega),\ a_{cy} = -2(\dot{z}\omega \sin \alpha + \dot{x}\Omega),\ a_{cz} = 2\omega(\dot{y} \sin \alpha + \dot{x} \cos \alpha). \tag{3}$$

After using Eq. (2) and Eq. (3) , taking into account the accepted assumptions, the equations of motion of the particle M of the suspension have the form:

$$\begin{aligned}
\ddot{x} + 6\pi\eta_1 r_0 \dot{x} &= -g \sin \alpha + (L_s \cos \alpha + x)\left(\omega^2 + \Omega^2\right) - \omega^2 \sin \alpha (x \sin \alpha - y \cos \alpha) \\
&\quad + z\dot{\omega} \cos \alpha - \dot{\Omega}(L_s \sin \alpha + y) + 2(\dot{z}\omega \cos \alpha - \dot{y}\Omega), \\
\ddot{y} + 6\pi\eta_1 r_0 \dot{y} &= g \cos \alpha + (L_s \sin \alpha + y)\left(\omega^2 + \Omega^2\right) + \omega^2 \cos \alpha (x \sin \alpha - y \cos \alpha) \\
&\quad - \dot{\Omega}(L_s \cos \alpha + x) + 2(\dot{z}\omega \sin \alpha - \dot{x}\Omega) + z\dot{\omega} \cos \alpha, \\
\ddot{z} + 6\pi\eta_1 r_0 \dot{z} &= \omega^2 z + 2\Omega\omega(x \sin \alpha - y \cos \alpha) - \dot{\omega}(L_s + y \sin \alpha + x \cos \alpha) \\
&\quad - 2\omega(\dot{y} \sin \alpha + \dot{x} \cos \alpha)
\end{aligned} \tag{4}$$

There r_0 – particle radius M of suspension $\Delta m = 4\pi {r_0}^3 \rho_p$, $\eta_1 = \eta/\Delta m = 3\eta/4\pi {r_0}^3 \rho_p = 3\rho_f \nu/4\pi {r_0}^3 \rho_p$, ρ_f – particle density, ρ_p – fluid density; ν – kinematic viscosity of a liquid (suspension); $\eta = \rho_f \nu$ – dynamic viscosity of the liquid (suspension); $6\pi\eta r_0$ – the coefficient of friction of a particle M (the coefficient of resistance force of a liquid of a medium) during its motion. The system of Eq. (4) is nonlinear, since $\omega = \omega(t)$, $\alpha = \alpha(t)$, $\Omega = \Omega(t)$, and the rotation angle α is an argument of trigonometric functions. In addition, the coefficients of the functions with coordinates and their derivatives are variables. In this regard, Eq. (4) has no exact solution. The rotor is rotated at a constant angular speed, $\omega = const$, that is,. At first we define the dependence of the angle of the angular speed of the rotor. For this purpose, we draw up the balance of forces acting on the container (see Fig. 2). Projecting force in the direction of $\tau\tau$ we get

$$g \sin \alpha = \omega^2 (L_{cm} + l \sin \alpha) \cos \alpha \tag{5}$$

Where l – distance from the axis of rotation to center of gravity of the container with a suspension of. From here it follows that

$$\omega = \sqrt{\frac{g \cdot tg\alpha}{L_s + l \sin \alpha}}. \tag{6}$$

If $\omega \to \infty, \alpha \to \pi/2\ or\ \alpha \to \pi/2, \omega \to \infty$ which follows from the physical meaning of the problem.

From Eq. (6) it is obvious that if the angular velocity ω of the rotor is constant, then the angle of rotation of the cups α is also constant. Each value of ω corresponds to a specific defined value of α (see Table 1).

Accordingly, the faster the angular velocity of rotation, the faster the process of sedimentation of particles. But since in the general case it is better to avoid acceleration of the rotor system to huge angular velocities, it is necessary to determine the optimal angle of inclination of the tubes and the corresponding rotation speeds that would satisfy

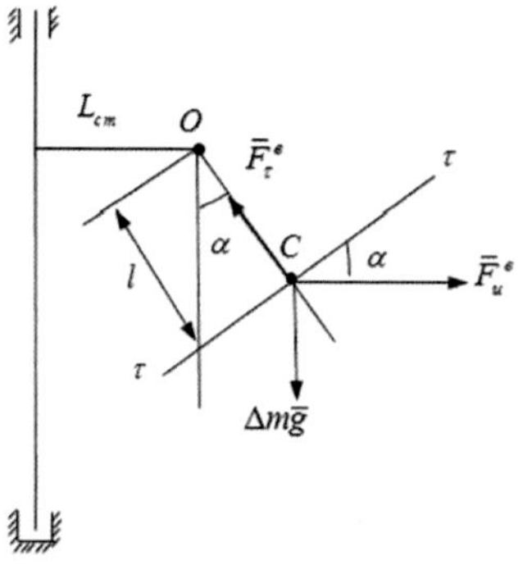

Fig. 2. Determination of forces

Table 1. The relationship between the angular velocity and the angle of the tubes.

Parameter	Value							
α, degree	15	30	45	60	75	80	85	88
ω, rpm	292	382	463	577	820	1005	1422	1640

the conditions of the problem of sedimentation of solid particles of the emulsion. Hence, each value of ω corresponds to the specific value of α. Then Eq. (4) as $\omega = const$, $\alpha = const$, $\Omega = 0$, $\varepsilon = 0$ we will have.

$$\begin{gathered} \ddot{x} + 2n\dot{x} - x\omega^2\cos^2\alpha = -g\sin\alpha + (L_s + y\sin\alpha)\omega^2\cos\alpha + 2\omega\cos\alpha\dot{z}, \\ \ddot{y} + 2n\dot{y} - y\omega^2\sin^2\alpha = g\cos\alpha + (L_s + x\cos\alpha)\omega^2\sin\alpha + 2\omega\sin\alpha\dot{z}, \\ \ddot{z} + 2n\dot{z} - \omega^2 z = -2\omega(\dot{y}\sin\alpha = \dot{x}\cos\alpha) \end{gathered} \tag{7}$$

where $2n = 6\pi\eta_1 r_0$ or $n = 9\nu\rho_f/4{r_0}^2\rho_p$ – friction coefficient. Equation (7) can be solved analytically, making some assumptions, which was done in [17]. In this case, taking into account the spatial motion of the deposited substance in the cavity, one can calculate the particle trajectory by a numerical method, estimate the time and trajectory of the sedimentation of particles in suspension for different values of the slope of the tube, particle size and resistance coefficient.

4 Results

In the calculation, three basic parameters vary, namely the particle radius, the angle of the tubes and the coefficient of resistance of the medium, which are the most important and fundamental in industrial production. Figure 3 shows the results of calculations for different tilt of the glasses (test tubes). As shown above, when the angular velocity of rotation of the rotor tends to an infinitely large value, the angle of inclination of the tubes tends to an absolutely horizontal position, i.e. to a value of 90°. It is impossible to achieve an absolutely horizontal position in industry, which is confirmed by a mathematical model. Therefore, it is necessary to determine the angle of inclination, which would provide a sufficient degree of cleaning and at the same time would be optimal from a structural point of view. For industrial centrifuges with operating speeds from 1500 rpm, angles from 80° are suitable. When tilted at 88°, particles settle faster (see Table 2) than at smaller bevel angles (tubes). We do not consider cases with a greater slope in connection with structural changes, which, for given parameters, can lead to a decrease in productivity, which follows from Eq. (6) . To achieve a tilt angle of 88°, it is necessary that the rotor rotates at a speed of 1640 rpm. Thus, if the centrifuge's working speed is 3000 rpm or more, as in many modern centrifuges of this class, the separation goal is achieved at much lower speeds. This will obviously reduce certain economic costs for centrifuges with a specified degree of purification. For an approximate estimate of the sedimentation time of particles T, one can use the expression obtained by the analytical method earlier in [17], i.e.

$$T = \ln\xi / b,$$

where

$$\xi = -\frac{p}{2} + \sqrt{\frac{p^2}{4} - q}, p = -\frac{2(L + A_0)b}{b(y_0 + A_0) + \dot{y}_0}, q = \frac{b(y_0 + A_0) - \dot{y}_0}{b(y_0 + A_0) + \dot{y}_0},$$

$$b_1 = b_2 = \sqrt{n^2 + \omega^2 \sin^2 \alpha} = b, b_1 + b_2 = 2b, \ A_0 = \frac{g \cos \alpha}{\omega^2 \sin^2 \alpha} + \frac{L_s}{\sin \alpha}.$$

In cases where the inclination angles are less than 80°, the sedimentation process proceeds more slowly and sedimentation occurs along spatial curves with increasing amplitudes due to an increase in the centrifugal force (see Fig. 3 and Table 2).

Table 2. The dependence of the separation time on the angle of the tube

Parameter	Value							
α, degree	15	30	45	60	75	80	85	88
T, s	87	61	49	37	28	25	23	19

The sedimentation rate of particles also depends on the nature of the multiphase fluid, and therefore it is advisable to consider the resistance of the medium as the next important parameter. For the calculation, drag coefficients for crude oil and some more viscous liquids were used. As expected, the higher the resistance index, the longer the separation time, respectively, the lower the resistance coefficient, the shorter the settling time of particles (see Fig. 4 and Table 3), which confirms the correctness of the proposed mathematical model.

The magnitude of the resulting resistance depends mainly on the mode of motion and the shape of the streamlined body. The mode of deposition of a solid phase particle can be taken as laminar until the condition Re $<$ 1–1.6 is fulfilled. In practical cases of centrifugation, the Reynolds number is less than the transition value from laminar

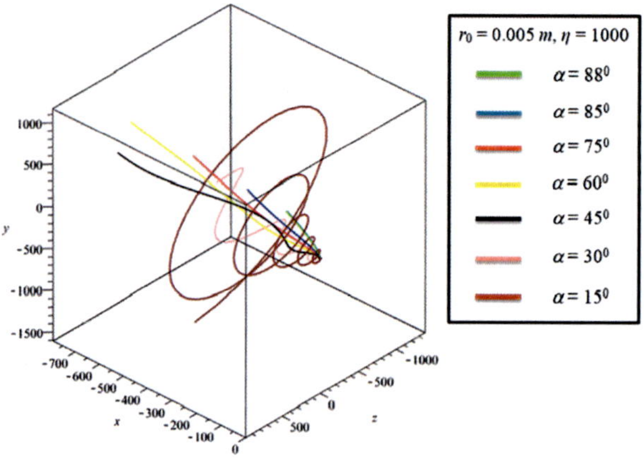

Fig. 3. Particle movement at different slope of the tube

to turbulent. During laminar motion, the body is surrounded by a boundary layer of fluid and flows smoothly around it. If the particle being deposited, having reached a certain distance from the axis of rotation, continued to experience the action of a constant centrifugal inertia force during further deposition, then the particle deposition rate would soon become constant. In this case, the fluid resistance would increase to the value of centrifugal force. The increase in resistance is caused by the fact that the liquid molecules in the layer adjacent to the body become denser, and with a decrease in the distance between them, the energy of mutual repulsion increases. But in reality, the centrifugal inertia force of a particle is always greater than the resistance force of a liquid medium due to its increase (force) as the particle moves away from the axis of rotation.

Table 3. The dependence of the separation time on the resistance of the medium

Parameter	Value at $\alpha = 88^0$, $r_0 = 0.005\ m$		
η, kg/s	1000	5000	10000
T, s	19	65	173

The third parameter is the dispersed composition of the particles, which must be separated. This characteristic is crucial when choosing a centrifuge. For a separator of this configuration, its required productivity is all the less, the smaller the particles of the solid phase in the treated suspension, which leads to a higher separation factor, which formally represents the Froude number. Since it is initially assumed that the particles have the shape of a sphere, their radius changes accordingly to consider different sizes of particles. It should be noted that in this model in one cycle of the system, the particle sizes were taken equal, i.e., in each case, all solid fractions in the suspension have the same size. Also, colloidal disperse systems are not taken into account, since it was found that for a certain range the particle size does not play the most important role in the separation process (see Fig. 5). For example, particles of size $r_0 = 0.01 - 0.05$ m settle along almost

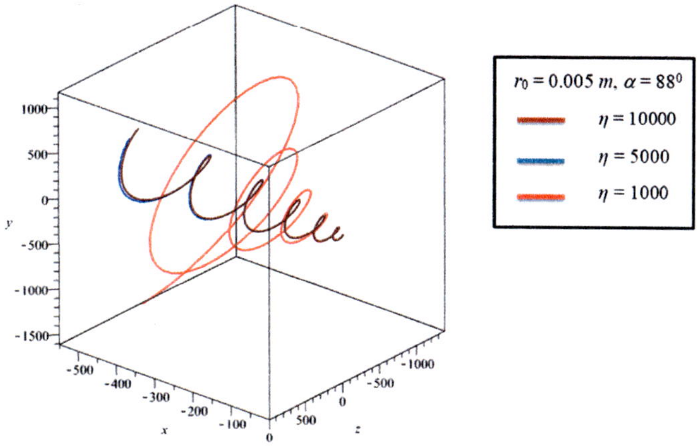

Fig. 4. Particle motion at different values of the drag coefficient of the medium

the same curved path, although here the particle size increases by 5 times (see Fig. 5). Initially, all particles up to the fourth second of the working regime of a fixed rotor have almost very close trajectories of subsidence. Up to the ninth second, particles of size $r_0 = 0.01 - 0.05$ m move along very close curves. Particles of size $r_0 = 5 \cdot 10^{-3} - 1 \cdot 10^{-6}$ m up to the ninth second settle along more curved paths, which further diverge even more (see Table 4).

Table 4. Separation time versus particle radius

Parameter	Value at $\alpha = 88^0$, $\eta = 1000$ kg/s			
r_0, m	0.000001	0.005	0.05	0.01
T, s	–	19	17	15

For these particles, these sedimentation curves are natural, since they are small in size, which means the need for higher operating speeds for micro- and nanoparticles. Since the separability of the suspension largely depends on the degree of dispersion, it becomes necessary to use the sedimentometric method to assess the nature of the particle distribution. When applying this method, the deposition time is plotted on the abscissa axis, and the total amount of solid phase deposited on the bottom of the centrifuge cup or passed through the reference level in a certain time is plotted along the ordinate axis. This amount consists of particles that have completely precipitated and are still precipitated [3]. This method requires the availability of experimental data, which is not always possible. At a certain stage of research, using the above sedimentation data for a qualitative assessment, the need for an experiment can be avoided.

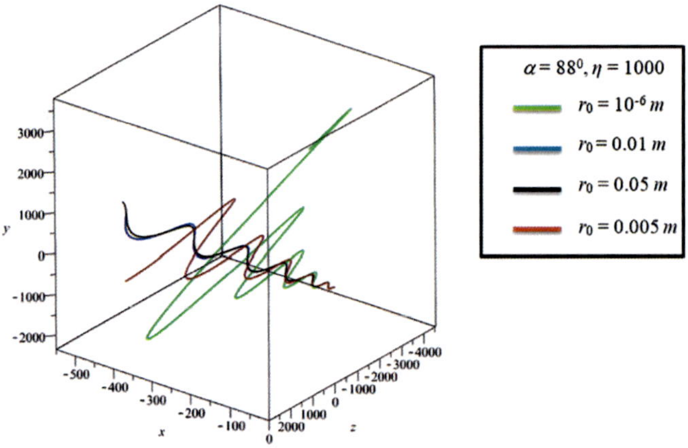

Fig. 5. Particle motion at different particle radius values

5 Conclusions

A qualitative analysis of the results based on classical numerical methods for solving differential equations is carried out. The time of particle deposition was estimated for different values of the angle of the tubes, the drag coefficients, and the size of these particles, taking into account spatial motion. The optimal parameters of the operating modes for the case of a fixed rotor, i.e. when each specific value of the angle of inclination corresponds to a specific angular speed of rotation of the disk on the flexible shaft. The results of the work performed confirm the physical meaning of the task, which can serve as a rationale for the use and implementation of this mathematical model in industrial production. The obtained dependences with sufficient accuracy for engineering practice allow us to determine such process characteristics as sedimentation time, deposition trajectory and the dependence of rotational speeds on inclination angles, as well as to predict the centrifuge operation efficiency. In addition, these results avoid the need to use the sedimentometric method and eliminate the need for additional costly experimental work.

References

1. Wilson, I.D.: Encyclopedia of separation science (2000)
2. Ruthven, M D.: Encyclopedia of Separation Technology. Wiley (1997)
3. Sokolov, V.I. Centrifugation. M.: Chemistry 18 (1976)
4. Grimwood, C.: Chapter 14 – Centrifugation. In: Chi Chou, C. (ed.) Handbook of Sugar Refining, pp. 203–244. Wiley, New York, USA (2000)
5. Lukyanenko, V.M., Taranets, A.V.: Industrial centrifuges, Moscow Chemistry, pp. 376 (1974)
6. Grimwood, G.C., Thewlis, M.J., Dean, A.J.: Efficient centrifugal operation. Int. Sugar J. **107**(1284), 680 (2005)
7. Swindells, R.J.: A mathematical model of a continuous sugar centrifuge (1982)
8. Splenter, L.E., Nirschl, H., Stickland, A.D., et al.: Pseudo two-dimensional modeling of sediment build-up in centrifuges: a comprotament approach using compressional rheology. AIChE J. **59**(10), 3843–3855 (2013)

9. Bell, G.R.A., Symons, D.D., Pearse, J.R.: Mathematical model for solids transport power in a decantor centrifuge. Chem. Eng. Sci. **107**, 114–122 (2014)
10. Bizard, A.F.M., et al.: Design guidelines for granular particles in a conical centrifugal filter. Chem. Eng. Res. Des. **91**(2), 348–360 (2013)
11. Semenov, E.V., Slavyanskii, A.A.: Calculation of emulsion separation process in a batch-operated centrifuge. Chem. Pet. Eng. **55**(9–10), 785–793 (2020). https://doi.org/10.1007/s10556-020-00694-y
12. Bullen, J., Bruijn, J.M.: Impact of the centrifugal speed of rotation on the quality of white sugar crystals. Zuckerindustrie **129**, 738–741 (2004)
13. Jullienne, L.M.S.A.: Washing sugar in batch A-centrifugals. Proc. SASTA, pp. 42–43 (1983)
14. Please, C.P., et al.: Extraction of molasses from sugar crystals in a centrifuge. Math. Model. Anal. **19**(3), 347–358 (2014)
15. Ligier, K.: Methods of diagnosing an acww 1000 sugar centrifuge with the use of vibration processes. Tech. Sci. **11**, 289–300 (2008)
16. Zachwieja, J., Ligier, K.: Numerical analysis of vertical rotor dynamics of ACWW 1000 centrifuge. J. Theor. Appl. Mech. **43**(2), 257–275 (2005)
17. Kydyrbekuly, A.B., Khajieva, L.A., Ybraev ,G.E.: Researching of the method of Separation of Fine-Grain Particles by Centrifugation in a Liquid Medium. In: Proceedings 12th International Conference on the Theory of Machines and Mechanisms, ADVANCES IN MECHANISM DESIGN II, Mechanisms and Machine Science. Vol. 44. pp. 105–116 (2017)
18. Lobanoff, V.S., Ross, R.R.: Centrifugal pumps design and application (2013)
19. Ungarish, M.: Fundamentals of centrifugal and gravity separation. In: Hydrodynamics of Suspensions. Springer (2013)
20. Dik, I.G., Minkov, L.L., Picuschak, E.V.: The influence of the particle size distribution function in a polydisperse suspension on the separation process in the classification apparatus. J. Tomsk State Univ. Math. Mech. **1**, 63–71 (2007)
21. Klasson, K.T., Taylor, P.A., Walker, J.F., Jr., Jones, S.A., Cummins, R.L., Richardson, S.A.: Modification of a centrifugal separator for in-well oil-water separation. Sep. Sci. Technol. **40**(1–3), 453–462 (2005)
22. Krebs, T., Schroën, C.G.P.H., Boom, R.M.: Separation kinetics of an oil-in-water emulsion under enhanced gravity. Chem. Eng. Sci. **71**, 118–125 (2012)

Analysis of the Stress-Strain State of Rotating Drill Strings with a Drilling Mud

Askar K. Kudaibergenov(✉), Askat K. Kudaibergenov, and L. A. Khajiyeva

Al-Farabi Kazakh National University, Almaty, Kazakhstan

Abstract. This paper studies the stress-strain state of a rotating drill string complicated by the effect of a drilling mud flow and external compressive and twisting forces. The drill string is considered in the form of a homogeneous isotropic elastic rod with constant cross-section. A nonlinear mathematical model of the drill string spatial lateral vibrations based on Novozhilov's nonlinear theory of elasticity is utilized. The Bubnov-Galerkin approach that allows reducing the given PDEs to ODEs and the numerical stiffness-switching method are applied to obtain a solution of the model. To analyze the stress-strain state of the drill string, we use the maximum stress intensity criterion and construct the graphs demonstrating changes of strain and stress in the chosen drill string cross-section with time, and stress distribution over the drill string length with time for nonlinear and linear cases. The research results indicate the significance of accounting for geometric nonlinearity to examine the drill string stress-strain state under the influence of the drilling mud.

Keywords: Drill string · Nonlinearity · Lateral vibrations · Stress-Strain state · Drilling mud

1 Introduction

The problem of strength of structures, parts of machines and mechanisms is of great importance in engineering practice. It is inextricably related to the determination of the magnitudes of stress and strain arising in deformable solids. Neglecting the requirements of physical and mathematical consistency, also associated with the use of basic stress-strain relationships, may result in large errors in calculations and serious ensuing consequences [1], especially when modelling the drill string dynamics complicated by factors of various nature.

The main reason of premature failure of drill string components, deterioration of the well trajectory and other undesirable phenomena emerging during the drilling process is the drill string vibrations, amongst which the coupled transverse mode is considered to be the most dangerous one [2]. Many drill string failures are preceded by fatigue fractures of metal, which are formed at multiple changes of magnitude and direction of loading at the sites of stress concentrations; however, they do not have a significant effect on the deformation of the drill string elements. At the same time, bending is a primary cause of residual stresses occurring during the drill string rotation.

© The Author(s), under exclusive license to Springer Nature Switzerland AG 2022
J. Beran et al. (Eds.): TMM 2020, MMS 85, pp. 114–122, 2022.
https://doi.org/10.1007/978-3-030-83594-1_12

The effect of the drilling mud (fluid) on the drill string motion was investigated in a number of works. In [3], the authors studied the impact of drilling hydraulics on drill stem vibrations using the Euler-Bernoulli beam theory and a finite element formulation for the drill stem discretization. The Herschel-Bulkley and power law fluid models were utilized to characterize the rheology of the drilling fluid. The results showed that the fluid dynamic pressure had a significant influence on the drill stem lateral frequencies. The effect of the drilling mud on the drill string dynamics based on the Timoshenko beam theory with the use of a simplified fluid-structure interaction model was studied in [4]. The impact of internal and external fluid flows on vibrations of the flexible cantilever pipe, which is similar in nature to the drill string, was investigated in [5]. It was obtained that the external fluid flow might cause a loss of the system stability by divergence. Analysis of coupled nonlinear lateral vibrations of a drill string under the effect of external supersonic gas and internal fluid flows was also carried out [6]. The research results indicated the presence of nonlinear effects and need for further study of the drill string dynamics in the gas and fluid flows in nonlinear formulation. For these reasons, this paper aims at studying the stress-strain state of the drill string taking into account geometric nonlinearity, the internal drilling mud flow and external loadings.

2 Governing Equations

Consider spatial lateral vibrations of a rotating drill string modelled in the form of a homogeneous isotropic elastic rod of length l with constant cross-section (Fig. 1).

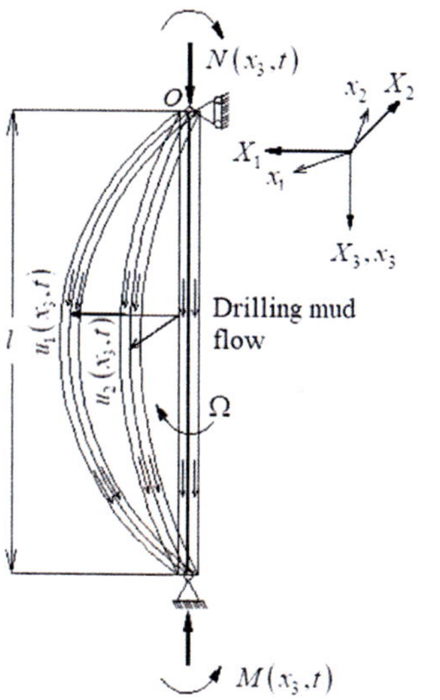

Fig. 1. Scheme of a drill string with a drilling mud.

The drill string is subject to the effect of an axial compressive force $N(x_3, t)$ and a torque $M(x_3, t)$. A drilling mud flow moves along the inner tube of the drill string in a positive direction of the vertical drill string axis x_3. Two Cartesian coordinate systems $OX_1X_2X_3$ and $Ox_1x_2x_3$ associated with lateral displacements of the drill string and accounting for its rotation with angular speed Ω, respectively, are utilized.

Since the drill string spatial lateral vibrations are studied, the displacement components of any point of the rod are given by

$$\begin{aligned} U_1(x_1, x_2, x_3, t) &= u_1(x_3, t), \\ U_2(x_1, x_2, x_3, t) &= u_2(x_3, t), \\ U_3(x_1, x_2, x_3, t) &= -\frac{\partial u_1(x_3,t)}{\partial x_3}x_1 - \frac{\partial u_2(x_3,t)}{\partial x_3}x_2, \end{aligned} \tag{1}$$

where the lateral displacements $u_1(x_3, t)$ and $u_2(x_3, t)$ are determined by solving the following nonlinear mathematical model including the effect of external compressive and twisting loadings and the drilling fluid flow pressure (see [7] for details):

$$\begin{aligned} &EI_{x_2}\frac{\partial^4 u_1}{\partial x_3^4} - \rho I_{x_2}\frac{\partial^4 u_1}{\partial x_3^2\partial t^2} + \frac{\partial^2}{\partial x_3^2}\left(M(x_3, t)\frac{\partial u_2}{\partial x_3}\right) + \frac{\partial}{\partial x_3}\left(N(x_3, t)\frac{\partial u_1}{\partial x_3}\right) \\ &-\frac{EA}{1-\nu}\frac{\partial}{\partial x_3}\left(\frac{\partial u_1}{\partial x_3}\right)^3 - \frac{EA(5-6\nu)}{2(1-\nu)}\frac{\partial}{\partial x_3}\left(\frac{\partial u_1}{\partial x_3}\left(\frac{\partial u_2}{\partial x_3}\right)^2\right) \\ &+(\rho A + \rho_f A_f)\left(\frac{\partial^2 u_1}{\partial t^2} - 2\Omega\frac{\partial u_2}{\partial t} - \Omega^2 u_1\right) - \rho_f I_{x_2}\left(\frac{\partial^4 u_1}{\partial x_3^4} + 2\frac{\partial^4 u_1}{\partial x_3^3\partial t} + \frac{\partial^4 u_1}{\partial x_3^2\partial t^2}\right) \\ &+\rho_f A_f\left(V_f^2\frac{\partial^2 u_1}{\partial x_3^2} + 2V_f\frac{\partial^2 u_1}{\partial x_3\partial t} - 2V_f\Omega\frac{\partial u_2}{\partial x_3}\right) \\ &+(\rho A + \rho_f A_f)g\left(\frac{\partial u_1}{\partial x_3} - (l - x_3)\frac{\partial^2 u_1}{\partial x_3^2}\right) = 0, \end{aligned} \tag{2.1}$$

$$\begin{aligned} &EI_{x_1}\frac{\partial^4 u_2}{\partial x_3^4} - \rho I_{x_1}\frac{\partial^4 u_2}{\partial x_3^2\partial t^2} - \frac{\partial^2}{\partial x_3^2}\left(M(x_3, t)\frac{\partial u_1}{\partial x_3}\right) + \frac{\partial}{\partial x_3}\left(N(x_3, t)\frac{\partial u_2}{\partial x_3}\right) \\ &-\frac{EA}{1-\nu}\frac{\partial}{\partial x_3}\left(\frac{\partial u_2}{\partial x_3}\right)^3 - \frac{EA(5-6\nu)}{2(1-\nu)}\frac{\partial}{\partial x_3}\left(\frac{\partial u_2}{\partial x_3}\left(\frac{\partial u_1}{\partial x_3}\right)^2\right) \\ &+(\rho A + \rho_f A_f)\left(\frac{\partial^2 u_2}{\partial t^2} + 2\Omega\frac{\partial u_1}{\partial t} - \Omega^2 u_2\right) - \rho_f I_{x_1}\left(\frac{\partial^4 u_2}{\partial x_3^4} + 2\frac{\partial^4 u_2}{\partial x_3^3\partial t} + \frac{\partial^4 u_2}{\partial x_3^2\partial t^2}\right) \\ &+\rho_f A_f\left(V_f^2\frac{\partial^2 u_2}{\partial x_3^2} + 2V_f\frac{\partial^2 u_2}{\partial x_3\partial t} + 2V_f\Omega\frac{\partial u_1}{\partial x_3}\right) \\ &+(\rho A + \rho_f A_f)g\left(\frac{\partial u_2}{\partial x_3} - (l - x_3)\frac{\partial^2 u_2}{\partial x_3^2}\right) = 0. \end{aligned} \tag{2.2}$$

Here E is Young's modulus, I_{x_1}, I_{x_2} axial inertia moments with respect to the x_2- and x_1-axes, ρ the drill string density, ν Poisson's ratio, A the cross-sectional area of the drill string, ρ_f the drilling mud density, A_f the internal cross-sectional area of the drill string, V_f the drilling mud flow speed.

Governing Eqs. (2.1), (2.2) meet the boundary conditions corresponding to the case of a simply supported rod:

$$\begin{aligned} &u_1(x_3, t) = u_2(x_3, t) = 0 \quad (x_3 = 0,\ x_3 = l), \\ &EI_{x_2}\frac{\partial^2 u_1(x_3,t)}{\partial x_3^2} = EI_{x_1}\frac{\partial^2 u_2(x_3,t)}{\partial x_3^2} = 0 \quad (x_3 = 0,\ x_3 = l). \end{aligned} \tag{3}$$

The initial conditions are written as

$$\begin{aligned} &u_1(x_3, t) = u_2(x_3, t) = 0 \quad (t = 0), \\ &\tfrac{\partial u_1(x_3,t)}{\partial t} = C_1, \quad \tfrac{\partial u_2(x_3,t)}{\partial t} = C_2 \quad (t = 0), \end{aligned} \tag{4}$$

where $C_1, \ C_2$ are constants defining the displacement rates of the rod cross-section from initial position in the Ox_1x_3- and Ox_2x_3-planes, respectively, at the initial time moment.

3 Determination of the Stress-Strain State

According to Novozhilov's nonlinear theory of elasticity [8], the strain tensor components ε_{ij} are determined through the displacement projections given by Eq. (1) as follows:

$$\begin{aligned} \varepsilon_{ii} &= \tfrac{\partial U_i}{\partial x_i} + \tfrac{1}{2}\left(\tfrac{\partial U_j}{\partial x_j}\right)^2, \quad i, j = \overline{1,3}, \\ \varepsilon_{ij} &= \tfrac{\partial U_i}{\partial x_j} + \tfrac{\partial U_j}{\partial x_i} + \tfrac{\partial U_k}{\partial x_i}\tfrac{\partial U_k}{\partial x_j}, \quad i, j, k = \overline{1,3} \quad (i \neq j). \end{aligned} \tag{5}$$

To find the stress tensor components σ_{ij}, the equations of generalized Hooke's law for a linear elastic homogeneous isotropic solid body in terms of Lame parameters $G, \ \lambda$ are applied:

$$\sigma_{ij} = 2G\varepsilon_{ij} + \lambda\delta_{ij}\varepsilon_{kk}, \quad i, j, k = \overline{1,3}, \tag{6}$$

where $G = \frac{E}{2(1+\nu)}, \ \lambda = \frac{E\nu}{(1+\nu)(1-2\nu)}$.

The maximum stress intensity criterion describing the transition to plastic deformation in metals with sufficient accuracy and proved by experiments [9] is used for determining the stress-strain state of the rotating drill string:

$$\sigma_* = \sqrt{\frac{1}{2}((\sigma_1 - \sigma_2)^2 + (\sigma_2 - \sigma_3)^2 + (\sigma_3 - \sigma_1)^2)}, \tag{7}$$

where σ_* is the critical stress, $\sigma_1, \ \sigma_2, \sigma_3$ principal stresses.

4 Numerical Results

To obtain a solution of the nonlinear mathematical model of drill string vibrations (2.1), (2.2) with boundary and initial conditions (3), (4), we utilize the Bubnov-Galerkin method that allows reducing the given PDEs to ODEs with further application of the numerical stiffness-switching method, as in [6]. The axial compressive force and torque are assumed to be distributed over the drill string length, i.e. $N(x_3, t) = N$, $M(x_3, t) = M$. All the computations are conducted in the Wolfram Mathematica package.

The initial values of the drilling system parameters used for numerical analysis are presented in Table 1.

Table 1. Drilling system parameters.

System parameter	Value
Drill string length, l	500 m
Angular speed of rotation, Ω	$2\,\mathrm{rad/s}$
Young's modulus, E	2.1×10^{11} Pa
Drill string material density, ρ	$7800\,\mathrm{kg/m^3}$
Poisson's ratio, ν	0.28
Outer diameter of the drill string, D	63.5×10^{-3} m
Wall thickness, h	4.5×10^{-3} m
Drill string cross-sectional area, A	$0.834 \times 10^{-3}\,\mathrm{m^2}$
Axial compressive load, N	3.5×10^{3} N
Torque, M	$10^4\,\mathrm{N \cdot m}$
Drilling mud density, ρ_f	$1120\,\mathrm{kg/m^3}$
Internal cross-sectional area, A_f	$2.33 \times 10^{-3}\,\mathrm{m^2}$
Drilling mud flow speed, V_f	$3.57\,\mathrm{m/s}$

Figure 2 illustrates the strain change in the drill string cross-section $x_3 = 0.8l$, Fig. 3 – the stress change in the considered cross-section with time (in seconds) at the initial values of the drilling system parameters. The cross-section point with coordinates $(x_1, x_2) = (0.03,\ 0.03)$ is chosen while drawing the graphs for conducting the comparative analysis of the application of nonlinear and linear mathematical models.

It is worth noting that the strain graphs are constructed for mean strain ε_0, which is calculated by the formula:

$$\varepsilon_0 = \frac{1}{3}(\varepsilon_{11} + \varepsilon_{22} + \varepsilon_{33}). \tag{8}$$

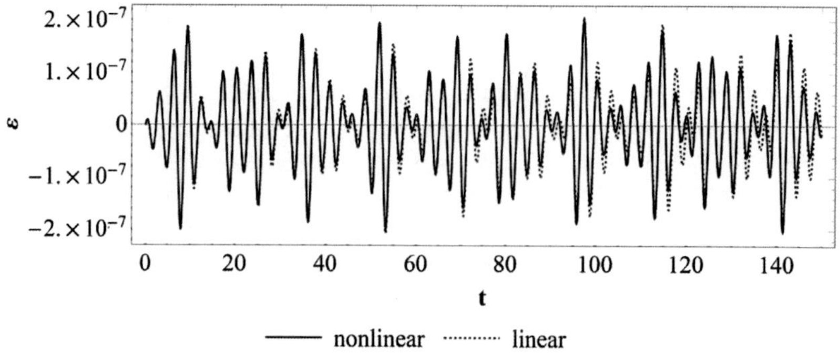

Fig. 2. Drill string strains at the initial parameter values.

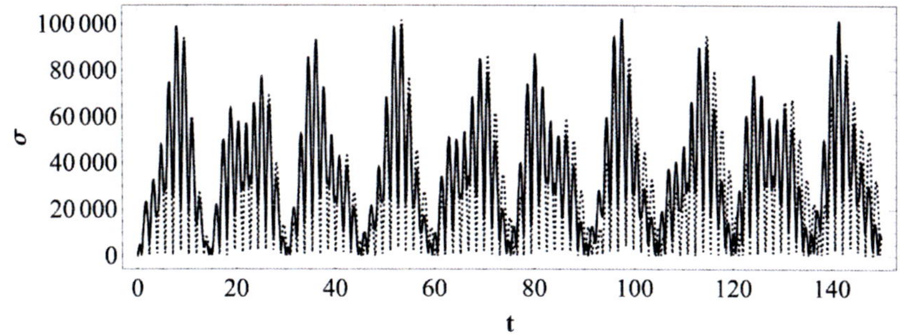

Fig. 3. Drill string stresses at the initial parameter values.

The linear model used for the comparative analysis is obtained on the basis of classical linear elasticity theory, according to which the strain components ε_{ij} linearly depend on the displacement projections U_i, and represents the linearized version of the governing Eqs. (2.1), (2.2). Hence, it does not involve the terms bringing the influence of geometric nonlinearity, namely the fifth and sixth terms in both Eqs. (2.1) and (2.2).

It follows from Figs. 2 and 3 that there is no significant difference in maximum values of strain and stress when applying the nonlinear model and its linear analogue at the initial system parameters.

Figures 4, 5 demonstrate the strain and stress changes in the selected cross-section of the rod $x_3 = 0.8l$, respectively, and Figs. 6, 7 give the stress distribution over the drill string length (in metres) with time when increasing the axial compressive load up to $N = 8 \times 10^3$N.

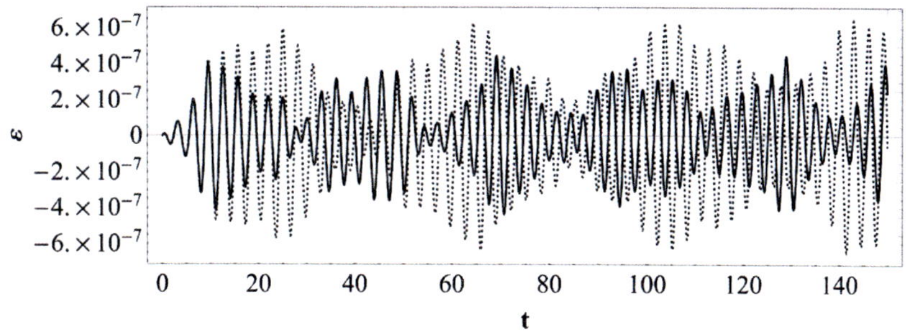

Fig. 4. Drill string strains under the axial force $N = 8 \times 10^3$ N.

As can be seen from Figs. 4 and 5, the increase of the axial force and accounting for geometric nonlinearity in the model results in considerable decrease of the mean strain and stress magnitudes in the given cross-section of the rotating drill string under the effect of the drilling mud flow. The stress distribution over the drill string length shows the sharp rise of stress in the lower part of the drill string (near 500 m) for nonlinear

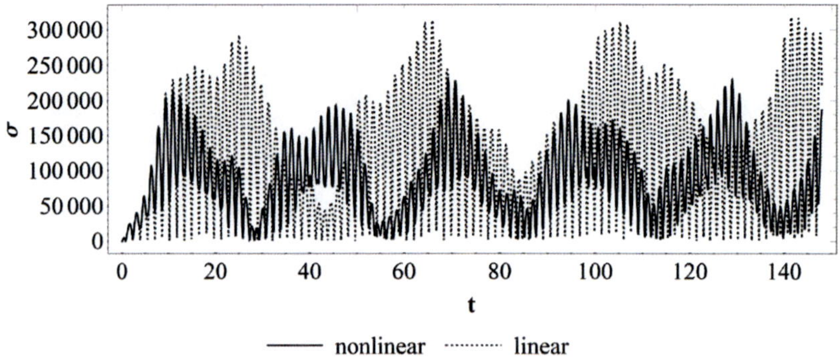

Fig. 5. Drill string stresses under the axial force $N = 8 \times 10^3$ N.

model (Fig. 6), which brings a correction to the results obtained in linear case when the most critical stresses are observed at a depth of 350–450 m (Fig. 7).

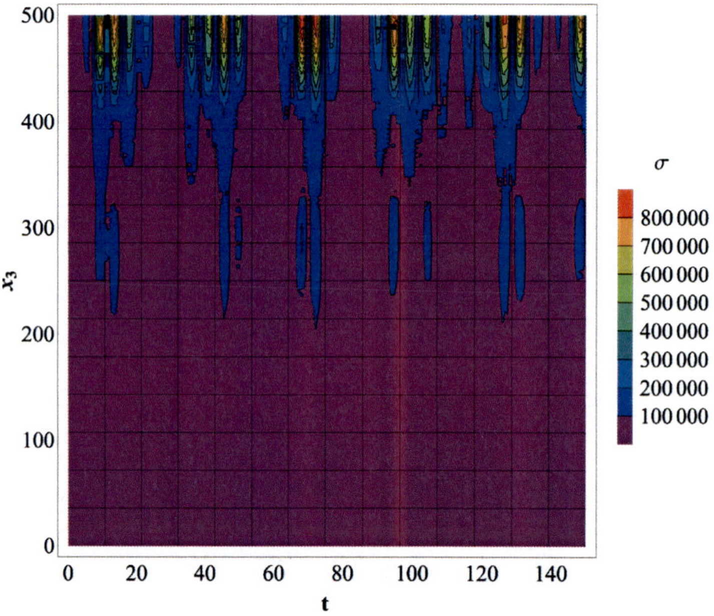

Fig. 6. Stress distribution over the drill string length at $N = 8 \times 10^3$ N for the nonlinear model.

The obtained results indicate the importance of accounting for geometric nonlinearity in mathematical models when increasing the values of the drilling system parameters to analyze the drill string stress-strain state under the influence of the drilling mud and complicated by external loadings.

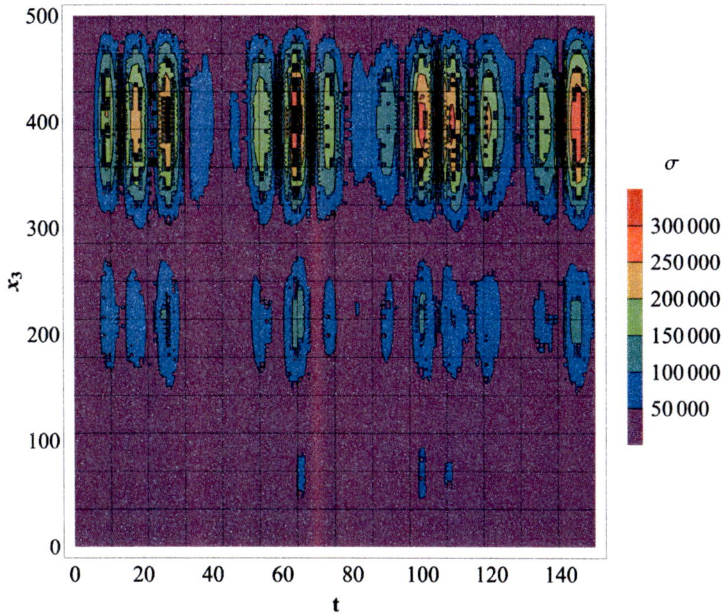

Fig. 7. Stress distribution over the drill string length at $N = 8 \times 10^3$ N for the linear model.

5 Conclusion

In this work, the analysis of the stress-strain state of a rotating drill string taking into account a drilling mud flow, an axial compressive load and a torque was conducted. A nonlinear mathematical model of the rod lateral vibrations, based on Novozhilov's nonlinear theory of elasticity, was utilized to describe the complex oscillatory process of the drill string and bring it closer to a real drilling process.

The graphs of the strain and stress change with time in a particular drill string cross-section, constructed on the basis of the nonlinear model, showed the significant decrease of the strain and stress magnitudes compared to its linear analogue when the compressive force was increased. The important result was obtained for stress distribution over the drill string length with time in nonlinear case. It revealed a shift of critical stresses to the bottom of the drill string compared to the results of linear model.

To further our research, the obtained results can be extended to a more complex problem taking into account the intermittent contact of the drill string with the borehole wall and damping effects.

Acknowledgements. This research is funded by the Science Committee of the Ministry of Education and Science of the Republic of Kazakhstan (Grant No. AP09261135).

References

1. Klyushnikov, V.D.: Physical and Mathematical Foundations of Strength and Plasticity. Moscow State Univ., Moscow (1994).(in Russian)
2. Ghasemloonia, A., Rideout, D.G., Butt, S.D.: Analysis of multi-mode nonlinear coupled axial-transverse drill string vibration in vibration assisted rotary drilling. J. Petrol. Sci. Eng. **116**, 36–49 (2014)
3. Al Dushaishi, M.F., Nygaard, R., Stutts, D.S.: Effect of drilling fluid hydraulics on drill stem vibrations. J. Nat. Gas Sci. Eng. **35**, 1059–1069 (2016)
4. Ritto, T.G., Sampaio, R., Soize, C.: Drill-string nonlinear dynamics accounting for the drilling fluid. In: 30° CILAMCE – Iberian-Latin-American Congress on Computational Methods in Engineering, Armação dos Búzios, RJ, Brazil, pp. 1–25 (2009)
5. Moditis, K., Paidoussis, M., Ratigan, J.: Dynamics of a partially confined, discharging, cantilever pipe with reverse external flow. J. Fluids Struct. **63**, 120–139 (2016)
6. Khajiyeva, L., Kudaibergenov, A., Kudaibergenov, A.: The effect of gas and fluid flows on nonlinear lateral vibrations of rotating drill strings. Commun. Nonlinear Sci. **59**, 565–579 (2018)
7. Kudaibergenov, A.K., Khajiyeva, L.A.: Modelling of drill string nonlinear dynamics with a drilling fluid flow. J. Math. Mech. Comp. Sci. **1**(97), 91–100 (2018)
8. Novozhilov, V.V.: Foundations of the Nonlinear Theory of Elasticity. Dover Publications, New York (1999)
9. Kachanov, L.M.: Foundations of Fracture Mechanics. Nauka, Moscow (1974).(in Russian)

Applications of the Modular Modelling Methodology to the Dynamic Analysis of Parallel Manipulators with Common Subsystems

Renato Maia Matarazzo Orsino[1], Tarcisio Antonio Hess-Coelho[2], and Fernando Malvezzi[1](✉)

[1] Mechanical Engineering Department, Maua Institute of Technology, São Caetano do Sul, SP, Brazil
{renato.orsino,fernando.malvezzi}@maua.br
[2] Department of Mechatronics and Mechanical Systems Engineering, Escola Politécnica, University of São Paulo, São Paulo, SP, Brazil
tarchess@usp.br

Abstract. Parallel manipulators offer advantages of high speed, precision and rigidity in comparison to serial robots, and these prospective benefits have attracted the attention of many researchers. One of the challenges related to the development of this class of manipulators is the derivation of the dynamic model. In fact, dynamic formulation used in parallel mechanisms is usually complex because of several independent kinematic chains which connect the end-effector to the base. The Modular Modelling Methodology (MMM) offers some advantages for modelling complex mechanical systems in comparison to the traditional ones. Basically MMM takes available models of subsystems as a starting point. Such models could have been derived from different types of formulations, being the result of prior investigations, testing and even validations.

The purpose of this work is to highlight the benefits of MMM from the perspective of the user/analyst by applying the methodology to parallel mechanisms with similar topologies.

Keywords: Parallel mechanism · Modular modelling methodology · Dynamic modelling · Multibody systems

1 Introduction

Dynamic formulation used in parallel mechanisms is usually complex because of several independent kinematic chains which connect the end-effector to the base. Implicit methods, such as Newton-Euler approach, and the Principle of Virtual Work are employed to generate and simulate dynamic models of complex systems [1–3]. However, these methods are not adequate for real-time computing,

© The Author(s), under exclusive license to Springer Nature Switzerland AG 2022
J. Beran et al. (Eds.): TMM 2020, MMS 85, pp. 123–133, 2022.
https://doi.org/10.1007/978-3-030-83594-1_13

and they are not recommended for co-simulation along with numerical optimization algorithms. Regarding explicit methods, Lagrangian formulation [4] is suitable for obtaining equations of motion in the explicit form. Nevertheless, this approach might become prohibitive for closed-loop mechanisms [5]. Orsino and Coelho have proposed recently a new methodology for modelling of multibody systems [6], namely Modular Modelling Methodology (MMM) which offers some advantages for modelling complex mechanical systems in comparison to the traditional ones. The main purpose of MMM is an improvement of the modelling process by integrating the use of libraries of mathematical models already available along with general-purpose computational tools. Then, according to the constraints that are imposed to those subsystems, a proper set of projection operators can be defined and the dependence among the kinematic variables can be adequately described, leading to the derivation of a consistent set of dynamic equations of motion for the complete system. The theoretical foundations of MMM approach are presented in [6], while in [7] it was presented as a particularized and detailed form of MMM to be applied to parallel mechanisms[1].

Kane's approach [3] requires an independent set of generalized speeds to be selected *a priori* according to the topology of the mechanical system in order to diminish the number of terms in the dynamical equations. However, if the topology changes, the whole process must be repeated in explicit form, not being a straightforward procedure to reuse any of the available previous information regarding either the previous structure or the lately attached one. The MMM, on the other hand, can be employed with any redundant selection of generalized speeds, in accordance with the preference of the analyst and, additionally, takes advantage of already available subsystems equations [7]. Hence, the derivation of the dynamic equations of a mechanical system with modified topology becomes a more direct procedure in either explicit or implicit form.

The purpose of this work is to highlight the benefits of MMM from the perspective of the user/analyst by applying the methodology to parallel mechanisms with similar topologies.

2 Modular Modelling Methodology (MMM) for Parallel Mechanisms: A Brief Overview

Hess-Coelho *et al.* [7] propose to represent a generic parallel mechanism $\mathcal{M}$ as a 3-level hierarchy (see Fig. 1), in which the end-effector E and the open-loop kinematic chains $\mathcal{K}_j$ $(j = 1, \ldots, n)$ that connect E to the fixed platform as considered as primary subsystems. Each kinematic chain $\mathcal{K}_j$, in turn, is composed of multiple rigid or flexible links, connected by joints which, in a first approach are modeled as ideal holonomic constraints. Denote by $\boldsymbol{\xi}_j$ the κ_j-vector of joint

[1] Regarding real-time applications, Coutinho et al. [8] presented experimental results for a planar 5R parallel mechanism for which a MMM based control system was designed. Good numerical performance was achieved since the numerical values of the matrices of the model were previously computed in accordance with the desired end-effector paths.

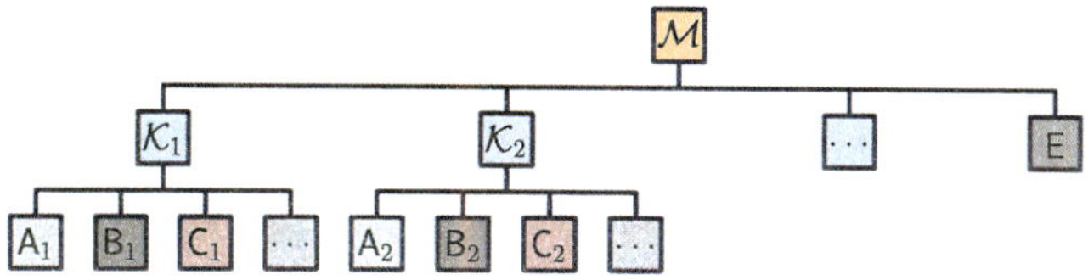

Fig. 1. Hierarchical conception of a parallel mechanism (adapted from [7]).

coordinates of the kinematic chain $\mathcal{K}_j$ $(j = 1, \ldots, n)$, with κ_j being its connectivity. Also, if the mechanism $\mathcal{M}$ has mobility μ, then a μ-vector $\mathbf{x}$ of generalized coordinates can be defined to describe the configuration (position and orientation) of the end-effector. For the sake of simplicity in notation, throughout this section, the end-effector will be alternatively denoted as subsystem $\mathcal{K}_{n+1} = \{\mathsf{E}\}$, constituted by a single body E, and $\boldsymbol{\xi}_{n+1} = \mathbf{x}$. In this case the expression of the Principle of Virtual Power for this system can be written as follows:

$$\sum_{j=1}^{n+1} \left[\delta \dot{\boldsymbol{\xi}}_j^{\mathsf{T}} \boldsymbol{\tau}_j + \sum_{\mathsf{L}_j \in \mathcal{K}_j} \delta \dot{W}_{\mathsf{L}_j} \right] = 0 \tag{1}$$

with:

- $\boldsymbol{\tau}_j$ denoting the generalized active forces provided by actuators and due to friction at the joints or some disturbance effects applied either to the links of the chains or to the end-effector[2].
- $\delta \dot{W}_{\mathsf{L}_j}$ denoting the virtual power associated with inertial and conservative forces acting in a given link $\mathsf{L}_j \in \mathcal{K}_j$.

Equation (1) can alternatively be expressed in terms of the variations of the generalized velocities $\delta \dot{\boldsymbol{\xi}}_j$ $(j = 1, \ldots, n, n+1)$:

$$\sum_{j=1}^{n+1} \delta \dot{\boldsymbol{\xi}}_j^{\mathsf{T}} \left[\mathbf{M}_j(\boldsymbol{\xi}_j) \ddot{\boldsymbol{\xi}}_j + \mathbf{h}_j(\boldsymbol{\xi}_j, \dot{\boldsymbol{\xi}}_j) + \mathbf{g}_j(\boldsymbol{\xi}_j) - \boldsymbol{\tau}_j \right] = 0 \tag{2}$$

In this last equation, the expressions between square brackets[3] would correspond to the dynamic models of subsystems in the absence of any constraints among the kinematic chains $\mathcal{K}_j$ and the end-effector E since, in this case, all $\delta \dot{\boldsymbol{\xi}}_j$ would be independent. These models for subsystems obtained under "relaxed constraints" can be considered the starting point for the application of MMM

[2] In typical parallel mechanisms forces or torques provided by the actuators are applied to the first joint of each kinematic chain. In this case, if friction and other disturbances are neglected, $\boldsymbol{\tau}_j = \left[\tau_j, 0, \ldots, 0\right]^{\mathsf{T}}$, for $j = 1, \ldots, n$, and $\boldsymbol{\tau}_{n+1} = \mathbf{0}$.

[3] In these expressions, $\mathbf{M}_j(\boldsymbol{\xi}_j)$ is the generalized inertia matrix of the kinematic chain $\mathcal{K}_j$, $\mathbf{h}_j(\boldsymbol{\xi}_j, \dot{\boldsymbol{\xi}}_j)$ stands for the corresponding terms related to generalized gyroscopic inertial forces, while $\mathbf{g}_j(\boldsymbol{\xi}_j)$ are related to generalized gravitational forces.

in the derivation of equations of motion for the parallel mechanism $\mathcal{M}$. As discussed in [7], mathematical models for typical kinematic chains adopted in the design of parallel mechanisms can be organized in libraries from which expressions for matrices $\mathbf{M}_j(\boldsymbol{\xi}_j)$, $\mathbf{h}_j(\boldsymbol{\xi}_j, \dot{\boldsymbol{\xi}}_j)$ and $\mathbf{g}_j(\boldsymbol{\xi}_j)$ would be readily available for the analyst or designer.

Even though the expressions for these matrices could be obtained from the application of the Analytical Mechanics formalism preferred by the analyst or even include some heuristic terms, Hess-Coelho *et al.* [7] proposed a direct approach, which resembles Kane's method [3], and is applicable whenever all the $\mathsf{L}_j \in \mathcal{K}_j$ could be modeled as rigid bodies. Let m_{L_j} be the total mass of L_j and $\mathbf{I}_{\mathsf{L}_j}$ be the matrix representation, in a given coordinate system, of the inertia tensor of this body with respect to its own center of mass L_j^*. For the sake of having a constant expression for $\mathbf{I}_{\mathsf{L}_j}$, typically, a body-fixed coordinate system is chosen. Also, denote by $\boldsymbol{\omega}_{\mathsf{L}_j}$ and $\boldsymbol{\alpha}_{\mathsf{L}_j}$ the column-vectors with the components, in the same coordinate system chosen for expressing $\mathbf{I}_{\mathsf{L}_j}$, of the angular velocity and angular acceleration of body L_j. Finally, let $\mathbf{v}_{\mathrm{L}_j^*}$, $\mathbf{a}_{\mathrm{L}_j^*}$ and $\mathbf{a}_{\mathrm{g}}$ be the column-vectors of components, in a same coordinate system (which typically is either body-fixed or defined in the fixed platform of the mechanism), of the velocity of L_j^*, the acceleration of L_j^* and the acceleration of gravity, respectively. All the aforementioned kinematic quantities must be measured with respect to an inertial reference frame, even though the coordinate systems chosen for representing their components could be defined in a non-inertial frame. In this case, it can be stated that:

$$\delta \dot{W}_{\mathsf{L}_j} = \delta \mathbf{v}_{\mathrm{L}_j^*}^{\mathsf{T}} m_{\mathsf{L}_j} \left(\mathbf{a}_{\mathrm{L}_j^*} - \mathbf{a}_{\mathrm{g}}\right) + \delta \boldsymbol{\omega}_{\mathsf{L}_j}^{\mathsf{T}} \left(\mathbf{I}_{\mathsf{L}_j} \boldsymbol{\alpha}_{\mathsf{L}_j} + \boldsymbol{\omega}_{\mathsf{L}_j}^{\times} \mathbf{I}_{\mathsf{L}_j} \boldsymbol{\omega}_{\mathsf{L}_j}\right) \tag{3}$$

with $\boldsymbol{\omega}_{\mathsf{L}_j}^{\times}$ denoting the skew-symmetric matrix representation of $\boldsymbol{\omega}_{\mathsf{L}_j}$. Notice that, for a planar system, $\mathbf{I}_{\mathsf{L}_j}$ is a scalar quantity, and $\boldsymbol{\omega}_{\mathsf{L}_j}^{\times} \mathbf{I}_{\mathsf{L}_j} \boldsymbol{\omega}_{\mathsf{L}_j} = \mathbf{0}$.

Also, since:

$$\mathbf{v}_{\mathsf{L}_j} = \mathbf{V}_{\mathsf{L}_j} \dot{\boldsymbol{\xi}}_j \quad \Rightarrow \quad \mathbf{a}_{\mathrm{L}_j^*} = \mathbf{V}_{\mathrm{L}_j^*} \ddot{\boldsymbol{\xi}}_j + \mathbf{e}_{\mathrm{L}_j^*} \tag{4}$$

$$\boldsymbol{\omega}_{\mathsf{L}_j} = \boldsymbol{\Omega}_{\mathsf{L}_j} \dot{\boldsymbol{\xi}}_j \quad \Rightarrow \quad \boldsymbol{\alpha}_{\mathsf{L}_j} = \boldsymbol{\Omega}_{\mathsf{L}_j} \ddot{\boldsymbol{\xi}}_j + \boldsymbol{\epsilon}_{\mathsf{L}_j} \tag{5}$$

with $\mathbf{e}_{\mathrm{L}_j^*}$ and $\boldsymbol{\epsilon}_{\mathsf{L}_j}$ expressing column-vectors of terms in $\mathbf{a}_{\mathrm{L}_j^*}$ and $\boldsymbol{\alpha}_{\mathsf{L}_j}$ that do not depend explicitly on $\ddot{\boldsymbol{\xi}}_j$, the following expressions are obtained for the matrices $\mathbf{M}_j(\boldsymbol{\xi}_j)$, $\mathbf{h}_j(\boldsymbol{\xi}_j, \dot{\boldsymbol{\xi}}_j)$ and $\mathbf{g}_j(\boldsymbol{\xi}_j)$:

$$\mathbf{M}_j(\boldsymbol{\xi}_j) = \sum_{\mathsf{L}_j \in \mathcal{K}_j} \left[m_{\mathsf{L}_j} \mathbf{V}_{\mathrm{L}_j^*}^{\mathsf{T}} \mathbf{V}_{\mathrm{L}_j^*} + \boldsymbol{\Omega}_{\mathsf{L}_j}^{\mathsf{T}} \mathbf{I}_{\mathsf{L}_j} \boldsymbol{\Omega}_{\mathsf{L}_j}\right] \tag{6}$$

$$\mathbf{h}_j(\boldsymbol{\xi}_j, \dot{\boldsymbol{\xi}}_j) = \sum_{\mathsf{L}_j \in \mathcal{K}_j} \left[m_{\mathsf{L}_j} \mathbf{V}_{\mathrm{L}_j^*}^{\mathsf{T}} \mathbf{e}_{\mathrm{L}_j^*} + \boldsymbol{\Omega}_{\mathsf{L}_j}^{\mathsf{T}} \left(\mathbf{I}_{\mathsf{L}_j} \boldsymbol{\epsilon}_{\mathsf{L}_j} + \boldsymbol{\omega}_{\mathsf{L}_j}^{\times} \mathbf{I}_{\mathsf{L}_j} \boldsymbol{\omega}_{\mathsf{L}_j}\right)\right] \tag{7}$$

$$\mathbf{g}_j(\boldsymbol{\xi}_j) = -\sum_{\mathsf{L}_j \in \mathcal{K}_j} m_{\mathsf{L}_j} \mathbf{V}_{\mathrm{L}_j^*}^{\mathsf{T}} \mathbf{a}_{\mathrm{g}} \tag{8}$$

Since joint coordinates $\boldsymbol{\xi}_j$ are adopted, all the existing internal constraints of kinematic chain $\mathcal{K}_j$ are already considered in the derivation of Eq. (2). In order to obtain the equations of motion for $\mathcal{M}$ from this weak formulation, it is still necessary to enforce the constraints between each chain $\mathcal{K}_j$ and the end-effector E, which can be expressed by the following constraint equations:

$$\mathbf{c}_j(\boldsymbol{\xi}_j, \mathbf{x}) = \mathbf{0} \quad \Rightarrow \quad \mathbf{J}_{\boldsymbol{\xi},j}(\boldsymbol{\xi}_j, \mathbf{x})\dot{\boldsymbol{\xi}}_j = \mathbf{J}_{\mathbf{x},j}(\boldsymbol{\xi}_j, \mathbf{x})\dot{\mathbf{x}} \tag{9}$$

with $\mathbf{J}_{\boldsymbol{\xi},j} = \partial\mathbf{c}_j/\partial\boldsymbol{\xi}_j$ and $\mathbf{J}_{\mathbf{x},j} = -\partial\mathbf{c}_j/\partial\mathbf{x}$. Thus, it can be stated that:

$$\dot{\boldsymbol{\xi}}_j = \mathbf{J}_{\boldsymbol{\xi},j}^{-1}\mathbf{J}_{\mathbf{x},j}\dot{\mathbf{x}} = \mathbf{C}_j(\boldsymbol{\xi}_j, \mathbf{x})\dot{\mathbf{x}} \quad \Rightarrow \quad \delta\dot{\boldsymbol{\xi}}_j = \mathbf{C}_j(\boldsymbol{\xi}_j, \mathbf{x})\delta\dot{\mathbf{x}} \tag{10}$$

The second time derivative of the constraint equations, in turn, leads to:

$$\ddot{\boldsymbol{\xi}}_j = \mathbf{C}_j(\boldsymbol{\xi}_j, \mathbf{x})\ddot{\mathbf{x}} + \mathbf{d}_j(\boldsymbol{\xi}_j, \mathbf{x}, \dot{\boldsymbol{\xi}}_j, \dot{\mathbf{x}}) \quad \text{with} \quad \mathbf{d}_j = \mathbf{J}_{\boldsymbol{\xi},j}^{-1}(\dot{\mathbf{J}}_{\mathbf{x},j}\dot{\mathbf{x}} - \dot{\mathbf{J}}_{\boldsymbol{\xi},j}\dot{\boldsymbol{\xi}}_j) \tag{11}$$

Notice that, since $\boldsymbol{\xi}_{n+1} = \mathbf{x}$, then: $\mathbf{C}_{n+1} = \mathbf{I}$ and $\mathbf{d}_{n+1} = \mathbf{0}$.

Finally, substituting Eqs. (10) and (11) in Eq. (2) and considering that $\delta\dot{\mathbf{x}}$ is a column-vector of arbitrary variations, the following equations of motion are obtained for $\mathcal{M}$:

$$\mathbf{M}(\boldsymbol{\xi}, \mathbf{x})\ddot{\mathbf{x}} + \mathbf{h}(\boldsymbol{\xi}, \mathbf{x}, \dot{\mathbf{x}}) + \mathbf{g}(\boldsymbol{\xi}, \mathbf{x}) = \mathbf{f}(\boldsymbol{\xi}, \mathbf{x}, \boldsymbol{\tau}) \tag{12}$$

with $\boldsymbol{\xi}$ being a column-vector constituted by the joint coordinates of all the kinematic chains $\mathcal{K}_j$ ($j = 1, \ldots, n$), and:

$$\begin{aligned} \mathbf{M} &= \sum_{j=1}^{n+1} \mathbf{C}_j^\mathsf{T}\mathbf{M}_j\mathbf{C}_j, & \mathbf{h} &= \sum_{j=1}^{n+1} \mathbf{C}_j^\mathsf{T}(\mathbf{M}_j\mathbf{d}_j + \mathbf{h}_j), \\ \mathbf{g} &= \sum_{j=1}^{n+1} \mathbf{C}_j^\mathsf{T}\mathbf{g}_j, & \mathbf{f} &= \sum_{j=1}^{n+1} \mathbf{C}_j^\mathsf{T}\boldsymbol{\tau}_j \end{aligned} \tag{13}$$

3 Case Study: Exploring the Common Modularity Between Two Planar Parallel Manipulators Through MMM

This case study explores the advantages of applying MMM for the derivation of equations of motion for parallel manipulators with similar topologies. In this present context, "topological similarity" should be understood as an interchangeability of one or more kinematic chains linking the fixed platform to the moving one, without affecting the mobility of the mechanism (i.e., keeping the same degrees of freedom of the end-effector).

Consider the topologically symmetric 3$\underline{\text{P}}$RR and the topologically asymmetric 2$\underline{\text{P}}$RR + R$\underline{\text{P}}$R manipulators illustrated in Figs. 2 and 3. In both of them, the position and orientation of the end-effector E, represented by the bar P_1P_3, in a vertical plane is controlled through actuators in the prismatic joints of the $\underline{\text{P}}$RR or R$\underline{\text{P}}$R kinematic chains. The corresponding active degrees-of-freedom

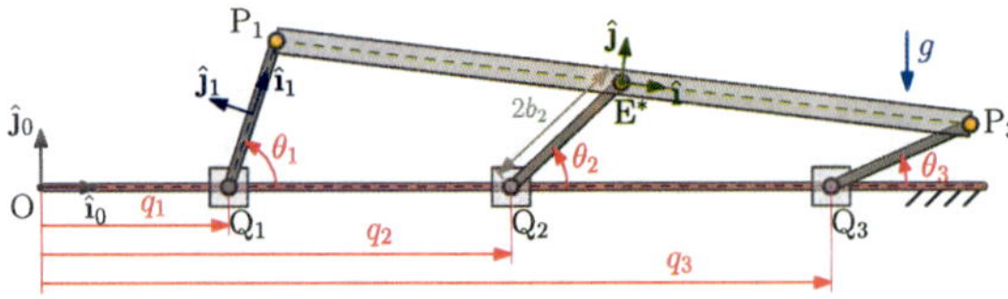

Fig. 2. 3PRR: kinematic diagram, joint coordinates and vector bases.

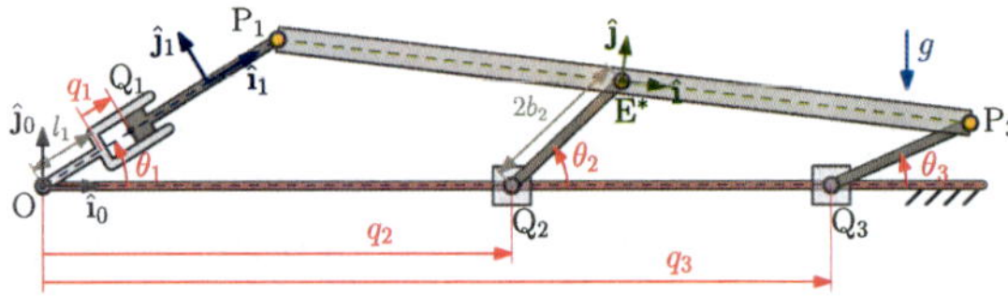

Fig. 3. 2PRR + RPR: kinematic diagram, joint coordinates and vector bases.

are described by the coordinates $\mathbf{q} = [q_1,\ q_2,\ q_3]^\mathsf{T}$. The 3-vector $\mathbf{x} = [x,\ y,\ \phi]^\mathsf{T}$, describes the position and orientation of the end-effector, with (x, y) being the Cartesian coordinates of the center of mass E^* in the coordinate system defined in the fixed platform, with origin O, and ϕ measuring the counterclockwise rotation between the longitudinal direction of the end-effector and the horizontal (notice that, in the configurations shown in Figs. 2 and 3, ϕ is negative). Also, the joint-coordinates of each kinematic chain are described by the 2-vectors $\boldsymbol{\xi}_j = [q_j,\ \theta_j]$, $j = 1, 2, 3$, with θ_j measuring the counterclockwise rotation between the segment $\mathrm{Q}_j\mathrm{P}_j$ and the horizontal, which are passive joint variables.

Define the orthonormal vector bases $(\hat{\mathbf{i}}_0, \hat{\mathbf{j}}_0)$, in the fixed platform, $(\hat{\mathbf{i}}, \hat{\mathbf{j}})$, attached to the end-effector E, and $(\hat{\mathbf{i}}_j, \hat{\mathbf{j}}_j)$, with $\hat{\mathbf{i}}_j$ being the unit vector directed along the segments $\mathrm{Q}_j\mathrm{P}_j$, $j = 1, 2, 3$. Also, let $\hat{\mathbf{k}} = \hat{\mathbf{i}} \times \hat{\mathbf{j}} = \hat{\mathbf{i}}_0 \times \hat{\mathbf{j}}_0 = \hat{\mathbf{i}}_j \times \hat{\mathbf{j}}_j$.

3.1 Position Kinematics

Four relevant points of the end-effector E must be highlighted: the center of mass E^* and the revolute joint centers P_1, $\mathrm{P}_2 = \mathrm{E}^*$ and P_3. Let $\vec{\mathbf{r}}_{\mathrm{P}_j|\mathrm{E}^*} = \sigma_j l_\mathsf{E} \hat{\mathbf{i}}$ be the position vector of P_j relative to E^*, with $\sigma_1 = -1$, $\sigma_2 = 0$ and $\sigma_3 = +1$. Also, by the definition of (x, y), $\vec{\mathbf{r}}_{\mathrm{E}^*|\mathrm{O}} = x\hat{\mathbf{i}}_0 + y\hat{\mathbf{j}}_0$. It can be stated that for each kinematic chain of these manipulators the following constraint equation must hold:

$$\vec{\mathbf{r}}_{\mathrm{P}_j|\mathrm{Q}_j} + \vec{\mathbf{r}}_{\mathrm{Q}_j|\mathrm{O}} = \vec{\mathbf{r}}_{\mathrm{P}_j|\mathrm{E}^*} + \vec{\mathbf{r}}_{\mathrm{E}^*|\mathrm{O}} \tag{14}$$

Particularly, for a PRR kinematic chain, $\vec{\mathbf{r}}_{\mathrm{Q}_j|\mathrm{O}} = q_j\hat{\mathbf{i}}_0$ and $\vec{\mathbf{r}}_{\mathrm{P}_j|\mathrm{Q}_j} = 2b_j\hat{\mathbf{i}}_j$, in which case the constraints among each chain and the end-effector E can be expressed as:

$$\mathbf{c}_j(\boldsymbol{\xi}_j, \mathbf{x}) = \begin{bmatrix} q_j + 2b_j\cos\theta_j - x - \sigma_j l_\mathsf{E}\cos\phi \\ 2b_j\sin\theta_j - y - \sigma_j l_\mathsf{E}\sin\phi \end{bmatrix} = \mathbf{0} \tag{15}$$

For a R<u>P</u>R kinematic chain, in turn[4], $\vec{\mathbf{r}}_{\mathsf{Q}_j|\mathsf{O}} = h_j\hat{\mathbf{\imath}}_0 + (l_j+q_j)\hat{\mathbf{\imath}}_j$ and $\vec{\mathbf{r}}_{\mathsf{P}_j|\mathsf{Q}_j} = 2b_j\hat{\mathbf{\imath}}_j$, which lead to the following constraint equations:

$$\mathbf{c}_j(\boldsymbol{\xi}_j, \mathbf{x}) = \begin{bmatrix} h_j + (q_j + l_j + 2b_j)\cos\theta_j - x - \sigma_j l_{\mathsf{E}}\cos\phi \\ (q_j + l_j + 2b_j)\sin\theta_j - y - \sigma_j l_{\mathsf{E}}\sin\phi \end{bmatrix} = \mathbf{0} \tag{16}$$

At each configuration, given the values of $\mathbf{x}$, the corresponding values of the $\boldsymbol{\xi}_j$, $j = 1, 2, 3$, can be determined from the numerical solution of the constraint equations of the associated kinematic chain.

3.2 Velocity and Acceleration Kinematics

<u>P</u>RR **Chain.** Let A_j denote the sliding block with center Q_j whose translation along the fixed guide is controlled and B_j denote the bar $\mathrm{Q}_j\mathrm{P}_j$. Assuming that the center of mass of A_j is $\mathrm{A}_j^* = \mathrm{Q}_j$ and the center of mass of B_j is the midpoint of the segment $\mathrm{Q}_j\mathrm{P}_j$, it can be stated that: $\vec{\mathbf{v}}_{\mathrm{A}_j^*} = \dot{q}_j\hat{\mathbf{\imath}}_0$, $\vec{\mathbf{v}}_{\mathrm{B}_j^*} = \dot{q}_j\hat{\mathbf{\imath}}_0 + b_j\dot{\theta}_j\hat{\mathbf{\jmath}}_j$, $\vec{\boldsymbol{\omega}}_{\mathsf{A}_j} = \vec{\mathbf{0}}$ and $\vec{\boldsymbol{\omega}}_{\mathsf{B}_j} = \dot{\theta}_j\hat{\mathbf{k}}$. Choosing the vector bases $(\hat{\mathbf{\imath}}_0, \hat{\mathbf{\jmath}}_0, \hat{\mathbf{k}})$ for representing $\vec{\mathbf{v}}_{\mathrm{A}_j^*}$ and $\vec{\boldsymbol{\omega}}_{\mathsf{A}_j}$, and $(\hat{\mathbf{\imath}}_j, \hat{\mathbf{\jmath}}_j, \hat{\mathbf{k}})$ for representing $\vec{\mathbf{v}}_{\mathrm{B}_j^*}$ and $\vec{\boldsymbol{\omega}}_{\mathsf{B}_j}$ in column-matrix forms:

$$\mathbf{V}_{\mathrm{A}_j^*} = \begin{bmatrix} 1 & 0 \\ 0 & 0 \end{bmatrix}, \quad \mathbf{V}_{\mathrm{B}_j^*} = \begin{bmatrix} \cos\theta_j & 0 \\ -\sin\theta_j & b_j \end{bmatrix}, \quad \boldsymbol{\Omega}_{\mathsf{A}_j} = \begin{bmatrix} 0 & 0 \end{bmatrix}, \quad \boldsymbol{\Omega}_{\mathsf{B}_j} = \begin{bmatrix} 0 & 1 \end{bmatrix} \tag{17}$$

For the corresponding expressions of accelerations and angular accelerations:

$$\mathbf{e}_{\mathrm{A}_j^*} = \begin{bmatrix} 0 \\ 0 \end{bmatrix}, \quad \mathbf{e}_{\mathrm{B}_j^*} = \begin{bmatrix} -b_j\dot{\theta}_j^2 \\ 0 \end{bmatrix}, \quad \boldsymbol{\epsilon}_{\mathsf{A}_j} = \boldsymbol{\epsilon}_{\mathsf{B}_j} = 0 \tag{18}$$

Also, for a <u>P</u>RR chain, relations among the $\dot{\boldsymbol{\xi}}_j$ and $\dot{\mathbf{x}}$ can be obtained from the time derivative of Eq. (15), which can be written in the form of Eq. (9), with:

$$\mathbf{J}_{\boldsymbol{\xi},j} = \begin{bmatrix} 1 & -2b_j\sin\theta_j \\ 0 & 2b_j\cos\theta_j \end{bmatrix} \quad \text{and} \quad \mathbf{J}_{\mathbf{x},j} = \begin{bmatrix} 1 & 0 & -\sigma_j l_{\mathsf{E}}\sin\phi \\ 0 & 1 & \sigma_j l_{\mathsf{E}}\cos\phi \end{bmatrix} \tag{19}$$

Therefore, by defining the matrices $\mathbf{C}_j = \mathbf{J}_{\boldsymbol{\xi},j}^{-1}\mathbf{J}_{\mathbf{x},j}$, the vectors $\boldsymbol{\xi}_j$, for $j = 1, 2, 3$, it can be stated that $\dot{\boldsymbol{\xi}}_j = \mathbf{C}_j\dot{\mathbf{x}}$, with:

$$\mathbf{C}_j = \frac{1}{\cos\theta_j}\begin{bmatrix} \cos\theta_j & \sin\theta_j & \sigma_j l_{\mathsf{E}}\sin(\theta_j - \phi) \\ 0 & 1/(2b_j) & (\sigma_j l_{\mathsf{E}}\cos\phi)/(2b_j) \end{bmatrix} \tag{20}$$

Finally, according to Eq. (11), $\ddot{\boldsymbol{\xi}}_j = \mathbf{C}_j\ddot{\mathbf{x}} + \mathbf{d}_j$, with:

$$\mathbf{d}_j = \frac{1}{\cos\theta_j}\begin{bmatrix} 2b_j\dot{\theta}_j^2 - \sigma_j l_{\mathsf{E}}\dot{\phi}^2\cos(\theta_j - \phi) \\ \dot{\theta}_j^2\sin\theta_j - \sigma_j l_{\mathsf{E}}\dot{\phi}^2\sin\phi/(2b_j) \end{bmatrix} \tag{21}$$

[4] In Fig. 3, since the position of the revolute joint between this chain and the fixed platform coincides with the origin O, then $h_1 = 0$.

R$\underline{P}$R **Chain.** Let A_j denote the link articulated to fixed platform at O_j (in Fig. 3, $O_1 = O$) and B_j denote the corresponding bar Q_jP_j. Also, assuming that the center of mass of A_j is $A_j^* = O_j$ and the center of mass of B_j is the midpoint of the segment Q_jP_j, it can be stated that: $\vec{\mathbf{v}}_{A_j^*} = \vec{\mathbf{0}}$, $\vec{\mathbf{v}}_{B_j^*} = \dot{q}_j\hat{\mathbf{i}}_j + (q_j + l_j + b_j)\dot{\theta}_j\hat{\mathbf{j}}_j$ and $\vec{\boldsymbol{\omega}}_{A_j} = \vec{\boldsymbol{\omega}}_{B_j} = \dot{\theta}_j\hat{\mathbf{k}}$. Choosing the vector basis $(\hat{\mathbf{i}}_j, \hat{\mathbf{j}}_j, \hat{\mathbf{k}})$ for the column-matrix representation of $\vec{\mathbf{v}}_{A_j^*}$, $\vec{\mathbf{v}}_{B_j^*}$, $\vec{\boldsymbol{\omega}}_{A_j}$ and $\vec{\boldsymbol{\omega}}_{B_j}$:

$$\mathbf{V}_{A_j^*} = \mathbf{0}, \quad \mathbf{V}_{B_j^*} = \begin{bmatrix} 1 & 0 \\ 0 & (q_j + l_j + b_j) \end{bmatrix}, \quad \boldsymbol{\Omega}_{A_j} = \boldsymbol{\Omega}_{B_j} = \begin{bmatrix} 0 & 1 \end{bmatrix} \tag{22}$$

For the corresponding expressions of accelerations and angular accelerations:

$$\mathbf{e}_{A_j^*} = \begin{bmatrix} 0 \\ 0 \end{bmatrix}, \quad \mathbf{e}_{B_j^*} = \begin{bmatrix} -(q_j + l_j + b_j)\dot{\theta}_j^2 \\ 2\dot{q}_j\dot{\theta}_j \end{bmatrix}, \quad \boldsymbol{\epsilon}_{A_j} = \boldsymbol{\epsilon}_{B_j} = 0 \tag{23}$$

Also, for a R$\underline{P}$R chain, relations among the $\dot{\boldsymbol{\xi}}_j$ and $\dot{\mathbf{x}}$ can be obtained from the time derivative of Eq. (16), which leads to:

$$\mathbf{J}_{\boldsymbol{\xi},j} = \begin{bmatrix} \cos\theta_j & -(q_j + l_j + 2b_j)\sin\theta_j \\ \sin\theta_j & (q_j + l_j + 2b_j)\cos\theta_j \end{bmatrix} \quad \text{and} \quad \mathbf{J}_{\mathbf{x},j} = \begin{bmatrix} 1 & 0 & -\sigma_j l_E \sin\phi \\ 0 & 1 & \sigma_j l_E \cos\phi \end{bmatrix} \tag{24}$$

Thus, $\dot{\boldsymbol{\xi}}_j = \mathbf{C}_j\dot{\mathbf{x}}$ and $\ddot{\boldsymbol{\xi}}_j = \mathbf{C}_j\ddot{\mathbf{x}} + \mathbf{d}_j$, with:

$$\mathbf{C}_j = \begin{bmatrix} \cos\theta_j & \sin\theta_j & \sigma_j l_E \sin(\theta_j - \phi) \\ \frac{-\sin\theta_j}{(q_j+l_j+2b_j)} & \frac{\cos\theta_j}{(q_j+l_j+2b_j)} & \frac{\sigma_j l_E \cos(\theta_j - \phi)}{(q_j+l_j+2b_j)} \end{bmatrix} \tag{25}$$

$$\mathbf{d}_j = \begin{bmatrix} (q_j + l_j + 2b_j)\dot{\theta}_j^2 - \sigma_j l_E \dot{\phi}^2 \cos(\theta_j - \phi) \\ \frac{-2\dot{q}_j\dot{\theta}_j + \sigma_j l_E \dot{\phi}^2 \sin(\theta_j - \phi)}{(q_j+l_j+2b_j)} \end{bmatrix} \tag{26}$$

End-Effector. Analogously, since $\vec{\mathbf{v}}_{E^*} = \dot{x}\hat{\mathbf{i}}_0 + \dot{y}\hat{\mathbf{j}}_0$ and $\vec{\boldsymbol{\omega}}_E = \dot{\phi}\hat{\mathbf{k}}$, and choosing the vector basis $(\hat{\mathbf{i}}_0, \hat{\mathbf{j}}_0, \hat{\mathbf{k}})$ for the associated column-matrix representations, it can be stated that:

$$\mathbf{V}_{E^*} = \begin{bmatrix} 1 & 0 & 0 \\ 0 & 1 & 0 \end{bmatrix} \quad \text{and} \quad \boldsymbol{\Omega}_E = \begin{bmatrix} 0 & 0 & 1 \end{bmatrix} \tag{27}$$

Also notice that, since $\vec{\mathbf{a}}_{E^*} = \ddot{x}\hat{\mathbf{i}}_0 + \ddot{y}\hat{\mathbf{j}}_0$ and $\vec{\boldsymbol{\alpha}}_E = \ddot{\phi}\hat{\mathbf{k}}$, then $\mathbf{e}_{E^*} = \mathbf{0}$ and $\boldsymbol{\epsilon}_E = 0$.

3.3 Active Generalized Forces

In order to apply the methodology described in Sect. 2 to derive the dynamic equations of motion for this system, it is still necessary to describe the effects due to gravity and the presence of actuators.

Regarding the acceleration of gravity, it can be stated that $\mathbf{a}_g = [0, -g]^\mathsf{T}$, for the bodies whose motion is described in matrix form in terms of components in the vector basis $(\hat{\mathbf{i}}_0, \hat{\mathbf{j}}_0, \hat{\mathbf{k}})$, which is the case of the end-effector E and bodies A_j of the PRR chains, or $\mathbf{a}_g = [-g \sin\theta_j, -g\cos\theta_j]^\mathsf{T}$, if the vector basis $(\hat{\mathbf{i}}_j, \hat{\mathbf{j}}_j, \hat{\mathbf{k}})$ is adopted instead, which is the case of bodies B_j of the PRR chains and bodies A_j and B_j of the A_j of the RPR chains.

Regarding the generalized forces in the joints of each kinematic chain, it can be stated that, since q_j are actuated degrees-of-freedom and θ_j are passive ones: $\boldsymbol{\tau}_j = [\tau_j, 0]^\mathsf{T}$.

Now, the equations of motion of this planar parallel manipulator can be obtained by computing the expressions for $\mathbf{M}(\boldsymbol{\xi}, \mathbf{x})$, $\mathbf{h}(\boldsymbol{\xi}, \mathbf{x}, \dot{\boldsymbol{\xi}}, \dot{\mathbf{x}})$, $\mathbf{g}(\boldsymbol{\xi}, \mathbf{x})$ and $\mathbf{f}(\boldsymbol{\xi}, \mathbf{x}, \boldsymbol{\tau})$ according to Eqs. (6)–(8) and Eq. (13).

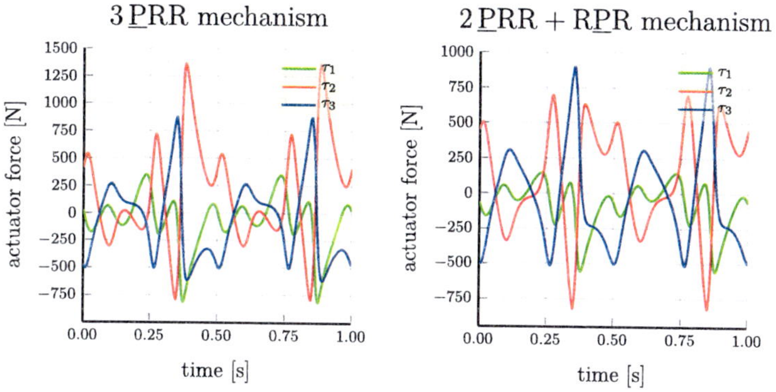

Fig. 4. Actuator torques τ_1, τ_2 and τ_3.

3.4 Simulation and Results

Inverse dynamic simulations are proposed in order to illustrate the use of the obtained equations of motion for each mechanism. In these simulations, assume that the center of mass E* describes uniformly, with a frequency of 2 Hz, a circular trajectory centered at (2.0 m, 0.3 m) whose radius is equal to 0.10 m. Simultaneously, the angle ϕ describes an in-phase oscillation whose amplitude is equal to 20° and whose frequency is 4 Hz.

Consider the following values for the simulation parameters:

- PRR chains: $b_1 = b_3 = 0.5$ m and $b_2 = 0.4$ m; also, consider $m_{\mathsf{A}_j} = 1.0$ kg, $m_{\mathsf{B}_j}/b_j = 1.0$ kg/m and $I_{\mathsf{B}_j} = m_{\mathsf{B}_j} b_j^2/3$, $j = 1, 2, 3$.
- RPR chain: $l_1 = 0.25$ m, $b_1 = 0.5$ m, $m_{\mathsf{A}_1} = 1.0$ kg, $I_{\mathsf{A}_j} = m_{\mathsf{A}_j} l_j^2/3$, $m_{\mathsf{B}_1} = 0.5$ kg and $I_{\mathsf{B}_j} = m_{\mathsf{B}_j} b_j^2/3$.
- End-effector: $l_\mathsf{E} = 0.5$ m, $m_\mathsf{E} = 2.0$ kg and $I_\mathsf{E} = m_\mathsf{E} l_\mathsf{E}^2/3$.

Figure 4 shows the computed actuator forces τ_1, τ_2 and τ_3 that should be provided by the mechanisms 3PRR (Fig. 2) and 2PRR + RPR (Fig. 3) so that the end-effector could describe, in the absence of any disturbance, the desired motion. Notice that, since for all the PRR chains as well as the RPR chain, $\boldsymbol{\tau}_j = [\tau_j, 0]^\mathsf{T}$ then, for both mechanisms:

$$\mathbf{f} = \sum_{j=1}^{3} \mathbf{C}^\mathsf{T}\boldsymbol{\tau}_j = \boldsymbol{\Gamma}^\mathsf{T}\boldsymbol{\tau}, \quad \text{with } \boldsymbol{\Gamma} = \begin{bmatrix} \mathbf{C}_{1,1} \\ \mathbf{C}_{2,1} \\ \mathbf{C}_{3,1} \end{bmatrix} \quad \text{and } \boldsymbol{\tau} = \begin{bmatrix} \tau_1 \\ \tau_2 \\ \tau_3 \end{bmatrix} \tag{28}$$

where $\mathbf{C}_{j,1}$ denotes the first row of $\mathbf{C}_j$, $j = 1, 2, 3$. Therefore, since $\boldsymbol{\Gamma}^\mathsf{T}$ has an inverse, which is true except from singular configurations:

$$\boldsymbol{\tau} = (\boldsymbol{\Gamma}^\mathsf{T})^{-1}\left[\mathbf{M}(\boldsymbol{\xi}, \mathbf{x})\ddot{\mathbf{x}} + \mathbf{h}(\boldsymbol{\xi}, \mathbf{x}, \dot{\mathbf{x}}) + \mathbf{g}(\boldsymbol{\xi}, \mathbf{x})\right] \tag{29}$$

The results shown in Fig. 4 not only reflect the expected periodicity in the time histories of forces (since the desired motion is periodic), but also highlight the differences in how the distribution of these forces among the actuators vary when the topology of the system is modified (i.e. when one of the PRR chains is replaced by a RPR chain).

4 Conclusions

This work dealt with the Modular Modelling Methodology from the perspective of the user/analyst by applying the methodology to a planar 3-dof parallel mechanism. Additionally, one can notice the suitability of MMM for the derivation of equations of motion for parallel manipulators with similar topologies.

In the case study, regarding the replacement of the first PRR kinematic chain by the RPR chain, both the mechanism mobility and the connectivity of the end-effector with respect to the fixed link are preserved. With respect to the actuator forces, the simulation results reveal that the chain replacement reduces the maximum absolute values of the necessary actuator forces for the end-effector motion and, consequently, actuators with lower capacity might be employed. Therefore, MMM can be adequate for performing the dynamic synthesis of a parallel mechanism.

Acknowlegement. The authors acknowledge grant #2018/12087-7, São Paulo Research Foundation (FAPESP)

References

1. Altuzarra, O., Eggers, P.M., Campa, F.J., Paraponiaris, C.R., Pinto, C.: Dynamic modelling of lower-mobility parallel manipulators using the boltzmann-hamel equations. In: Corves B., Lovasz, E.C., Husing, M. (eds.) Mechanisms, Transmissions and Applications. Mechanisms and Machine Science, vol. 31. Springer, Cham (2015). https://doi.org/10.1007/978-3-319-17067-1_17
2. Fontes, J.V.C., Santos, J.C., da Silva, M.M.: Numerical and experimental evaluation of the dynamic performance of kinematically redundant parallel manipulators. J. Braz. Soc. Mech. Sci. Eng. **40**, 142 (2018). https://doi.org/10.1007/s40430-018-1072-1

3. Mitiguy, P.C., Kane, T.R.: Motion variables leading to efficient equations of motion. Int. J. Robot. Res. **15**(5), 522–532 (1996)
4. Pakzad, S., Akhbari, S., Mahboubkhah, M.: Kinematic and dynamic analyses of a novel 4-DOF parallel mechanism. J. Braz. Soc. Mech. Sci. Eng. **41**(12), 1–13 (2019). https://doi.org/10.1007/s40430-019-2058-3
5. Potosakis, N., Paraskevopoulos, E., Natsiavas, S.: Application of an augmented Lagrangian approach to multibody systems with equality motion constraints. Nonlinear Dyn. **99**(1), 753–776 (2019). https://doi.org/10.1007/s11071-019-05059-6
6. Orsino, R.M.M., Hess-Coelho, T.A.: A contribution on the modular modelling of multibody systems. Proc. R. Soc. A **471**, (2015). https://doi.org/10.1098/rspa.2015.0080
7. Hess-Coelho, T.A., Orsino, R.M.M., Malvezzi, F.: Modular modelling methodology applied to the dynamic analysis of parallel mechanisms. Mech. Machine Theor., **161**, (2021). https://doi.org/10.1016/j.mechmachtheory.2021.104332
8. Coutinho, A.G., Bartholomeu, V.P., Stevanni, I., Oliveira-Fuess, J.M., Hess-Coelho, T.A., Colon, D.: Design and control of 2-dof parallel mechanism, In: 25th ABCM International Congress of Mechanical Engineering, October 20–25, Uberlandia, Brazil (2019)

Tuning the Dynamics of Bistable Mechanisms by Introducing Travel Limits

Kylian van Puffelen, Thijs W. A. Blad(✉), and Ron van Ostayen

Department Precision and Microsystems Engineering,
Delft University of Technology, Delft, The Netherlands
t.w.a.blad@tudelft.nl

Abstract. Vibrations are a promising source for powering wireless sensors, for example in low-frequency environments like human motion. These environments suffer from unpredictable vibration spectra and their low-frequency and large amplitude characteristics offer great possibilities for mechanisms with double well potential energy characteristics. The dynamical output performance of a bistable mechanism depends on the oscillation in the large amplitude trajectory between the two potential wells. However, requires enough force to overcome potential energy barrier. This work aims to improve the occurrence of interwell oscillation by lowering the potential energy barrier between the two potential wells by the influence of hard mechanical travel limits. A bistable mechanism is numerically modelled and experimentally tested to investigate the influence of the mechanical travel limits for low-frequency excitations. An axial loaded buckling beam was used to introduce bistability and combined with a parallel guidance mechanism to compensate for the strong negative stiffness. A single-degree of freedom model is used to model the bistable characteristics and is expanded with a coefficient of restitution model to represent the mechanical characterization of the travel limits. This combination resulted in a decrease in required force for the oscillation in the desired large amplitude trajectory by constraining the oscillator motion with travel limits. Furthermore, the results from the numerical bistable model in combination with mechanical characteristics of the travel limits at impact, proves to be in good agreement with the experimentally obtained results.

Keywords: Vibration energy harvesting · Bistability · Low-frequency · Mechanical travel limits · Impact · Coefficient of restitution

1 Introduction

Buckled bistable oscillators are mechanisms, consisting of flexible elements and allow to travel with different trajectories between the stable equilibrium positions [1]. Their difference in comparison with conventional monostable mechanisms, are the potential energy characteristics. Monostable oscillators consist of a single potential energy well and are mostly designed to work as a resonant oscillator, resulting in high output performance on a specific frequency [2, 3]. However, if the oscillator is not accurately tuned to the input signal of the real world, poor output performance can be expected [4].

© The Author(s), under exclusive license to Springer Nature Switzerland AG 2022
J. Beran et al. (Eds.): TMM 2020, MMS 85, pp. 134–143, 2022.
https://doi.org/10.1007/978-3-030-83594-1_14

The potential energy characteristics of bistable structures consist of double potential energy wells, which are segregated by an energy barrier. The barrier between the two wells allows the oscillator to travel with two distinct trajectories, intra- or interwell motion, respectively confined in one well, or between the wells [5]. The energy capture mechanism depends on the ability of the oscillator to move between the stable equilibria, however this oscillation can not always be guaranteed and depends on the height of the potential energy barrier. If the bistable oscillator is not designed properly, large amount of input energy is consumed to overcome the potential energy barrier. As a consequence, low range of motion as well as poor output performance can be expected, resulting in the requirement for high input acceleration signals [6].

In order to overcome this problem a positive stiffness can be used to counteract the strong negative stiffness of the bistable mechanism and reduces the height of the potential energy barrier. In the work of Blad et al. [7] the principle of static balancing is used to compensate for the negative stiffness. With this principle the potential energy barrier can be flattened, which reduces the energy consumption by the jump over the energy barrier. In the work of Zhu et al. [6] a midpoint magnetic force is added, to accomplish the same principle by actively tuning the height of the potential energy barrier with magnets.

However, both stiffness compensation mechanisms do not only reduce the potential energy barrier, but also reduces the advantage of the dynamical nonlinearities. Since the stiffness compensation counteracts the negative stiffness in the bistable mechanism, which is connected to the nonlinear snap-through motion [8]. In this work, a method is studied in which the travel limit is manipulated as a design parameter. This proposed design parameter reduces the height of the barrier by relocating the stable equilibrium positions in the mechanical domain with travel limits. This allows the oscillator to be tuned to the input excitation without having to compensate for the original stiffness characteristics between the stable equilibria positions.

In Sect. 2, the working principle and design of the bistable oscillator is presented, followed by a mechanical and dynamical analysis. Section 3 describes the results and are discussed in Sect. 4. The most significant conclusions are described in Sect. 5.

2 Method

2.1 Mechanical Design

The mechanical design and working principle of the planar bistable mechanism is shown in Fig. 1a–c. Figure 1a shows the mechanism after fabrication which consist of a shuttle suspended by two parallel flexures on one end, and a single flexure at the other end. In Fig. 1b the preloading process is shown, in which the initially flat mechanism is buckled by compressing it over a distance of *dl*. Figure 1c shows the post-buckled mechanism with the identified degree-of-freedom (DoF). In this mechanism, the boundary condition of the flexures are fixed.

The force-deflection relation of the motion in this DoF has the following interesting properties. The uncompressed monostable configuration in Fig. 1a, consist only of positive stiffness and has a straight line relation as force-deflection characteristics. As soon as bistability is introduced, the single flexure is buckled and the mechanism consist of two stable equilibrium positions, as can be seen in Fig. 1b. Between the stable equilibria the positive stiffness is altered with a negative stiffness slope and exhibits nonlinear phenomena. The mechanism losses stability as it acquire enough input force, this will result in an attractive force to the other stable equilibria position, and is called snap-through motion. This nonlinearity improves the ability of bistable oscillators to outperform their performance in comparison with monostable oscillators [9].

From theory, we know that this mechanism will deflect as it acquire enough force [10]. In this work, it is expected that the force required for action is influenced by hard mechanical travel limits as can be seen fin Fig. 1c. Because of this limitation, the motion of this DoF is reduced and the system will come at rest at an other stable equilibrium position. This position is much closer to the local maximum of the potential energy curve and therefore requires less input force for interwell motion. To investigate this phenomena a contactless force will be exerted on the mechanism with the use of a mass and an acceleration signal, to examine the influence of the travel limits on the required force for interwell motion. Moreover, the dependency of the stable equilibria positions is uncoupled from the preload by the integration of the new design parameter. In this configuration the height of the potential energy barrier can be set with the preload and influenced with the design parameter by relocating the stable equilibrium positions.

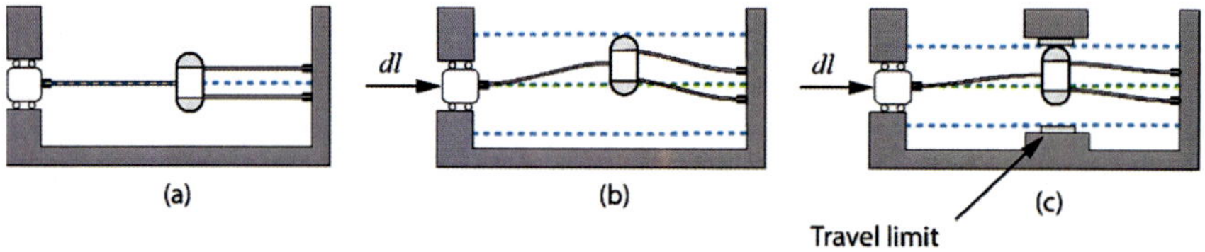

Fig. 1. Schematic working principle of the bistable mechanism, in blue stable and in green unstable equilibrium positions. a.) Monostable structure, b.) Unconstrained bistable structure, c.) Constrained bistable structure with travel limits

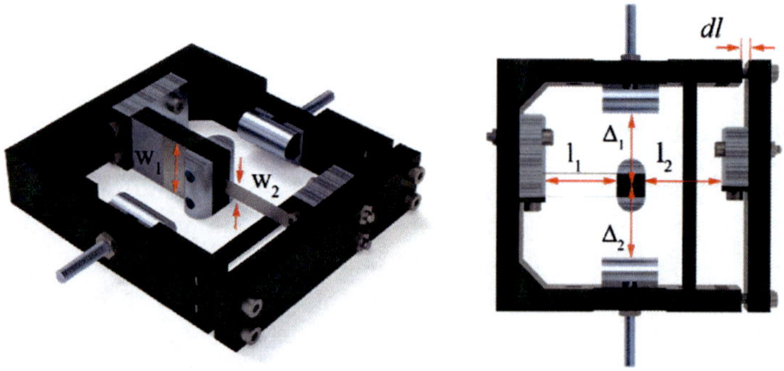

(a) (b)

Fig. 2. The unloaded planar bistable mechanism design with indicated parameters, a) Isometric view, b) top view

2.2 Fabrication

The bistable mechanism prototype can be seen in Fig. 2. The flexural beams in this design are made from 0.1 mm thick spring steel with the properties: ($E_f = 190$ GPa, $v_f = 0.34$ $\rho_f = 7.89 \frac{kg}{m^3}$) and are manufactured using a micro laser cutting machine. The proofmass was constructed out of a stack of steel shims and squeezed by two cylindrical shaped aluminium components with the following properties: ($E_s = 69$ GPa, $v_s = 0.34$, $\rho_s = 2.70 \frac{kg}{m^3}$). The same shape of the aluminium components were used for the travel limits, however placed perpendicular to the orientation of the proofmass to create a Hertz contact point. Both cylinders of the proofmass and travel limits have corresponding dimensions for the radius (5 mm) and length (20 mm). The frame was 3D printed in PLA at 100% infill, in order to give the system its maximum rigidity and hinder the influence of its flexibility during dynamical loading.

An important design aspect in bistable mechanisms are the clamping conditions of the suspensension, slight differences in alignment tightening torque and material stiffness, result in different behaviour. Therefore the assembly procedure was according to the following order. A spacer with thickness *dl* were used to fill the gap between the main frame and the tightening bracket, then the bistable suspension were installed in the unloaded configuration as shown in Fig. 1a. As the suspension was installed, the spacer was removed and the bracket was tightened. In this order, the bistability could be introduced and guaranteed with the correct predefined axial load displacement. The tuning parameter in this design is defined in Fig. 2 with the travel limit (Δ), which defines the locations of the new stable equilibria positions. The variation of this parameter is done with the use of 0.5 mm thick spacers and placed between the frame and the backside of the travel limits, which were kept in place with a screw thread. Furthermore, the parameters used for this prototype are summarized in Table 1.

Table 1. Planar bistable mechanism design parameters

Parameter	Symbol	Value
Unloaded parallel flexure length	l_1	26 mm
Unloaded buckling flexure length	l_2	27.7 mm
Parallel flexure width	w_1	20 mm
Buckling flexure width	w_2	6.8 mm
Axial displacement	*dl*	2.7 mm
Proofmass weight	m	0.0152 kg

2.3 Numerical Model

Finite Element Analysis

To model the dynamical performance, the force-deflection relation of the motion in the DoF in this bistable mechanism was obtained first. This mechanical analysis was performed in a finite element (FEM) analysis in ANSYS Mechanical APDL. The model was constructed out of BEAM 188 elements based upon Timoshenko's beam theory. To model the bistable characteristics, both ends of the structure were constrained and the in plane longitudinal degree-of-freedom of the single buckling flexure were relieved. On this flexure, a small force was applied in the middle, to help the system to bifurcate to the first buckling mode when an axial load displacement (*dl*) was applied on the relieved DoF. After the bistability was introduced, a displacement were applied on the middle of the proofmass, where simultaneous the reaction forces were recorded. The simulated force deflection behaviour can be obtained in Fig. 3.

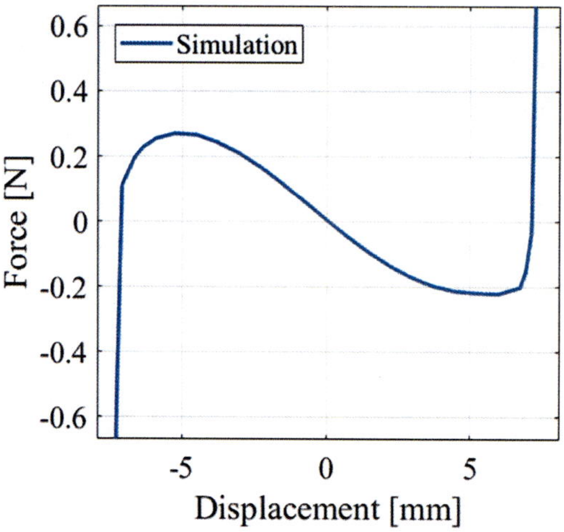

Fig. 3. Simulated force-deflection behaviour of the bistable mechanism

Dynamical Analysis

The dynamical model of the bistable system is described using a lumped-parameter single-degree-of freedom model. Without the interaction of the travel limits, the system can be described using the mass-spring-damper model, represented by equation Eq. (1):

$$m\ddot{x} + c\dot{x} + F_{Mechanism}(x) = F(x) \tag{1}$$

In this equations x, $\dot{x}$ and $\ddot{x}$ represent the oscillator displacement, velocity and acceleration relative to its base excitations. The mass (m) of the system is defined by the weight of the proofmass and damping coefficient (c) is calculated with the observed damping ratio ($\xi = 0.023$) during a logarithmic decrement measurement, where the system at rest is excited with an plucking based motion. The nonlinear force of the bistable oscillator suspension is defined as $F_{Mechanism}$ and is extracted locally from the force-deflection simulation. The locations of the stable equilibria of the unconstrained configuration are defined by the amount of pretension, however the introduction of mechanical travel limits results in an expansion of the equation of motion with an extra nonlinear force. This force is defined as the parameter, responsible for the new locations of the stable equilibria, when the motion is constrained by the mechanical travel limits as can be observed from Fig. 1c.

When the stable equilibria are shifted, the stiffness and damping of the system increases significantly due to the rigid characteristics of the travel limits. The contact mechanics of this additional force of the travel limits is simplified using a coefficient of restitution (CoR) model, where the incoming velocity is reduced with an reduction factor of CoR = 0.36. This value represent the amount of kinetic energy loss at the collision with the travel limit. This value was observed during impact measurement, where the prototype was excited with a sinusoidal input excitation and the bouncing behavior of the impact were recorded with an laser sensor. The expression of Eq. (2) is used in the model [11].

$$CoR = \frac{V2}{V1} \tag{2}$$

The synthesis of the models are integrated as follows: the lumped-parameter model is calculated using the ODE solver in MATLAB. As soon as the oscillator reaches the travel limit, a collision is detected by an Event function in the ODE solver and changes the direction and reduces the incoming velocity according to the relation of Eq. (2). This procedure is repeated until all the energy is dissipated by the travel limit. The point of interest is the point at which the system loses stability and deflects between the stable equilibria positions, therefore an acceleration input signal is used to excite the model and repeated for varies travel limit positions. To represent an linear increasing acceleration sweep, a time series of 10 s is used for each separate acceleration signals between 0 g and 2 g, with an incremental increase of 0.02 g. From each signal the RMS velocity was calculated when the oscillator reached steady state conditions.

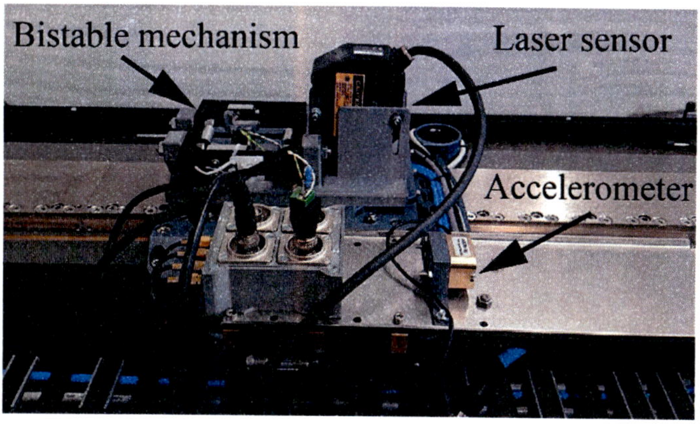

Fig. 4. Overview of the experimental setup on the linear air bearing stage

2.4 Experimental Characterization

Experimental Setup. The validation of this numerical model is performed on a custom made linear air bearing stage, shown in Fig. 4. This stage consist of a range of motion of 500 mm and is able to perform experiments for low frequency and large amplitude excitations. A DaQ chassis consisting of a NI-9215 and NI-9263 module were used. The

NI-9263 output module were used to provide the controller with input signals to actuate the air bearing stage. Although the controller is feedback controlled, an external AS28e 5g accelerometer is used to check the consistency of the input/output relation of the stage. A Keyence LK-H052 laser sensor were used to record the dynamical performance of the bistable mechanism. By placing the laser sensor on top of the base, relative displacement of the proofmass with respect to the base excitation could be recorded and simultaneously measured with the acceleration data by the NI-9215 input module.

Experimental Measurement. In order to rely on the numerical outcomes of the simulation, the dynamical performance is experimentally tested to validate the correctness. The measurement were executed as follows. The bistable mechanism is excited with a sinusoidal acceleration sweep, in order to provide a contactless force and investigate the influence of the travel limits on the acquired force to deflect between the stable positions. The sweep is started from rest with zero initial conditions and swept linearly from 0 g to 2 g acceleration over a time span of 300 s to ensure that each acceleration signal reaches steady state conditions. The excitation were kept at a constant frequency of 3 Hz. During this measurement, the displacement of the proofmass and the acceleration profile of the base excitation were measured and used as a check for correctness of the numerical model.

3 Results

In Fig. 6 the calculated oscillator travel results from the numerical model are compared with the experimentally obtained results. From this figure it can be seen that both results yield towards the same deflection point, where the contactless force from the acceleration profile is enough to overcome the potential barrier and start to bifurcate from intrawell to steady state interwell oscillation. From the section view, minor differences are observable in the rebounce when the travel limits are triggered.

In Fig. 6. the numerical calculated RMS velocity results for different accelerations and travel limit distances are shown. From the diagonal line in the surf plot, it can identified that less force is required for oscillation in the desired interwell trajectory as the bistable oscillator motion is more constrained by the travel limits.

4 Discussion

The dynamical response of the numerical model and the experimentally investigated response of the prototype, observed in Fig. 5, yield both towards the same deflection point as the acceleration sweep is linearly increased for a frequency of 3 Hz. Though difference could be observed at the impact as can be seen in the section view. The rebounce of the experiment consist of a maximum error of 8%, which means a small inaccuracy and result in larger damping characteristics than experimentally measured. However, from this figure it can also be seen that the rebounce characteristics of the experiment is not the same in both stable equilibrium positions, the reason can be explained by the small errors

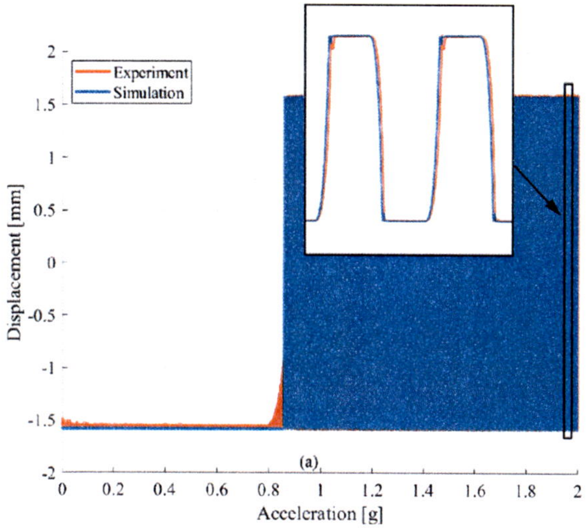

Fig. 5. Experimental validation of simulated dynamical response of an acceleration sweep and frequency of 3 Hz

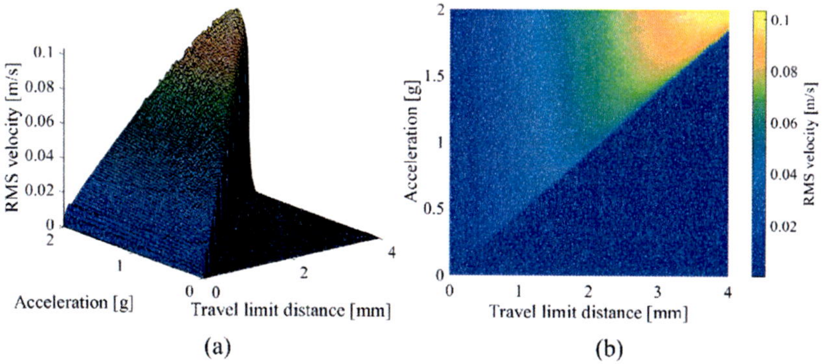

Fig. 6. Numerical calculated RMS velocity results for different accelerations and travel limit distances, a) Isometric view, b) Top view of the surf plot

in the location of the travel limits in the prototype, which causes this small asymmetrical behaviour of the rebounce at the collision in the experiment.

The surf plot of Fig. 6. confirms the research objective, from this plot it can be observed that less input force for actuation is needed for interwell oscillation as the motion is more constrained. However, this does not directly mean that constraining the oscillator motion will result in a better dynamical performance, but depends on a trade-off between maximum allowable decrease in RMS velocity and required force for actuation. In other words, as oscillation occur at for example an acceleration signal of 1 g and the travel limit distance is set to 4 mm, the bistable oscillator has the freedom to oscillate, but the input force is too low and poor dynamical performance can be expected. The method described in this paper recommends to tune the travel limit distance, however in case of excessive tuning by constraining the motion too much, it has the ability to oscillate but not the freedom to move due to the travel limits. Therefore a trade-of should be made first and the travel limit distance should be found with an optimisation process to estimate for each different environment the correct location for the travel limit distance.

5 Conclusion

In this work, the use of travel limits was studied as a method to reduce the energy barrier of bistable mechanisms such that interwell motion can be more easily achieved. The bistable mechanism was constructed from a flexure that was fixed at both ends and compressed beyond the point of buckling. A proofmass was mounted in the middle of the flexure and its motion is constrained by a mechanical stop. A numerical model was constructed to study the dynamics of this mechanism in which the travel limit is manipulated as a design parameter. The force-deflection relation of the bistable mechanism was obtained through finite element simulation and used as an input for the dynamical model. It was demonstrated that tuning the travel limit can result in an increased RMS velocity of the proofmass by allowing interwell motion. A prototype was constructed and tested experimentally under low-frequency vibrations to validate the results. It was found that travel limits can be used as a method to achieve interwell motion in bistable systems. It is recommended for future work to expand this research on bistable mechanisms by incorporating a transducer and comparing the performance for energy harvesting applications.

References

1. Masana, R., Daqaq, M.F.: Relative performance of a vibratory energy harvester in mono- and bi-stable potentials. J. Sound Vib. **330**(24), 6036–6052 (2011). https://doi.org/10.1016/j.jsv.2011.07.031
2. Blad, T.W.A., Tolou, N.: On the efficiency of energy harvesters: a classification of dynamics in miniaturized generators under low-frequency excitation. J. Intell. Mater. Syst. Struct. **30**(16), 2436–2446 (2019). https://doi.org/10.1177/1045389X19862621
3. Zhu, D., Harris, N., Beeby, S.: Performance of linear vibration energy harvesters under broadband vibrations with multiple frequency peaks. Procedia Eng. **47**, 5–8 (2012). https://doi.org/10.1016/j.proeng.2012.09.070
4. Blystad, L.-C.J., Halvorsen, E.: A piezoelectric energy harvester with a mechanical end stop on one side. Microsyst. Technol. **17**(4), 505–511 (2011). https://doi.org/10.1007/s00542-010-1163-0
5. Harne, R.L., Wang, K.W.: A review of the recent research on vibration energy harvesting via bistable systems. Smart Mater. Struct. **22**(2), 023001 (2013). https://doi.org/10.1088/0964-1726/22/2/023001
6. Zhu, Y., Zu, J.W.: Enhanced buckled-beam piezoelectric energy harvesting using midpoint magnetic force. Appl. Phys. Lett. **103**(4), 041905 (2013). https://doi.org/10.1063/1.4816518
7. Blad, T.W.A., van Ostayen, R.A.J., Tolou, N.: A method for tuning the stiffness of building blocks for statically balanced compliant ortho-planar mechanisms. Mech. Mach. Theory **162**, 104333 (2021). https://doi.org/10.1016/j.mechmachtheory.2021.104333

8. Ha, C.S., Lakes, R.S., Plesha, M.E.: Design, fabrication, and analysis of lattice exhibiting energy absorption via snap-through behavior. Mater. Des. **141**, 426–437 (2018). https://doi.org/10.1016/j.matdes.2017.12.050
9. Jiang, W.-A., Chen, L.-Q.: Snap-through piezoelectric energy harvesting. J. Sound Vib. **333**(18), 4314–4325 (2014). https://doi.org/10.1016/j.jsv.2014.04.035
10. Jia, Y.: Review of nonlinear vibration energy harvesting: duffing, bistability, parametric, stochastic and others. J. Intell. Mater. Syst. Struct. **31**(7), 921–944 (2020). https://doi.org/10.1177/1045389X20905989
11. Ashraf, K., Md Khir, M.H., Dennis, J.O., Baharudin, Z.: A wideband, frequency up-converting bounded vibration energy harvester for a low-frequency environment. Smart Mater. Struct. **22**(2), 025018 (2013). https://doi.org/10.1088/0964-1726/22/2/025018

Simulation Analysis of a Motorcycle with Passive, Idealized Semi-active and Active Suspension Systems

Stefan Segla(✉) and Sayantan Roy

Technical University of Košice, Košice, Slovakia
stefan.segla@tuke.sk

Abstract. The paper is devoted to simulation analysis of three different types of motorcycle suspension systems - passive, idealized semi-active and active suspensions. The equivalent motorcycle mathematical model is derived, in which the leaning front and rare spring-damper units are replaced by the equivalent vertical spring-damper units. The road roughness excitation is considered in the form of a deterministic bump and also stochastic excitation modelled by the Shinozuka method with a given power spectral density. The effective acceleration of the motorcycle body centre of mass, which significantly determines the rider comfort level, is used to compare the three different types of suspension systems. The simulation results for the stochastically uneven road show 25.4% improvement for the semi-active suspension and 31.2% improvement for the active suspension. For the deterministic "hat" bump the improvements are 33.2% for the semi-active suspension and 61.5% for the active suspension - compared to the passive motorcycle suspension system.

Keywords: Motorcycle · Passive suspension · Semi-active suspension · Active suspension · Comfort · Simulation

1 Introduction

A motorcycle dynamics model consists of a rigid body which is connected to the front and rear wheels with the front and rare suspension systems [1, 2]. The motorcycle body consists of the chassis, rider, steering head and engine. It represents the motorcycle sprung mass. The front and rare unsprung masses are represented by wheels and all other masses attached to them. The suspension system of a motorcycle is designed to meet the following two conflicting demands:

– a good level of the rider comfort as he drives along an uneven road,
– a good wheel grip on the road to maintain good tyre to road contact.

The conventional non-adjustable passive suspension systems involve passive springs and dampers with non-variable rates. The stiffness and damping parameters are chosen based on a compromise between the ride comfort and good tyre to road contact over a

© The Author(s), under exclusive license to Springer Nature Switzerland AG 2022
J. Beran et al. (Eds.): TMM 2020, MMS 85, pp. 144–151, 2022.
https://doi.org/10.1007/978-3-030-83594-1_15

wide range of road and speed conditions. Their characteristics of vibration isolation are rather limited. Low damping leads to good vibration isolation at high frequencies but poor resonance characteristics. Higher damping leads to good resonance characteristics but the performance at higher frequencies is poor.

The necessity of improving vibration isolation of passive suspension systems motivated investigation of controlled suspension systems [3, 4], where e.g., damping characteristics are controlled in closed loops. The control strategies can be divided into two categories: active and semi-active. In the fully active suspension system, the conventional suspension elements are replaced with electric or hydraulic actuators. The force demand signal is typically generated in a microprocessor governed by a control law (e.g., sky-hook or balance control). The optimum transmissibility has no resonance amplification. Their performance is superior to any conventional passive suspension system. But they are still costly, less reliable, having high energy consumption and possible instability.

A compromise between active and passive suspension systems are semi-active suspension systems. In automotive engineering, especially magnetorheological dam-pers attracted a lot of interest over the past decades for their quick time response and low energy needs.

2 Modelling of a Motorcycle

The motorcycle dynamics model, Fig. 1, is composed of the the sprung mass 1 (consisting of the motorcycle rigid body and the driver), the unsprung front mass 2 (the front wheel), the unsprung rare mass 3 (the rare wheel), the front spring-damper modul 4 and the rare spring-damper modul 5. Both the leaning spring-damper modu-les are placed between the sprung and unsprung masses [1, 2].

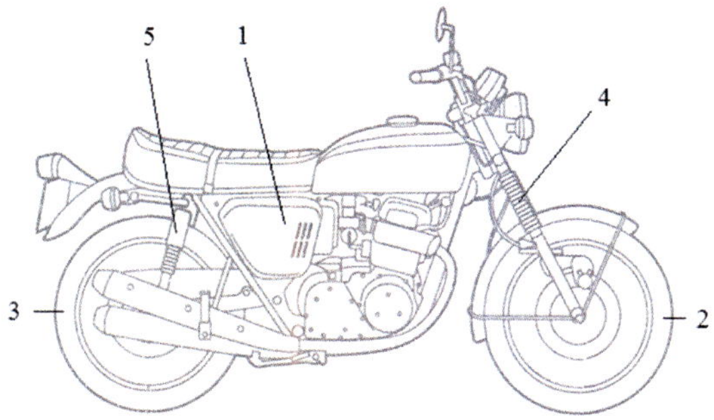

Fig. 1. Motorcycle parts.

Motion of motorcycle parts is described by four coordinates z_1,z_2, z_3,φ_3, Fig. 2, which are independent. The coordinate z_1 describes the vertical displacement of the front unsprung mass and the coordinate z_2 describes the vertical displacement of the rare unsprung mass. The coordinate z_3 describes the vertical motion of the sprung mass centre T_3 and finally the angular coordinate φ_3 describes the pitching angular motion of the sprung mass m_3 about the sprung mass centre T_3. All coordinates are measured from their equilibrium positions. The coordinates u_1 and u_2 describe the uneven road surface. The tyre stiffness coefficients are k_3 and k_4 and the equivalent coefficients of stiffness and damping $k_{1,eq}$, $k_{2,eq}$, $b_{1,eq}$ and $b_{2,eq}$(corresponding to the leaning spring-damper modules, Fig. 1) in the main suspension are given below, Eqs. [1] to [4]. The motion equations of a motorcycle with the leaning passive suspension systems are derived in the form of sa set of the second order ordinary differential equations, which were derived in [5].

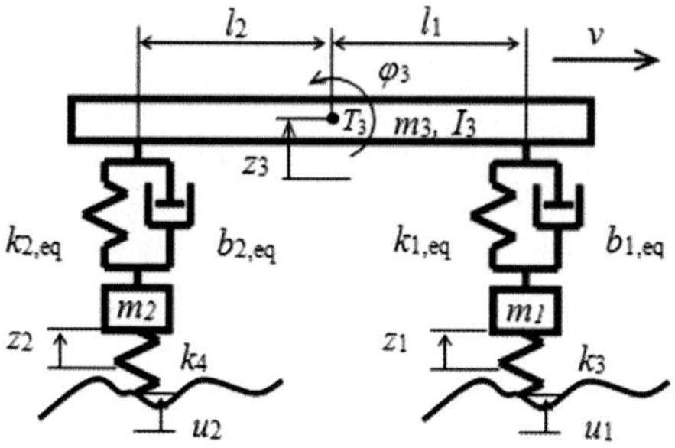

Fig. 2. Equivalent motorcycle model with vertical passive suspension systems.

Based on potential energy considerations (equality of potential energy of the original leaning spring-damper module and the substitute vertical spring-damper module), the equivalent stiffness coefficient $k_{1,eq}$ in the front suspension is determined by the following equation derived in [5]

$$k_{1,eq} = k_1 \cos^2 \varepsilon_1, \tag{1}$$

where k_1 is the spring stiffness coefficient of the original leaning spring-damper module and ε_1 is the lean angle. Similarly based on the Rayleigh dissipative function the equivalent damping coefficient can be derived

$$b_{1,eq} = b_1 \cos^2 \varepsilon_1. \tag{2}$$

Similarly for the rare suspension the equivalent spring and damping coefficients of the substitute vertical spring-damper module above the rare wheel are as follows [5]

$$k_{2,eq} = k_2 \left(\frac{L_1}{L}\right)^2 \cos^2 \varepsilon_2, \tag{3}$$

$$b_{2,eq} = b_2 \left(\frac{L_1}{L}\right)^2 \cos^2 \varepsilon_2, \tag{4}$$

where ε_2 is the lean angle of the rare spring-damper module, L is the length of the swinging arm and L_1 is the distance between the point at which the swinging arm is joined to the motorcycle body and the point at which the substitute vertical spring-damper module is joined to the swinging arm, Fig. 3.

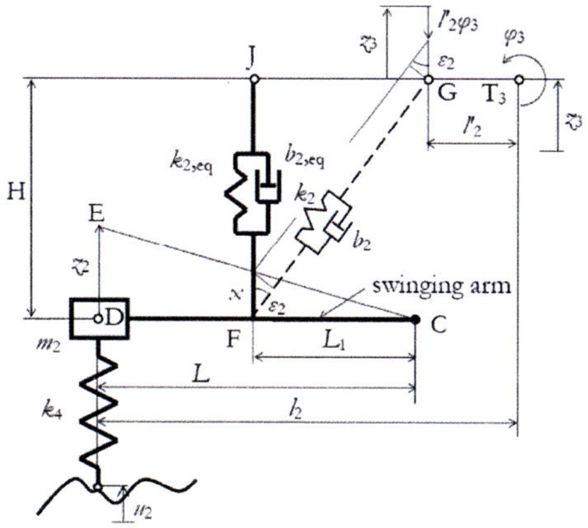

Fig. 3. Equivalent stiffness and damping coefficients of the rare suspension.

3 Semi-active Suspension System

The equations of motion of the ideal semi-active suspension, Fig. 4, have the same form as the equations of the passive system [5], but instead of the damping coefficients of the hydraulic dampers $b_{1,\mathrm{eq}}$ and $b_{2,\mathrm{eq}}$, the variable coefficients of the semi-active dampers $b_{\mathrm{sa},1,\mathrm{eq}}$ and $b_{\mathrm{sa},2,\mathrm{eq}}$ are used

$$m_1\ddot{z}_1 - b_{sa,1,eq}(\dot{z}_3 + l_1\dot{\varphi}_3 - \dot{z}_1) - k_{1,eq}(z_3 + l_1\varphi_3 - z_1) + k_3(z_1 - u_1) = 0, \tag{5}$$

$$m_2\ddot{z}_2 - b_{sa,2,eq}(\dot{z}_3 - l_2\dot{\varphi}_3 - \dot{z}_2) - k_{2,eq}(z_3 - l_2\varphi_3 - z_2) + k_4(z_2 - u_2) = 0, \tag{6}$$

$$\begin{aligned} m_3\ddot{z}_3 + b_{sa,1,eq}(\dot{z}_3 + l_1\dot{\varphi}_3 - \dot{z}_1) + b_{sa,2,eq}(\dot{z}_3 - l_2\dot{\varphi}_3 - \dot{z}_2) \\ + k_{1,eq}(z_3 + l_1\varphi_3 - z_1) + k_{2,eq}(z_3 - l_2\varphi_3 - z_2) = 0, \end{aligned} \tag{7}$$

$$\begin{aligned} I_3\ddot{\varphi}_3 + b_{sa,1,eq}l_1(\dot{z}_3 + l_1\dot{\varphi}_3 - \dot{z}_1) - b_{sa,2,eq}l_2(\dot{z}_3 - l_2\dot{\varphi}_3 - \dot{z}_2) \\ + k_{1,eq}l_1(z_3 + l_1\varphi_3 - z_1) - k_{2,eq}l_2(z_3 - l_2\varphi_3 - z_2) = 0, \end{aligned} \tag{8}$$

3.1 Sky-Hook Control of Semi-active Dampers

The idealized semi-active suspension systems allow us to determine the force intervals of real magnetorheological dampers and their efficiency in comparison with the idealized semi-active dampers. A conventional sky-hook control [4] is used.

The forces exerted by fictitious sky-hook dampers (imagined to be hooked in the sky) acting at points A and B, Fig. 4, are as follows

$$F_{sky,1} = b_{sky,1}(\dot{z}_3 + l_1\dot{\varphi}_3)\,, \tag{9}$$

$$F_{sky,2} = b_{sky,2}(\dot{z}_3 - l_2\dot{\varphi}_3)\,. \tag{10}$$

The damping forces exerted by the front and rare semi-active dampers $b_{sa,1,eq}$, $b_{sa,2,eq}$ are

$$F_{sa,1,eq} = b_{sa,1,eq}(\dot{z}_3 + l_1\dot{\varphi}_3 - \dot{z}_1), \tag{11}$$

$$F_{sa,2,eq} = b_{sa,2,eq}(\dot{z}_3 - l_2\dot{\varphi}_3 - \dot{z}_2). \tag{12}$$

Equalities of these forces

$$F_{sa,1,eq} = F_{sky,1}, \tag{13}$$

$$F_{sa,2,eq} = F_{sky,2}, \tag{14}$$

lead to equations determining the equivalent damping coefficients of the semi-active dampers

$$b_{sa,1,eq} = \frac{b_{sky,1}(\dot{z}_3 + l_1\dot{\varphi}_3)}{(\dot{z}_3 + l_1\dot{\varphi}_3 + \dot{z}_1)}, \tag{15}$$

$$b_{sa,2,eq} = \frac{b_{sky,2}(\dot{z}_3 - l_2\dot{\varphi}_3)}{(\dot{z}_3 - l_2\dot{\varphi}_3 - \dot{z}_2)}. \tag{16}$$

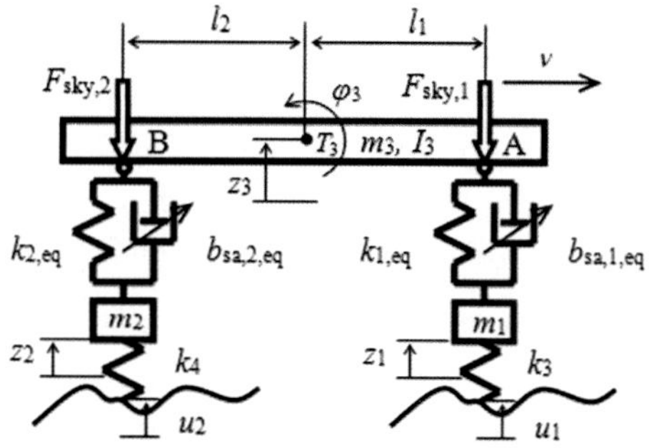

Fig. 4. Equivalent motorcycle model with semi-active suspension systems.

The desired damping forces $F_{sa,1,eq}$ and $F_{sa,2,eq}$ can be exerted only if the the relative velocities between the sprung and unsprung masses and the absolute velocities of points A and B are of the same sign. The on-off conditions of the front semi-active damper are as follows: if

$$(\dot{z}_3 + l_1\dot{\varphi}_3)\,(\dot{z}_3 + l_1\dot{\varphi}_3 - \dot{z}_1) > 0, \tag{17}$$

then the damper is on, otherwise is off. If

$$(\dot{z}_3 - l_2\dot{\varphi}_3)\,(\dot{z}_3 - l_2\dot{\varphi}_3 - \dot{z}_2) > 0, \tag{18}$$

then the rare damper is on, otherwise is off.

3.2 Active Suspension System

The equations of motion of the motorcycle with the active suspension system differ from the equations of motion of the motorcycle with the semi-active suspension system in the fact that instead of the semi-active dampers $b_{\mathrm{sa,1,eq}}$ and $b_{\mathrm{sa,2,eq}}$ the actuators are used. The actuators have to exert the forces $F_{sky,1}$ and $F_{sky,2}$. However, in this case the on-off conditions are not applied, because the actuators are able to exert the required forces in required directions at any time, unlike the semi-active dampers.

4 Simulations with Matlab and Results

The deterministic non-normalized bump of "hat" shape is shown in Fig. 5. It takes into account to some extent the actual tyre shape, because the road-tyre contact is in Matlab simulation considered as point contact.

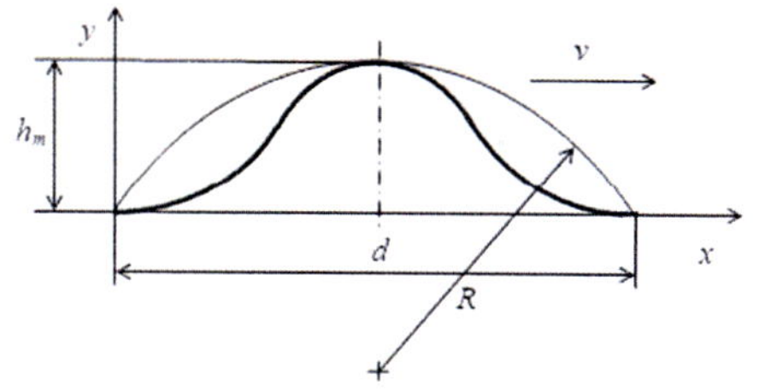

Fig. 5. Road bump of "hat" shape.

In Fig. 5 is: the circumscribed circle radius R, the bump length d, the bump height h_{m}. The stochastic road is represented by the sum of sine functions using the Shinozuka method [6].

Values of parameters of the motorcycle simulation model are: $m_1 = 15$ kg, $m_2 = 18$ kg, $m_3 = 194$ kg, $I_3 = 38\ \mathrm{kg \cdot m^2}$, $k_1 = 15000$ N/m, $k_2 = 24000$ N/m, $k_3 = 180000$ N/m, $k_4 = 180000$ N/m, $b_1 = 710\ \mathrm{N \cdot s/m}$, $b_2 = 1171$ N s/m, $l_1 = 0.64$ m, $l_2 = 0.7$ m, $\varepsilon_1 = 27°$, $\varepsilon_2 = 20°$, $L = 0.6$ m, $L_1 = 0.6$ m, $v = 11$ m/s, $R = 4.68$ m, $d = 2.68$ m, $h_m = 0.06$ m. Parameters of the stochastic road: $S_u(\Omega_0) = 22.10^{-6}\ \mathrm{m}^3$ (asphalt-concrete road of average quality), $v = 20\ \mathrm{m/s}$. The sky-hook damping coefficients are: $b_{sky,1} = b_{sky,2} = 2200\,\mathrm{kg/s}$.

In Table 1 values of the motorcycle body centre of mass effective acceleration and displacement for the ride over the deterministic „hat" bump and along the stochastic road are shown.

Dependence of the centre of mass T_3 vertical displacement z_3 and acceleration $\ddot{z}_3$ on time for the ride over the deterministic bump and along the stochastic road for the three different suspension systems is shown in Figs. 6 and 7.

Table 1. The motorcycle body centre of mass T3 effective acceleration and displacement.

Suspension system	Deterministic bump		Stochastic road	
	$\ddot{z}_{3,ef}$ [m/s^2]	$z_{3,ef}$ [m]	$\ddot{z}_{3,ef}$ [m/s^2]	$z_{3,ef}$ [m]
Passive	0.0498	0.00038	0.0725	0.0035
Semi-active	0.0332	0.00031	0.0541	0.0034
Active	0.0191	0.00023	0.0499	0.0034

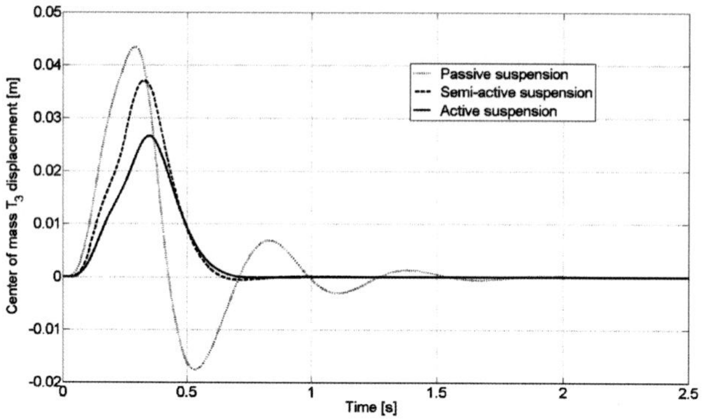

Fig. 6. The centre of mass T3 displacement z_3- „hat“ shaped bump.

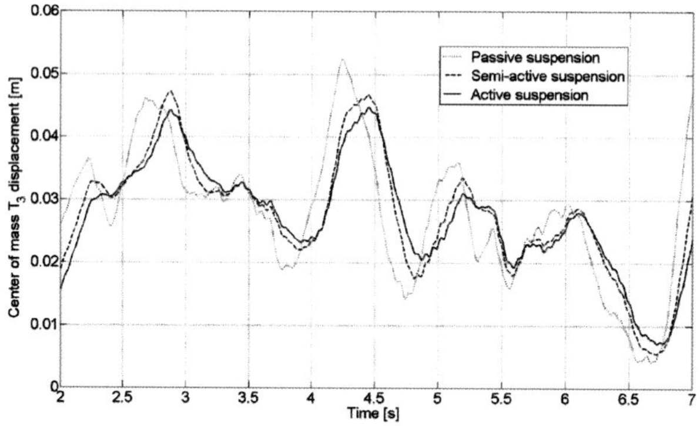

Fig. 7. The centre of mass T3 acceleration z_3- stochastic road.

5 Conclusions

The simulation results presented in Abstract show substantial reduction of the value of the motorcycle body centre of mass effective acceleration for both active and semi-active motorcycle suspension systems and both deterministic bump and stochastic road in comparison with the passive motorcycle suspension system. This fact is important for improving comfort of the motorcycle rider because the motorcycle body centre of mass effective acceleration is a decisive factor determining his comfort level.

Based on experience with other vehicle and seat suspension systems it is possible to expect significant improvements also in the case of using real magnetorheological dampers in the motorcycle suspension systems. Their use will be investigated in further research.

Acknowledgements. This work was funded by project grant VEGA No. 1/0500/20.

References

1. Jazar, R.N.: Vehicle Dynamics: Theory and Application. Springer, Melbourne (2008)
2. Cossalter, V.: Motorcycle Dynamics. 2nd edn. Vittore Cossalter, Padova (2002)
3. Venhovens, P.J.Th.: Optimal control of vehicle suspensions. Doctoral Thesis, Delft University of Technology. Fac. Mech. Eng. (1993)
4. Guglielmino, E., Sireteanu, T., Stammers, C.W., Ghita, G., Giuclea, M.: Semi-active Suspension Control. Improved Vehicle Ride and Road Friendliness. Springer, London (2008). https://doi.org/10.1007/978-1-84800-231-9
5. Segla, S., Roy, S.: Dynamic simulation analysis of a motorcycle suspension system – assessment of comfort. Manuf. Technol. **20**(3), 373–377 (2020)
6. Nigam, N.C., Narayanan, S.: Applications of Random Vibration. Springer-Verlag, Berlin (1994)

Computational Analysis of Cracks in Hardfacings of Wedges of Quick-Acting Main Steam Isolation Valves

Martin Sivý(✉) and Róbert Berky

ROEZ R&D, Leškova 3/A, Bratislava 811 04, Slovakia
{martin.sivy,robert.berky}@roez.sk

Abstract. The main steam isolation valves are safety valves installed in nuclear power plants designed for high energy line break of the steam pipeline, the main steam collector or the steam generator feedwater pipeline. During the regular shutdowns, the radial cracks regularly situated around the circumference in hardfacings of wedges were observed some of which were assessed as non-permissible. The primary objective of the solution is to develop and verify a suitable methodology, based on which it will be possible to decide on the operability of wedges. However, the solution aims at the potential risk prediction of brittle fracture or fatigue cracks growth under the operational loading during the period between regular diagnostics or until the end of the planned service life of the valves. The paper deals with the computational analysis of cracks focused mainly on the abovementioned fracture failures. It is also dedicated to comparing the acquired results from the analyses computed by finite element method (ANSYS) with classical analytical models and experimental testing, respectively.

Keywords: Crack · Brittle fracture · Fatigue crack growth · Wedge · Hardfacing · Gate valve

1 Introduction

The main steam isolation valves (MSIV) are safety valves installed in a nuclear power plant (NPP) which are designed for a high energy line breaks (HELB), the main steam collector or the steam generator feedwater pipeline. Their task is to separate the broken part of the pipeline from the secondary circuit in a maximum of 5 s. MSIVs are, from a structural point of view, gate valves operated by a pneumatic actuator fed from the air tank. The individual components of the MSIV are presented in Fig. 1.

During the regular shutdowns of units, radial cracks around the circumference were observed in the hardfacings of wedges (Fig. 1a, pos. 3). The presence of these cracks has so far been tolerated since the cracks will not influence the tightness of the MSIVs in a closed position, only if they began to propagate to the interface between hardfacing (Deloro 40) and base material (SA 105).

© The Author(s), under exclusive license to Springer Nature Switzerland AG 2022
J. Beran et al. (Eds.): TMM 2020, MMS 85, pp. 152–162, 2022.
https://doi.org/10.1007/978-3-030-83594-1_16

The operator's attitude to radial cracks of hardfacing changed when a radial crack (Fig. 1b) was identified by a penetration test and was declared as non-permissible by the test report. Based on this finding, the operator decided to develop a permanent solution for transverse cracks in hardfacings of wedges.

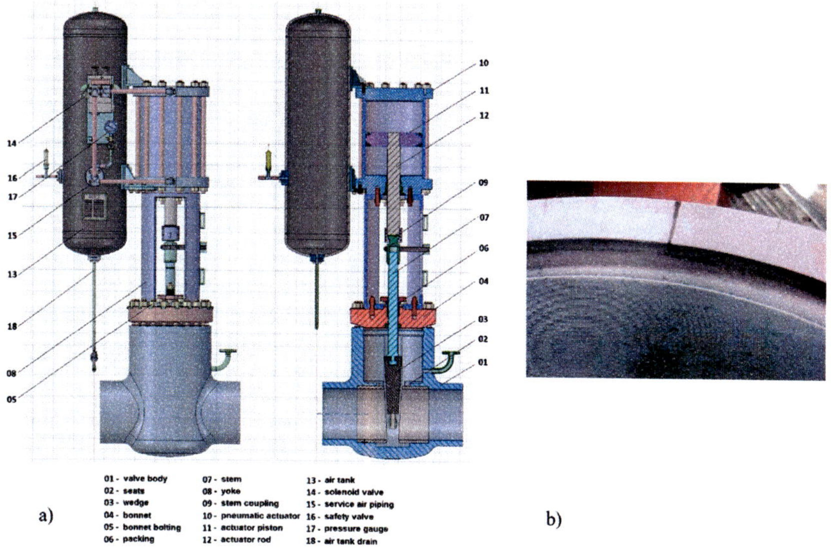

Fig. 1. Main steam isolation valve. Radial cracks in the hardfacing of the wedge.

The aim of the investigation is to develop a methodology for the diagnostics of the cracks in the hardfacing of wedges and subsequent evaluation of their admissibility. This investigation represents one partial task of the huge project dealt with the wedges of MSIVs and their renovation.

2 Basic Concept

The paper is focused mainly on the development of the methodology for the prediction of fatigue crack growth and potential brittle fracture failure under operational loadings using computational analyses. This requires combining experimental testing of specimens prepared from the real wedge with analytical crack growth theories and calculation of corresponding numerical models.

2.1 Experimental Measurements

The experimental program is divided two parts. First part aims at testing of the base material as well as the hardfacing from which mechanical properties were used as input data for the computational analyses. The fracture mechanical material properties were gained according to standards [1–3]. It was desirable to compare data with the data from operation to make sure the results from the experimental program would be transferable to the operated wedges.

Experimental measurements within second part served for comparison and accuracy verification of used analytical and numerical calculation methods.

2.2 Analytical Approach

By the crack propagating, new free surfaces, i.e. fracture surfaces are formed in the initially coherent materials. This is an inhomogeneous process over time and space. There are many publications dedicated to basic crack theories and approaches (LELM, EPLM) [4, 5], mechanisms and conceptions of brittle or fatigue fracture failure [6], factors and conditions which support the crack formation and propagation [7, 8], aspects and criteria of resistance and safety evaluation [9].

For the brittle fracture, it is necessary to follow the stress intensity factor (SIF) which must be lower than its limit value – fracture toughness. It can be determined as follows

$$K = Y\sigma\sqrt{\pi a} \tag{1}$$

in which Y represents a numerical correction factor, σ is a stress load and a is a crack length. In publications (such as [10]), there are derived numerical approximate values of correction factors or respective formulas for specific specimens and types of loading.

The fatigue crack growth rate can be expressed by functional dependence which considers factors such as cyclic load parameters, crack shape and length and material parameters that determine resistance to crack propagation. The process is interpreted by the dependence of the ratio of the differentials $\mathrm{d}a/\mathrm{d}N$ and the range of the stress intensity factors in the form of a sigmoidal curve, namely the crack initiation area, the slow subcritical and critical growth area and the final fracture. The middle part of the curve is linear in logarithmic coordinates and can be expressed analytically by the Paris relationship

$$\frac{\mathrm{d}a}{\mathrm{d}N} = C(\Delta K)^n \tag{2}$$

where C and n represent material parameters (obtained by material testing).

According to the standard [3], this range of SIF for a specific specimen shape and type of loading can be defined as follows

$$\Delta K = \frac{\Delta FL}{B\sqrt{W}} \frac{3\left(\frac{a_0}{W}\right)^{0,5}\left[1,99 - \left(\frac{a_0}{W}\right)\left(1 - \frac{a_0}{W}\right)\left(2,15 - 3,93\frac{a_0}{W} + 2,7\left(\frac{a_0}{W}\right)^2\right)\right]}{2\left(1 + 2\frac{a_0}{W}\right)\left(1 - \frac{a_0}{W}\right)^{1,5}} \tag{3}$$

where ΔF is a force range, L a distance between supports, a_0 initial crack length, B specimen thickness and W specimen width.

2.3 Numerical Computation

Verification of the Paris law analytical approach for fatigue crack growth for given input material parameters, loading conditions, number of cycles, etc. was performed in software ANSYS based on FEM. In addition to the classical static method, an integrated quasi-static SMART method was used to simulate crack growth for the purposes of numerical calculation. This method automatically updates the finite element mesh as the geometry of the crack changes due to the defect increase in each calculation step. All assumptions and limitations are introduced in [11].

Calculation of crack growth is performed in the solution phase after calculation of stress field and fracture characteristics, followed by calculation of crack extension according to Life-cycle method. The increment for a crack growth Δa in the model is calculated as the maximum size of the finite element at the crack face which is then compared with the minimum or maximum possible increment of a crack growth. The increment for the number of cycles ΔN is calculated considering the increment of the crack growth and the maximum value of the stress intensity factor $K_{\max}$ at the nodal points on the crack face

$$\Delta N = \Delta a / f(K_{\max}, R) \tag{4}$$

The crack growth at the *i*-th node is determined

$$\Delta a_i^{node} = \Delta N \bullet f(K_i, R) \tag{5}$$

3 The Experimental Program and Results Verification

The result from the first part of experimental program, i.e. material properties for assessment of cracks in wedges, such as transition temperature, fracture toughness and parameters for fatigue crack growth at ambient temperature are introduced in Table 1.

Table 1. Material properties for computational purposes

Material property	Value	Unit
Transition temperature	12	°C
Fracture toughness	314	$MPa.m^{0.5}$
Material parameter C for fatigue crack growth	4.54e-11	mm/(cycles $MPa^n m^n$)
Material parameter n for fatigue crack growth	4.5791	-

The purpose of the second part of the experimental program was to cycle specimens (Fig. 2) to find such a pulsating force at which fatigue crack growth occurs at a low number of cycles as required by the operator (approx. 500 cycles). The specimens (each of ca 20 × 23 × 140 mm) designed for three-point bending test were prepared from the wedge on which the hardfacing (ca 4 mm) was coated to simulate the real crack growth conditions. In the middle of the specimen length, the lateral crack with depth to the interface was introduced. The specimens were loaded by a pulsating force with stress ratio 0.1. The force amplitude was gradually increasing from 10 000 N to 20 000 N.

From the given measurements the dependence of the range of the stress intensity factor ΔK on the ratio $\mathrm{d}a/\mathrm{d}N$ in logarithmic coordinates was performed which was fitted with a curve according to principles of the Paris law. Figure 2 shows the curves obtained by individual measurements on experimental specimens. Based on the results, constants C and n obtained from the curves of the individual specimens are identical. The results were compared with the curve obtained from the fatigue crack growth measurement for the base material of the wedge. The test specimens were subsequently assessed by fractography, showing that all cracks grew evenly over their entire width.

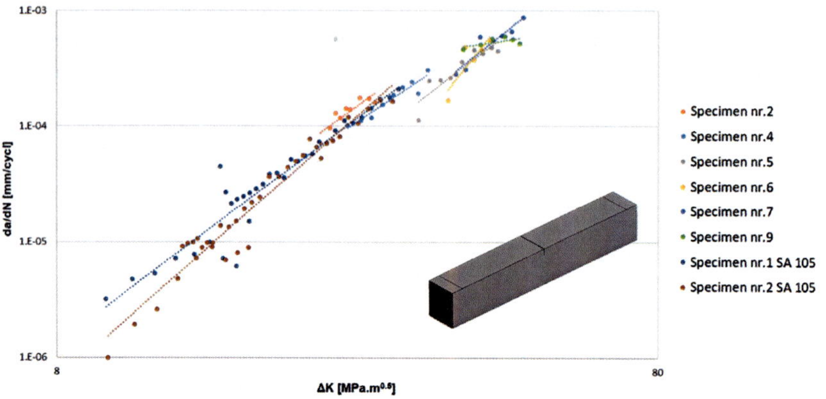

Fig. 2. Results comparison from testing of fatigue crack growth and specimens for verification

Verification calculations in the form of cycling of specimens with hardfacing and radial crack to the melting boundary serve as a basis for verifying the accuracy of the proposed methodology. The aim of the solution is to compare these results with two approaches, i.e. the analytical calculation based on the Paris law and the numerical calculation based on the FEM in ANSYS.

Specimen nr. 4 is a representative example illustrating analytical and numerical calculations which are compared with experimental results. Table 2 shows a comparison of the analytically determined crack growth based on the Paris law relationship with the experimentally determined crack growth for a given number of cycles and the experimentally determined crack growth for a given cyclic force with amplitude of 12 500 N. The force amplitude for the initial and final crack length within the respective experimental measurement is also calculated. Based on the achieved results and the calculated error, a good results agreement can be concluded.

Table 3 compares the analytically determined values of the range of the stress intensity factors with the values of the numerical calculation. For each numerical calculation, new models of the specimens were created with a crack depth corresponding to the increase by the experiment. The value of the range of the stress intensity factor was determined for the load without using the SMART crack growth method. Based on the achieved results, a good agreement can be concluded.

Table 2. Verification calculation. Results comparison. Crack extension.

Nr	Number of cycles N	Cum. number of cycles N	Crack extension *a* [mm]		Error [%]	Force [N]	Error [%]
			Experiment	Analytically			
1	8 000	08 000	0.815	0.804	1.37	12 542	0.34
2	4 000	12 000	1.260	1.321	4.62	12 367	1.06
3	3 000	15 000	1.615	1.777	9.12	12 249	2.01
4	2 500	17 500	2.000	2.212	9.58	12 250	2.00
5	2 000	19 500	2.375	2.604	8.79	12 283	1.73
6	1 500	21 000	2.700	2.928	7.79	12 317	1.46
7	1 300	22 300	3.015	3.231	6.69	12 349	1.21
8	1 100	23 400	3.225	3.507	8.04	12 323	1.42
9	1 100	24 500	3.560	3.801	6.34	12 366	1.07

Table 3. Verification calculation. Results comparison. Range of stress intensity factor.

Nr	Number of cycles N	Cum. number of cycles N	Range of SIF [MPa mm$^{0.5}$]		Error [%]
			Experiment	Analytically	
1	8 000	08 000	24.5	23.5	4.01
2	4 000	12 000	25.7	25.8	0.38
3	3 000	15 000	26.8	27.4	2.25
4	2 500	17 500	28.0	28.5	1.83
5	2 000	19 500	29.1	29.8	2.24
6	1 500	21 000	30.2	31.2	3.20
7	1 300	22 300	31.3	32.4	3.46
8	1 100	23 400	32.0	33.5	4.78
9	1 100	24 500	33.2	34.3	3.09

Figure 3 shows a summary of crack growth results for a given load obtained from the experiment, analytical and numerical calculation (using SMART crack growth) for specimen nr. 4. The crack growth results are compared with the coefficients C and n obtained for the base wedge material. The results are presented in a logarithmic scale. The sets of points were fitted with curves according to the Paris law to compare the coefficient values. Comparing the individual curves, the slopes of the curves, i.e. the corresponding coefficients n agree with each other.

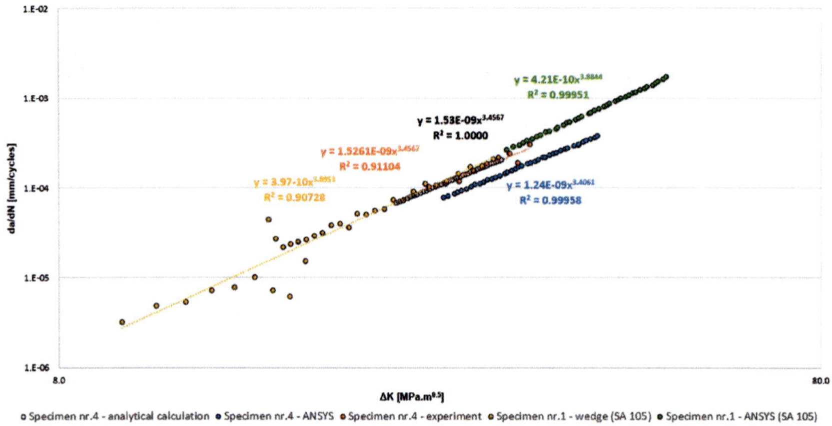

Fig. 3. Results comparison. Fatigue crack growth in specimen nr. 4.

Figure 4 shows a comparison of crack growth results as a function of the number of cycles. Based on the achieved results, a good agreement between the individual calculations and the experiment can be concluded.

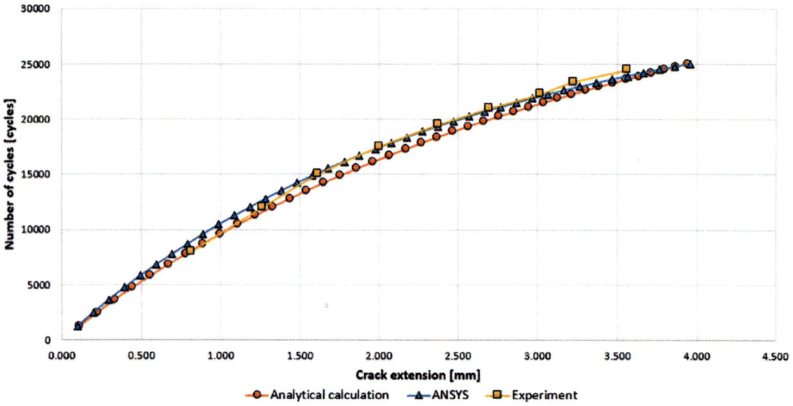

Fig. 4. Results comparison. Fatigue crack growth in specimen nr.4.

4 Wedge Assessment for the Brittle Fracture Mechanism

For the assessment of the wedge for unstable crack propagation in the MSIV's wedge, it was assumed the crack would grow when the value of the stress intensity factor is greater than the value of the fracture toughness K_{Ic} (K_{Jc}). The calculation for low-energy crack propagation was based on fracture toughness value K_{Ic} obtained from fracture toughness measurement. This value is relatively high since there was no sudden unstable fracture in the measurements which would correspond to the provisional fracture toughness K_Q but the specimen material began to deform plastically.

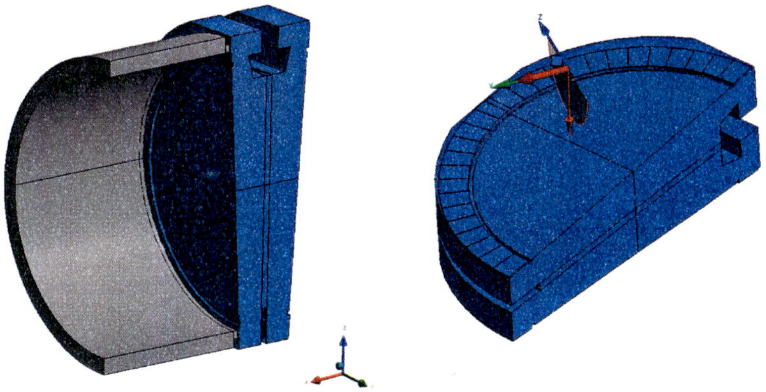

Fig. 5. Calculation model. Wedge with the radial oriented cracks settled in seats.

For SIF calculation, a set of cracks radially oriented around the circumference of the hardfacing of the wedge was regularly spaced 50 mm apart (Fig. 5). The crack depth was assumed to the melting boundary layer (ca 4 mm). For calculation, the symmetric model of the wedge embedded in the seats was assumed loaded by one-sided pressure of 4.6 MPa on the opposite side of the wedge representing the worst condition of the wedge by closing the flow. The calculation is performed at a conservative temperature of 20°C because the value of fracture toughness increases with increasing temperature.

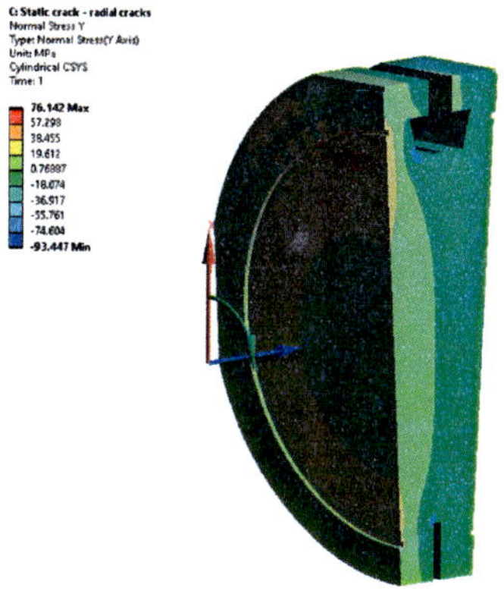

Fig. 6. Tangential stresses distributed in the wedge of MSIV

Figure 6 shows the state of the wedge stress in the form of tangential stresses. It is important to know the tangential stresses to assess crack propagation in the wedge of MSIV. These stresses are directly related to the first (tensile) mode of fracture which results in the crack opening and its further propagation.

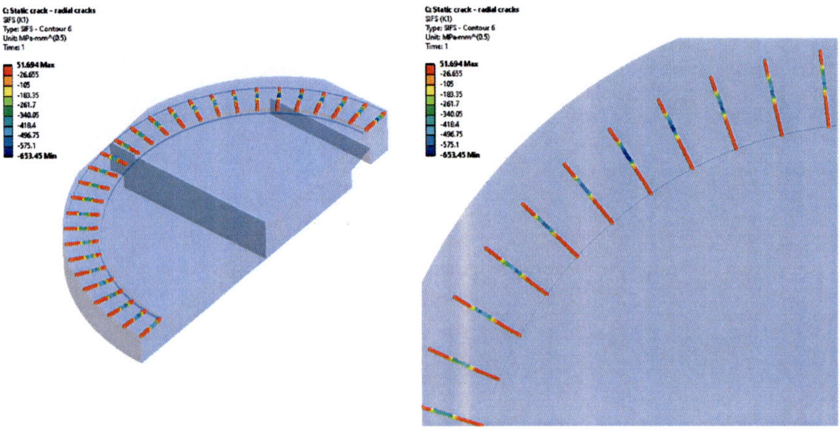

Fig. 7. Stress intensity factors of wedge cracks

Figure 7 shows the SIFs along the crack faces resulting from the pressure acting on the wedge. It follows the highest values are formed in the upper part of the wedge where the stress concentration occurs due to weakening of the wedge by the recess.

5 Wedge Assessment for the Fatigue Crack Growth

The assessment of the wedge for fatigue crack growth was based on an experimental program aimed at finding a cycling load with a positive pulsating cycle that, at a low number of cycles (500 cycles), would result in the fatigue crack opening associated with its growth. Based on the state of stress results of the wedge assessment for unstable crack propagation associated with low tangential stress values, the fatigue crack growth model was based on a non-cracked model. The purpose of the calculations was a more detailed assessment of the state of stress for a given load. The model and boundary conditions were taken from the previous analysis.

Figure 8a shows the distribution of tangential stresses on the wedge due to the one-sided pressure load. The figure shows that the highest stresses are concentrated in the wedge weakened region. Stress distribution at the bottom of the wedge without the influence of the recess is shown in Fig. 8b. In Fig. 8c, stresses are shown only in the hardfacing, whose values are low and do not affect the fatigue crack growth for the given loading conditions.

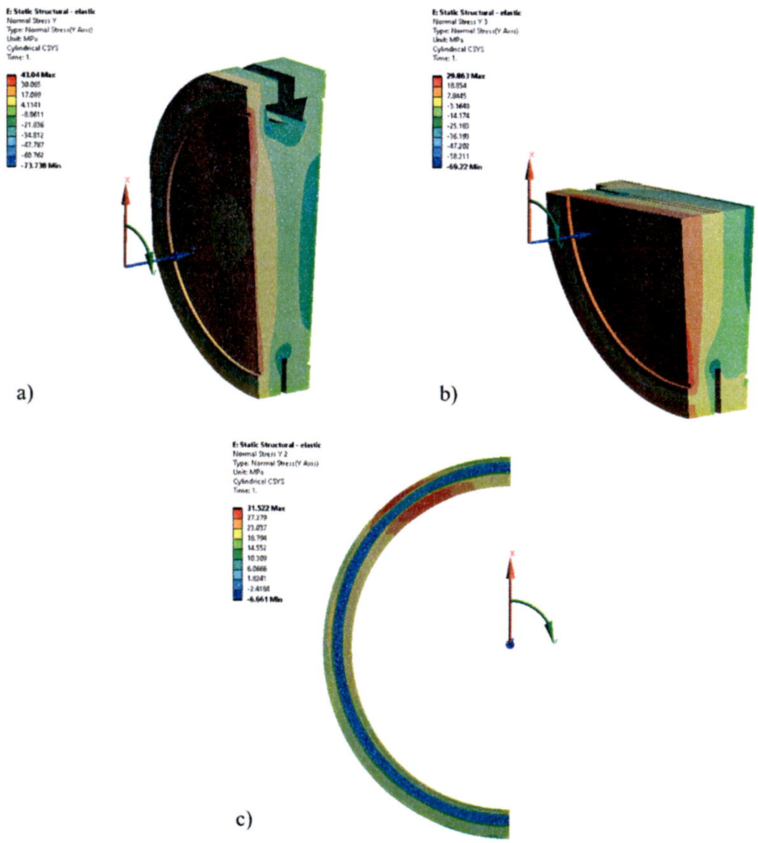

Fig. 8. Tangential stresses distributed in the wedge and in hardfacing

6 Conclusion

The development of this methodology represents only one part of the project solution. Within the scope of the project, the experimental program was performed to determine the relevant properties of the base material as well as the hardfacing material. The results of the experimental program served as input and for verification of calculation procedures as well as for the formulation of conclusions. The risk of crack propagation through brittle fracture or fatigue growth was assessed by computational analysis.

The main results of the solution can be summarized as follows:

- *Assessment of the risk of crack propagation through the brittle fracture mechanism:* Based on the results of the experimental program as well as the results of the computational analyses, the risk of uncontrolled crack propagation by the brittle fracture mechanism within the operating temperature range can be clearly excluded.
- *Assessment of the risk of crack growth by a fatigue mechanism:* The results of the experimental program show that macroscopic crack growth by fatigue mechanisms within 10^3 cycles requires significantly greater stress than is presented in hardfacing

at specified load (one-sided external pressure 4.6 MPa). Based on this result, the risk of crack growth by the fatigue mechanism is low.

To conclude, the achieved results correspond to the initial requirements and serve as an excellent input for other parts of the project.

References

1. ISO 148–1:2009, Metallic Materials – Charpy Pendulum Impact Test – Part 1: Test Method.
2. ISO 12135:2016, Metallic Materials – Unified Method of Test for the Determination of Quasistatic Fracture Toughness.
3. ISO 12108:2012, Metallic Materials – Fatigue Testing – Fatigue Crack Growth Method.
4. Anderson, T.L.: Fracture Mechanics – Fundamentals and Applications. CRC Press, Boca Raton (2005)
5. Kunz, J.: Applied Fracture Mechanics (in Czech). Vydavatelství ČVUT, Praha (2005).
6. Klesnil M., Lukáš, P.: Fatigue of Metallic Materials. Elsevier (1992).
7. Broek, D.: Elementary Engineering Fracture Mechanics. Hingham, Kluwer Boston (1982).
8. Veles, P.: Mechanical Properties and Testing of Metals (in Slovak). Alfa, Bratislava (1989).
9. Peciar, P. et al.: Process Engineering II (in Slovak). Vydavatelstvo STU, Bratislava (2016).
10. Ashby, M.F., Jones, D.: Engineering Materials 1: An Introduction to Properties. Applications and Design. Butterworth-Heinemann, Boston (2012)
11. ANSYS tutorials, https://ansyshelp.ansys.com, last accessed 2019/12/16.

A Study on Stress Relaxation Behavior of Isotropic Magnetorheological Elastomeric Composite

Tran Huu Nam(✉), Iva Petríková, and Bohdana Marvalová

Faculty of Mechanical Engineering, Technical University of Liberec, Studentská 1402/2, 461 17 Liberec 1, Czech Republic
{nam.tran.huu,iva.petrikova,bohda.marvalova}@tul.cz

Abstract. Experimental study and numerical investigation of stress relaxation behavior of isotropic magnetorheological elastomeric composite (MEC) were carried out in this article. The isotropic MEC was produced from silicone rubber reinforced with micro-sized carbonyl iron particles. The stress relaxation response of the isotropic MEC was investigated at different loading rates, constant strain levels, and under various electromagnetic fields through the single relaxation test with double-lap shear specimens. Research results indicated that the stress relaxation of the isotropic MEC depended slightly on the loading rate, but it was considerably dependent on the constant strain and the electromagnetic field. The shear stress and modulus of the MEC in the relaxation period enhanced with increasing the constant strain and electromagnetic field intensity as well. The stress relaxation of the isotropic MEC was examined numerically using the four-parameter fractional derivative viscoelastic Zener model. The studied fractional derivative viscoelastic model was fitted well to the measured relaxation modulus of the isotropic MEC. The calculated shear stresses of the isotropic MEC with long-term predictions agreed well with the measured ones. Therefore, the investigated fractional derivative viscoelastic model can apply to predict the long-term stress relaxation behavior of the isotropic MEC.

Keywords: Magnetorheological elastomer · Stress relaxation · Relaxation test · Viscoelastic modeling · Fractional calculus

1 Introduction

Magnetorheological elastomeric composites (MECs) have been prepared by dispersing ferromagnetic powder into a non-magnetic elastomeric matrix. MECs are considered to be intelligent materials due to controllable rheological and mechanical properties under an external magnetic field [1]. The typical behavior of MECs is changing their stiffness and damping properties under the magnetic field. With controllable stiffness and damping properties, MECs have been used in a variety of engineering applications [2]. In addition, MECs are regarded to be viscoelastic composite materials. The mechanical behavior of viscoelastic materials is time-dependent. The time-dependent response of

© The Author(s), under exclusive license to Springer Nature Switzerland AG 2022
J. Beran et al. (Eds.): TMM 2020, MMS 85, pp. 163–172, 2022.
https://doi.org/10.1007/978-3-030-83594-1_17

MECs makes them prone to creep and stress relaxation. Therefore, studies of the stress relaxation behavior of MECs are necessary to investigate their viscoelastic properties for long-term applications.

The viscoelastic properties of MECs have been studied numerically using fractional derivative models [3–5]. Fractional derivative models based on classical viscoelastic models (Kelvin–Voigt, Maxwell, Zener, etc.) had been built by different combinations of elastic spring and fractional-order dashpot. The fractional derivative models were effective in describing the stress relaxation behavior of polymeric materials [6]. For this paper, the isotropic MEC was developed from silicone rubber and carbonyl iron powders (CIPs). The stress relaxation response of the isotropic MEC was investigated using the single relaxation test with double-lap shear samples. Influences of the loading rate, constant strain, and magnetic flux density (MFD) on the stress relaxation of the MEC were examined. The stress relaxation of the MEC was calculated numerically using a four-parameter fractional derivative viscoelastic Zener model. The model parameters fitting to measured data of the single relaxation test were used to predict the long-term stress relaxation behavior of the isotropic MEC.

2 Experimental Investigation

2.1 Materials

The isotropic MEC was developed using micro-sized CIPs, RTV silicone rubber ZA13, and its catalyst. The micro-sized CIPs (type: 44890), which were provided by Sigma-Aldrich (USA), had overall spherical shapes with 2–5 μm ($\geq$99.5%) in diameter. The RTV silicone rubber ZA13 and its catalyst were developed by Zhermack S.P.A (Italy) and were supplied by Havel Composites Ltd. (Czech Republic).

2.2 Fabrication of Isotropic MEC

The isotropic MEC samples were fabricated by mixing the silicone rubber ZA13, its catalyst, and 27 vol.% CIPs. The processing of the isotropic MEC specimens and their microstructural morphology can be found in detail in our earlier reports [3–5].

2.3 Single Relaxation Test

The stress relaxation behavior of the isotropic MEC was investigated via the single relaxation test with double-lap shear specimens. The double-lap shear samples were fabricated by sandwiching two isotropic MEC square blocks between the aluminum slabs. The single relaxation test for the isotropic MEC was performed in the Instron Electropuls testing system at different loading rates (0.01, 0.1, and 1.0/s) and various constant strains (5, 10, 15, and 20%). Besides, the single relaxation test for the MEC samples was conducted with the rise of the MFD up to 0.58 T using an electromagnet. The electromagnet was used to generate magnetic fields with the applied direction perpendicular to the sample shear force. Before each relaxation test, the MEC specimen was loaded cyclically to maximum strain amplitude to eliminate the Mullins effect.

The shear force and displacement were recorded for 1000 s in each single relaxation test. The double-lap shear test, the MEC specimen, and the electromagnet system were described in our previous articles [3–5].

2.4 Experimental Results

The single shear stress relaxation behavior of the isotropic MEC under a 20% constant strain at different loading rates is described in Fig. 1. The influences of applied constant strains on the shear stress relaxation response of the isotropic MEC are presented in Fig. 2. Shear stresses of the isotropic MEC in the relaxation period as functions of time and the MFD at various constant strain levels are depicted in Fig. 3. The influences of different MFDs on the relaxation modulus of the isotropic MEC at various constant strains are shown in Fig. 4.

The increase in the loading rate resulted in the rise of the shear stress at the same strain (Fig. 1a). Besides, the stress relaxation rate of the isotropic MEC enhances with increasing the loading rate. At the beginning of relaxation, the peak stress at the loading rate of 1.0/s is highest and higher than that at the lower loading rates. The isotropic MEC loaded at a faster rate has greater peak stress than that loaded at a lower rate, because a low rate affords a longer time for the isotropic MEC to relax during loading. Moreover, the modulus relaxation rate increases with the rise of the loading rate (Fig. 1b). However, the shear stress and modulus of the isotropic MEC in the relaxation period after the loading ramp at a lower rate are slightly greater than those at a higher rate. In general, the isotropic MEC exhibits that the higher the loading rate, the greater the stress relaxation rate, but its relaxation modulus does not depend much on the loading rate.

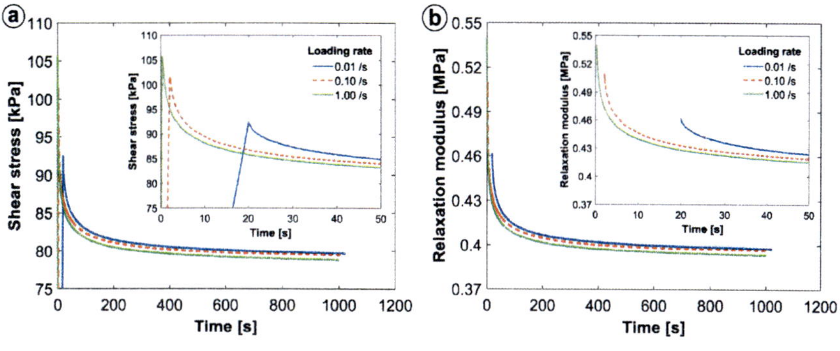

Fig. 1. Shear stress-time curves and relaxation modulus curves of the isotropic MEC under a 20% constant strain at different loading rates. The inset figures are the zooms of the first 50 s.

The rise of applied constant strains leads to a significant increase in the shear stress and a slight reduction of the modulus in the relaxation period (Fig. 2). In addition, the shear stress and modulus versus time curves show strongly stress relaxation during the initial 200 s and then indicate an extremely slow rate of relaxation that continues in an asymptotic sense, as reported in [9]. The stress relaxation rate is dependent on the overstress, which is defined as the difference between the current stress and the equilibrium stress. As Fig. 2a shows, the stress relaxation rate increases with raising the strain level. Therefore, the high strain level shows a larger overstress than the low strain level. Generally, the single relaxation test carried out at higher strain levels possessed greater overstresses and showed faster stress relaxation than those at lower strain levels with smaller overstresses, as presented by Amin et al. [10].

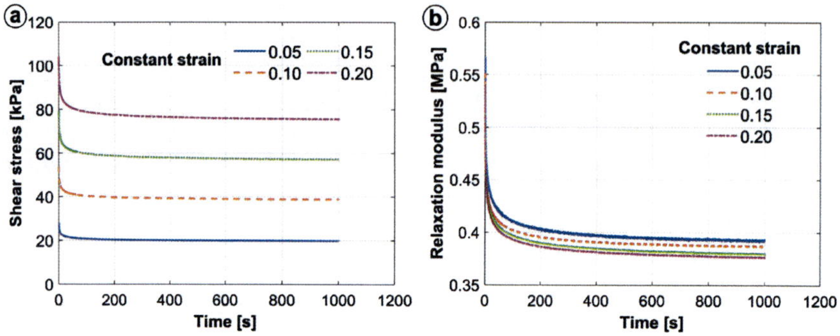

Fig. 2. Shear stress-time curves and relaxation modulus curves of the isotropic MEC under different constant strain levels.

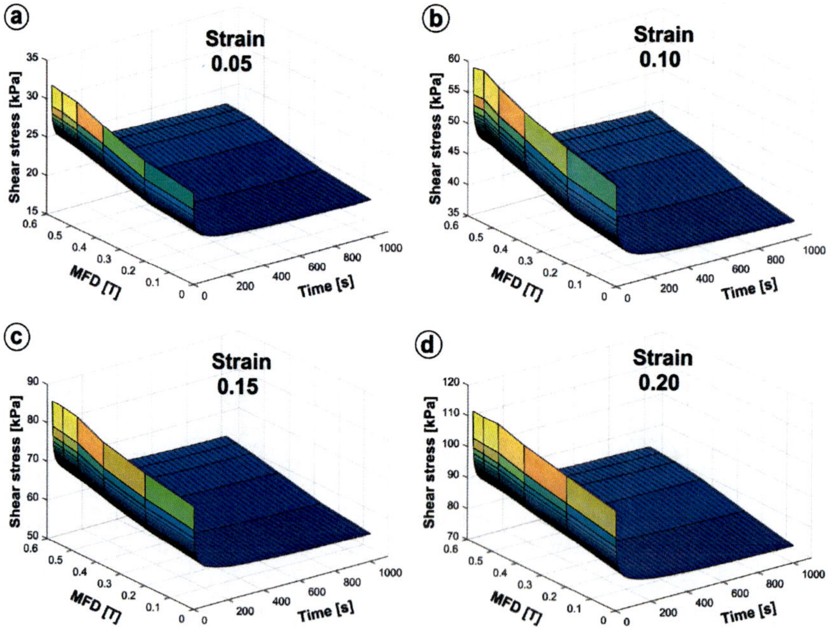

Fig. 3. Shear stress of the isotropic MEC in the relaxation period as a function of time and the MFD at different constant strain levels.

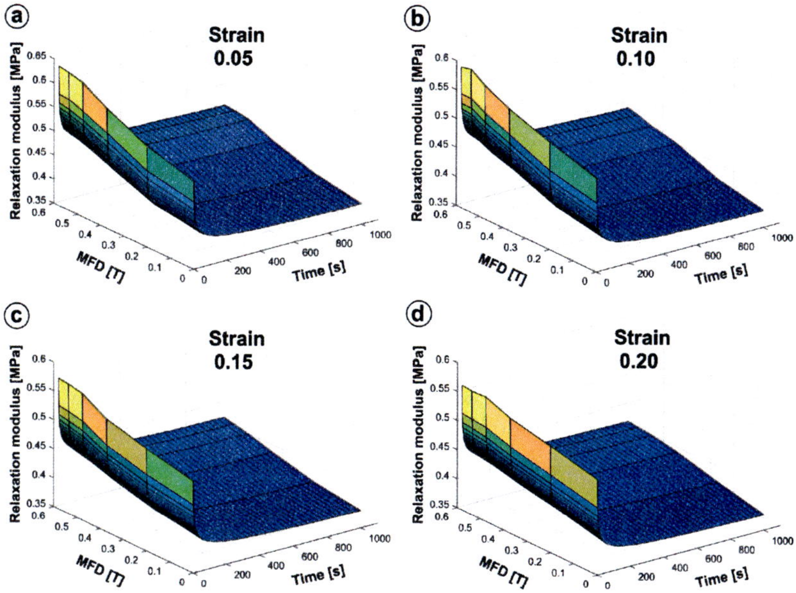

Fig. 4. Relaxation modulus of the isotropic MEC over time under various MFDs at different constant strain levels.

As observed in Figs. 3, 4, the shear stress and modulus increase with increasing the MFD. They boost significantly with raising the MFD to about 0.5 T, then enhance somewhat above 0.5 T. The stress relaxation of the MEC is more largely as the MFD rises, so it will take more time to reach the equilibrium state. The enhancement in the relaxation modulus with increasing the MFD is attributable to raising the MR effect of the MEC [4, 5]. The increase in the shear modulus with enhancing the MFD is ascribable to the rise in the magnetic attraction force between CIPs in the MEC. The augmentation of the magnetic force makes CIPs closer, resulting in an increase in the MEC stiffness. The variation in the shear stress and modulus of the MEC is related to the position alternation tendency of CIPs under an electromagnetic field. Once an electromagnetic field is applied to the MEC, the CIPs tend to reach the positions of minimum energy state [4]. The movement of CIPs introduces deformations in the rubber matrix, leading to the rise in the modulus of the MEC [5].

3 Numerical Modeling

3.1 A Fractional Derivative Viscoelastic Model for the MEC

The stress relaxation of the MEC was investigated using the four-parameter fractional derivative Zener model with a Mittag-Leffler function kernel. The model was composed of an elastic spring and a fractional Maxwell element in parallel [6]. The fractional derivative with the Mittag–Leffler function kernel is defined as [7]:

$$D^{\alpha} f(t) = \frac{1}{1-\alpha} \frac{\mathrm{d}}{\mathrm{d}t} \int_0^t f(x) M_{\alpha}\left[-\frac{\alpha(t-x)}{1-\alpha}\right] \mathrm{d}x \tag{1}$$

where α is the fractional parameter with value changing between 0 and 1 [7].

The definition of one-parameter Mittag–Leffler function $M_{\alpha}(x)$ is given as [8]:

$$M_{\alpha}(x) = \sum_{n=0}^{\infty} \frac{x^n}{\Gamma(1+\alpha n)} \tag{2}$$

where $\Gamma(1 + \alpha n)$ is the gamma function with the argument $(1 + \alpha n)$.

The constitutive equation for the four-parameter fractional derivative Zener model in the time domain is expressed as follows:

$$\sigma(t) + \tau^{\alpha} \frac{\mathrm{d}^{\alpha}\sigma}{\mathrm{d}t^{\alpha}} = E_0 \varepsilon(t) + (E_0 + E_1)\, \tau^{\alpha} \frac{\mathrm{d}^{\alpha}\varepsilon}{\mathrm{d}t^{\alpha}} \tag{3}$$

where E_0 and E_1 are the elastic moduli of the two springs of the model, and τ is the relaxation time of the fractional dashpot.

The relaxation modulus of the investigated model obtained by the application of the Laplace transform to Eq. (3) is expressed as follows:

$$G(t) = E_0 + E_1 M_{\alpha}\left[-\left(\frac{t}{\tau}\right)^{\alpha}\right] \tag{4}$$

3.2 Numerical Simulation Results and Comparison to Experimental Data

The stress relaxation of the isotropic MEC was simulated numerically using the presented model. Equation (4) was used to fit the measured relaxation modulus from the single relaxation test. Four parameters in the vector $\mathrm{x} = (E_0, E_1, \alpha, \tau)^{\mathrm{T}}$ of the investigated model were obtained by fitting Eq. (4) to the measured data. The least-squares fit of the relaxation modulus to the measured one was conducted by minimizing an objective function with the optimization parameters using the derivative-free method in Matlab.

The model parameters fitting to measured data by minimizing the objective function were given in Tables 1, 2. It is clear from Table 1 that the elastic modulus E_0 and the fractional parameter α decreased slightly with increasing the loading rate. As Table 2 shows, the parameter E_0 changed slightly with the rise of the applied constant strain. However, the parameter E_0 boosted rapidly with raising the MFD to approximately 0.5 T and grew slightly above 0.5 T. The enhancement in the shear modulus of the isotropic MEC with increasing the MFD is corresponding to the magnetic-controllable stiffness of the isotropic MEC. Besides, the fractional parameter α varied irregularly with the rise of the MFD and the constant strain as well (Table 2).

Table 1. The model parameters fitting to measured data at loading rates for the MEC.

Loading rate [/s]	Parameter			
	E_0 [MPa]	E_1 [MPa]	α	τ [s]
0.01	0.395	9.23	0.690	5.67E-02
0.10	0.385	31.9	0.376	1.60E-06
1.00	0.381	0.34	0.366	2.94E-01

Table 2. The model parameters fitting to measured data for the MEC under different constant strains and MFDs.

Strain	Parameter	MFD [T]					
		0	0.201	0.373	0.478	0.538	0.580
0.05	E_0 [MPa]	0.370	0.389	0.422	0.438	0.436	0.427
	E_1 [MPa]	0.520	0.449	0.381	0.545	1.162	4.082
	α	0.306	0.341	0.384	0.346	0.284	0.239
	τ [s]	7.1E-02	1.9E-01	5.6E-01	1.3E-01	3.6E-03	5.0E-06
0.10	E_0 [MPa]	0.355	0.380	0.408	0.415	0.420	0.419
	E_1 [MPa]	1.378	0.430	0.447	0.513	0.544	1.852
	α	0.262	0.352	0.363	0.338	0.325	0.268
	τ [s]	4.1E-04	1.6E-01	1.6E-01	1.1E-01	9.8E-02	2.4E-04
0.15	E_0 [MPa]	0.356	0.366	0.383	0.397	0.403	0.404
	E_1 [MPa]	1.264	3.628	3.783	1.156	0.888	1.229
	α	0.267	0.251	0.252	0.281	0.293	0.274
	τ [s]	6.8E-04	5.5E-06	4.7E-06	2.1E-03	7.7E-03	1.5E-03
0.20	E_0 [MPa]	0.355	0.374	0.401	0.406	0.408	0.409
	E_1 [MPa]	2.637	1.358	0.420	0.554	0.606	0.871
	α	0.270	0.281	0.386	0.353	0.344	0.323
	τ [s]	3.9E-05	7.3E-04	1.9E-01	6.7E-02	4.5E-02	9.5E-03

The model fittings of the relaxation modulus to measured data of the isotropic MEC in the single relaxation test at different loading rates, applied constant strains, and under various MFDs were shown in Fig. 5. Results showed the excellent fittings of relaxation modulus to the measured one of the isotropic MEC. In addition, the stress relaxation stress of the isotropic MEC as a function of the time at various loading rates, constant strains, and under different MFDs was calculated using the studied model with estimated parameters, with results described in Fig. 6. The long-term predictions of the shear stress of the isotropic MEC over a wide range of time using the presented model were shown in Fig. 6. Although the stress relaxation was measured in only 1000 s, the studied model can calculate the stress relaxation of the isotropic MEC for a longer time (Fig. 6). The maximal relative error between estimated and measured values within 1000 s for both the relaxation modulus and shear stress is less than 2%. Generally, the four-parameter fractional derivative Zener model was fitted well to experimental data for the isotropic MEC in the single relaxation test. The investigated fractional derivative viscoelastic model can be used for predicting the long-term stress relaxation of the isotropic MEC.

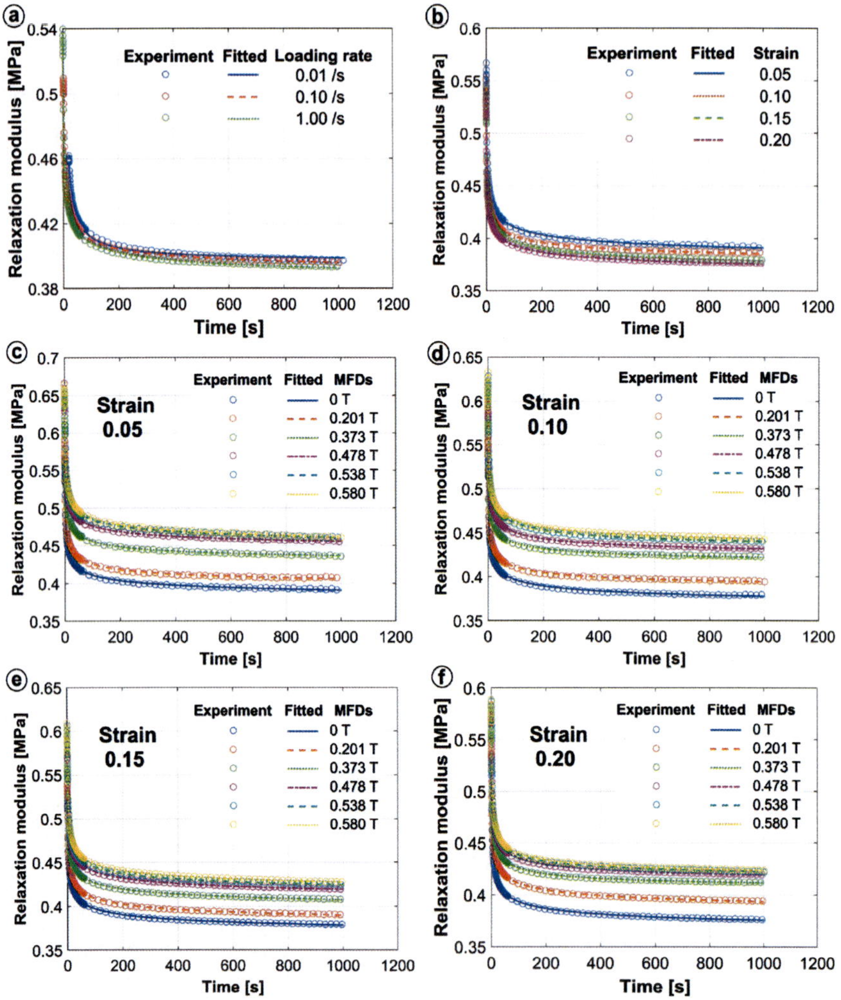

Fig. 5. Experimental and model fitted curves of the relaxation modulus of the isotropic MEC at various loading rates, applied constant strains, and under different MFDs.

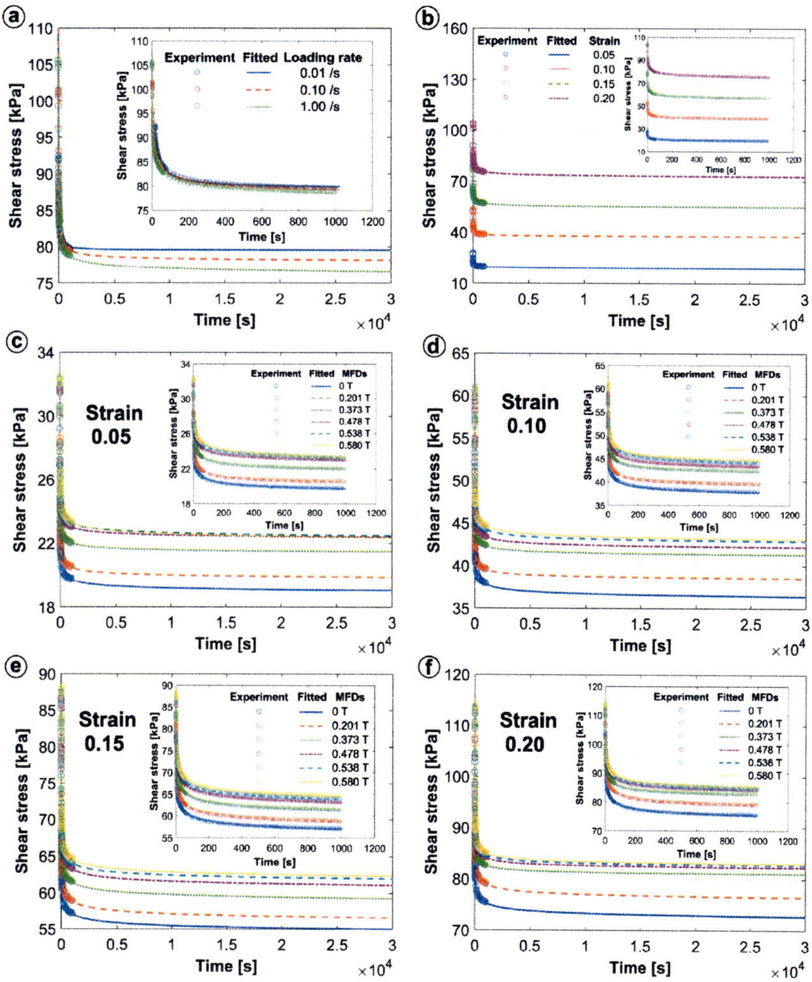

Fig. 6. Experimental and model fitted curves of the shear stress of the isotropic MEC at different loading rates, applied constant strains, and under various MFDs. The inset figures are the zooms of the first 1000 s.

4 Conclusions

Stress relaxation behavior of the isotropic MEC produced from silicone rubber and micro-sized CIPs was investigated experimentally and numerically in this study. Effects of loading rates, applied constant strains, and electromagnetic fields on the stress relaxation of the isotropic MEC were examined. Measured results indicated that the stress relaxation of the isotropic MEC depended slightly on the loading rate, but it was strongly dependent on the applied constant strain and the electromagnetic field. Although the shear stress of the isotropic MEC in the relaxation period enhanced with increasing the constant strain, the relaxation modulus reduced moderately. The shear stress and modulus of the isotropic MEC boosted rapidly with raising the MFD to 0.5 T and increased slightly over 0.5 T. The four-parameter fractional derivative viscoelastic Zener model was fitted well to measured data of the isotropic MEC. The estimated shear stresses of the isotropic MEC with long-term predictions agreed well with the measured ones. The maximal relative error between experimental and calculated values of both shear stress and relaxation modulus is less than 2.0%. In short, the investigated fractional derivative viscoelastic model can apply to predict the long-term stress relaxation behavior of the isotropic MEC.

Acknowledgments. This work was supported by the Ministry of Education, Youth and Sports of the Czech Republic and the European Union – European Structural and Investment Funds in the frames of Operational Program Research, Development and Education –project Hybrid Materials for Hierarchical Structures (HyHi, Reg. No. CZ.02.1.01/0.0/0.0/16_019/0000843).

References

1. Bastola, A.K., Hossain, M.: A review on magneto-mechanical characterizations of magnetorheological elastomers. Compos. B Eng. **200**, 108348 (2020)
2. Ahamed, R., Choi, C.B., Ferdaus, M.M.: A state of art on magneto-rheological materials and their potential applications. J. Intell. Mater. Syst. Struct. **29**(10), 2051–2095 (2018)
3. Nam, T.H., Petríková, I., Marvalová, B.: Experimental and numerical research of dynamical mechanical properties of magneto-sensitive elastomeric composites. In: Huneau, B., et al. (eds.) Constitutive Models for Rubbers XI, pp. 138–143. Taylor & Francis, UK (2019)
4. Nam, T.H., Petríková, I., Marvalová, B.: Experimental characterization and viscoelastic modeling of isotropic and anisotropic magnetorheological elastomers. Polym. Test. **81**, 106272 (2020)
5. Nam, T.H., Petríková, I., Marvalová, B.: Experimental and numerical research of stress relaxation behavior of magnetorheological elastomer. Polym. Test. **93**, 106886 (2021)
6. Guo, X., Yan, G., Benyahia, L., Sahraoui, S.: Fitting stress relaxation experiments with fractional Zener model to predict high frequency moduli of polymeric acoustic foams. Mech. Time Depend. Mater. **20**(4), 523–533 (2016). https://doi.org/10.1007/s11043-016-9310-3
7. Atangana, A., Baleanu, D.: New fractional derivatives with nonlocal and non-singular kernel: theory and application to heat transfer model. Therm. Sci. **20**, 763–769 (2016)
8. Niedziela, M., Wlazło, J.: Notes on computational aspects of the fractional-order viscoelastic model. J. Eng. Math. **108**(1), 91–105 (2017). https://doi.org/10.1007/s10665-017-9911-0
9. Haupt, P., Sedlan, K.: Viscoplasticity of elastomeric materials: experimental facts and constitutive modelling. Arch. Appl. Mech. **71**, 89–109 (2001)
10. Amin, A.F.M.S., Lion, A., Sekita, S., Okui, Y.: Nonlinear dependence of viscosity in modeling the rate-dependent response of natural and high damping rubbers in compression and shear: experimental identification and numerical verification. Int. J. Plast. **22**, 1610–1657 (2006)

A Simulation Stand for Human Limb Movements During Nordic Walking

Sławomir Wudarczyk, Bogusz Lewandowski(✉), Jarosław Szrek, and Jacek Bałchanowski

Wrocław University of Science and Technology, 50-371 Wrocław, Poland
Bogusz.Lewandowski@pwr.edu.pl

Abstract. According to physiotherapists and personal trainers, gait exercises with Nordic Walking poles (NW) are one of the most frequently used forms of physical rehabilitation in treating many diseases. The first stage of rehabilitation is teaching patients the correct gait technique. At least in the initial stage of the rehabilitation process, this involves the constant supervision of a physiotherapist. The development of a NW gait monitoring system, which will also provide information about errors in the performed exercises, will allow the patient to improve the NW gait technique independently. This paper presents the construction of the simulation stand for testing the components of the designed NW mechatronic poles system. The main task of this station is to simulate the movements of the lower and upper limbs in a repeatable manner, which are to be used to verify the operation of the developed mechatronic NW poles. The developed control and measurement system of the station allows to simulate the kinematics of the upper and lower limbs obtained from actual measurements on human limbs during the NW gait. The paper presents the research results on the accuracy of human limb movement during NW gait on the simulation stand.

Keywords: Gait research · Control system · Gait kinematics · Motor rehabilitation

1 Introduction

In modern medicine, rehabilitation is considered as part of the treatment process, which is an important adjunct to pharmacological and surgical treatments. Motor rehabilitation is a very common type of rehabilitation. It is dedicated to people with limited motor functions.

One of the methods for faster recovery is the use of Nordic Walking (NW) in rehabilitation. The NW gait is used to support the treatment of many diseases (strokes, myocardial infarction, limb injuries, intermittent claudication, diabetes, obesity) [1–5].

Compared to regular walking, Nordic Walking involves the human muscular system much more, especially the upper body [6]. The effect is an increased response of the cardiovascular and respiratory systems with less fatigue [7, 8]. A prerequisite for good rehabilitation results is the correct use of walking poles. It should be emphasized that

© The Author(s), under exclusive license to Springer Nature Switzerland AG 2022
J. Beran et al. (Eds.): TMM 2020, MMS 85, pp. 173–182, 2022.
https://doi.org/10.1007/978-3-030-83594-1_18

improper use of NW poles means that the NW march not only does not fulfil its task but may even cause an injury. It should be borne in mind that usually, patients have not experienced this form of activity before and do not have the correct technique of NW gait. Therefore, the first stage of rehabilitation is teaching patients the correct gait technique. To achieve this, the march must be supervised by a physiotherapist or trainer, limiting the availability and common use of this method [9].

Execution of NW poles supporting and supervising the development of proper march technique improve the effectiveness of rehabilitation. With an appropriate NW march monitoring system, the patient could perform exercises and improve the technique without constant supervision by a physiotherapist or trainer.

As part of the carried out research project, the authors created mechatronic poles for NW equipped with a measuring, diagnosing and monitoring system for the patient's gait [10, 11]. The final stage of the project is to equip the NW poles with a system that will allow the measurement of the following parameters while marching: the angles of the pole about the vertical in the sagittal plane, the force of the pole hitting the ground, the patient's grip force and the detection of the pole passes the leg event. Monitoring these parameters requires equipping poles with an appropriate electronic and IT system.

This paper presents the problems of research and construction of a simulation stand for human limb movements while walking with Nordic Walking poles. The main task of this stand is to simulate the movements of the lower and upper limbs in a reproducible manner, which are to be used to verify the operation of the NW mechatronic poles. The paper also presents the results of tests and measurements of the accuracy of performing limb movements on the developed test stand. The stand's developed control and measurement system allow simulating the upper and lower limb kinematics obtained from actual measurements on human limbs during the NW march.

2 Materials and Methods

In order to accurately reproduce human walking with NW poles, a test stand was developed (Fig. 1). The mechanism had 6 degrees of freedom, which reflect the movements of the main parts of the upper and lower limbs. It consists of a frame with three shelves as the base. The lowest shelf contains ProNet-04AMG controllers from ESTUN. The second shelf consists of a mechanism for simulating the movement of the lower limb, with mobility $W = 3$ consisting of three segments simulating the movement: thighs (h), calves (k) and feet (a). Attached to the third shelf is a mechanism for simulating the movement of the upper limb, also with mobility $W = 3$, consisting of the following segments: arm (s), elbow (e) and hand (w) [12]. Therefore the test stand consists of 6 independently controlled axes φ_{sg}, φ_{eg}, φ_{wg}, φ_{hg}, φ_{kg} and φ_{ag}, which determine the orientation of individual limb segments relative to the base. Every motor M_s, M_e, M_w, M_h, M_k and M_a drives a dedicated ball screw s_s, s_e, s_w, s_h, s_k, s_a and causes the toothed belt to move linearly. The further part of the modules consists of 6 belt transmissions with a ratio of $i_k = 1$, which excites the elements simulating upper and lower extremity movements (Fig. 1).

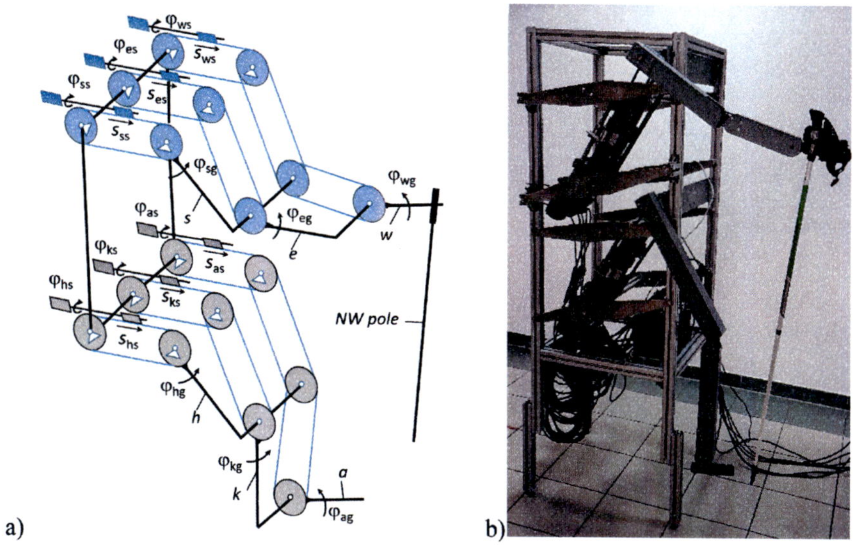

Fig. 1. A test stand for simulating limb movements during NW gait: a) kinematic diagram, b) view of the test stand.

A formula according to which the angular position of the test stand parts is calculated about the angular position of the drive shaft is as follows:

$$\varphi_{is} = \pi d / h \varphi_{ig} \tag{1}$$

where:

φ_{is} – the angle of drive shaft rotation ($i = s, e, w, h, k, a$),
φ_{ig} – the angle of toothed pulley wheel ($i = s, e, w, h, k, a$),
h – thread pitch,
d – pitch diameter of gears.

The stand was designed to simulate NW gait and calibrate mechatronic NW poles. The NW pole is attached to the hand w (Fig. 1). Calibration and experimental studies of the mechatronic NW poles measuring system require the cooperation of the upper and lower limbs. The upper part is responsible for the position and orientation of the pole. On the other hand, the lower limb allows you to check the correct position of the lower limb about the NW pole, the appropriate length and gait characteristics. The simulation stand is controlled via the dSPACE ControlDesk application [13]. In each driving axis of the system, there is a sensor responsible for the angular position measurement. Thus, the position of each segment of the device is known.

The signals of the set angular displacements are transferred to the real-time controller dSPACE, which generates control signals for each axis of the stand. This information is passed to ProNet controllers, which directly control each of the 6 drives. The current angular position, measured with 16-bit encoders, is transmitted to the dSPACE controller (Fig. 2) [13].

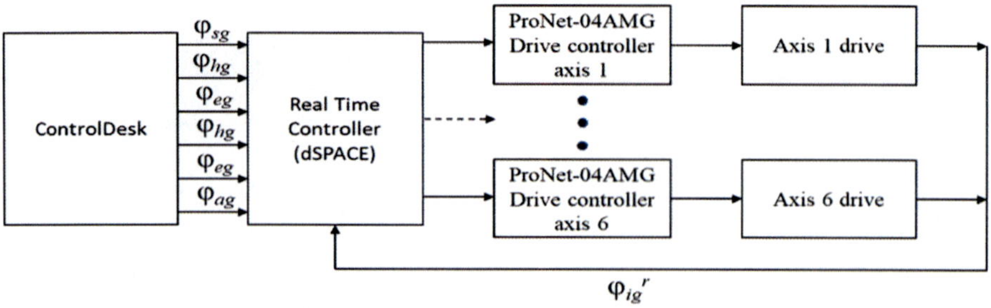

Fig. 2. Block diagram of the stand control system.

The control signal φ_{ig} individual axes should impose anatomical movement of the patient's limb segments during NW gait. The necessary parameters of this gait were determined based on measurements made with the use of the IMU STT-System inertial sensor system [14].

Measurements were carried out using a set of IMU sensors and iSEN software by the biomechanical protocol "Full Body T-Stance" with a frequency of 100 Hz [14]. The anatomical courses of the values of the rotation angles of the upper limb in the shoulder joint φ_s, elbow joint φ_e and wrist joint φ_w joint were obtained, and femoral joint φ_h, knee joint φ_k and ankle joint φ_a in lower limb movement. All angle values were measured in the sagittal plane (Fig. 3).

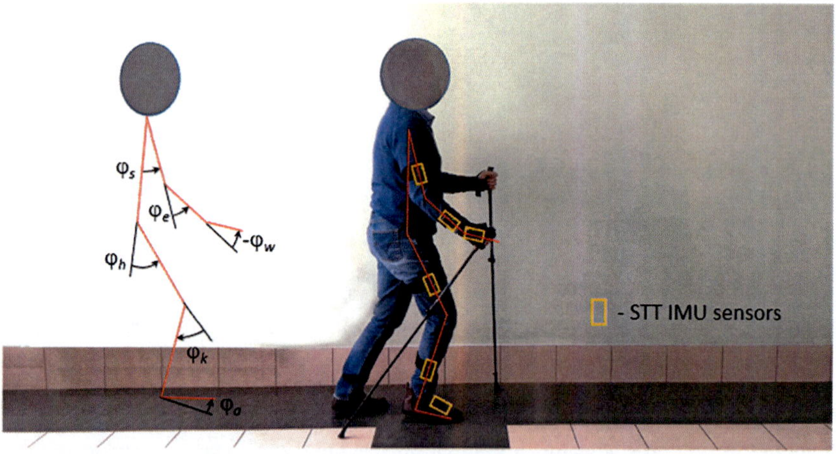

Fig. 3. View of a human during NW control gait with the STT IMU measurement system and a scheme for rotation angle measurement in the sagittal plane.

To control the stand, the angles φ_{ig} of the human limbs segments orientation concerning the bases are used. The anatomical angles φ_i obtained from the measurements with IMU STT-System sensors are converted into φ_{ig} angles of individual limb segments according to the formulas:

$$\begin{aligned} \varphi_{sg} &= \varphi_s \\ \varphi_{hg} &= \varphi_h \\ \varphi_{eg} &= \varphi_s + \varphi_e \\ \varphi_{kg} &= \varphi_h - \varphi_k \\ \varphi_{wg} &= \varphi_s + \varphi_e - \varphi_w \\ \varphi_{ag} &= \varphi_h - \varphi_k + \varphi_a \end{aligned} \tag{2}$$

where:

φ_i – anatomical angles ($i = s, e, w, h, k, a$),
φ_{ig} – stand angles relative to the base ($i = s, e, w, h, k, a$).

3 Results

The experimental research of the developed stand for simulating the movement of human limbs was divided into two stages:

(a) measuring the gait parameters with NW poles using IMU sensors,
(b) testing the implementation of limb movement on the simulation stand.

In the first stage, the required courses of changes in the limb rotation angles in the joints during NW gait were obtained. Measurements were made using a set of STT-System [14] sensors with a frequency of 100 Hz on the control path for 15 s in accordance with the diagram shown in Fig. 3. The graphs in Figs. 4, 5, 6, 7, 8 and 9 show the obtained values of the rotation angles of the upper limb in the shoulder joint φ_s, elbow joint φ_e, and wrist joint φ_w, as well as lower limb joints: thigh φ_h, knee φ_k, and ankle φ_a (Fig. 3). In the second stage of the experiment, tests were carried out on the accuracy of human limb movements on the developed stand. The limb movements obtained during the control passage were simulated on the stand. The set parameter values φ_{ig} for the control systems were converted in accordance with (2) on the basis of φ_i angles. During the tests, the actual angles of rotation of the drives φ_s^r, exciting limb movements with a frequency of 100 Hz were measured. The values of the angles obtained were converted according to the formulas (1) and (2) into the values of anatomical angles of rotation φ_s^r, φ_e^r, φ_w^r, φ_h^r, φ_k^r and φ_a^r. Examples of waveforms obtained from one representative study are shown in the graphs in Figs. 4, 5, 6, 7, 8 and 9.

Analyzing the obtained waveforms of rotation angles in human joints φ_s, φ_e, φ_w, φ_w, φ_k, φ_a and courses of simulated movements of the limbs on the station φ_s^r, φ_e^r, φ_w^r, φ_w^r, φ_k^r, φ_a^r high accuracy (convergence) of the results can be observed. The courses of set and obtained angles values presented in pairs in the diagrams assume similar values.

In order to quantify the accuracy of the movements, errors were determined $\Delta\varphi_i$ ($i = s, e, w, h, k, a$) in the implementation of simulated rotations in the joints according to the formula:

$$\Delta\varphi_i = \varphi_i - \varphi_i^r \tag{3}$$

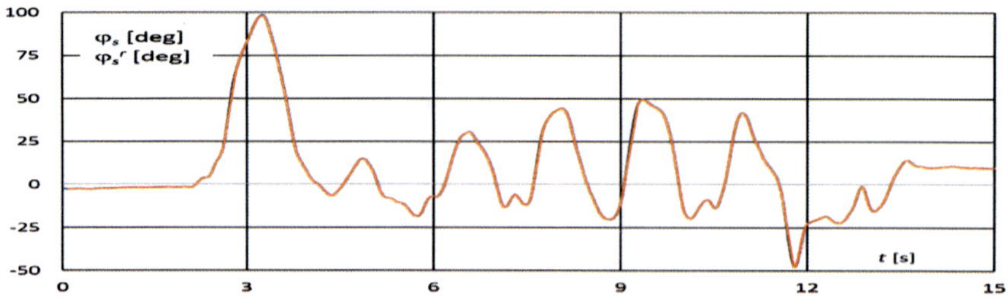

Fig. 4. The rotation angles in the shoulder joint: φ_s obtained from measurements with IMU sensors, φ_s^r obtained from measurements on the test stand.

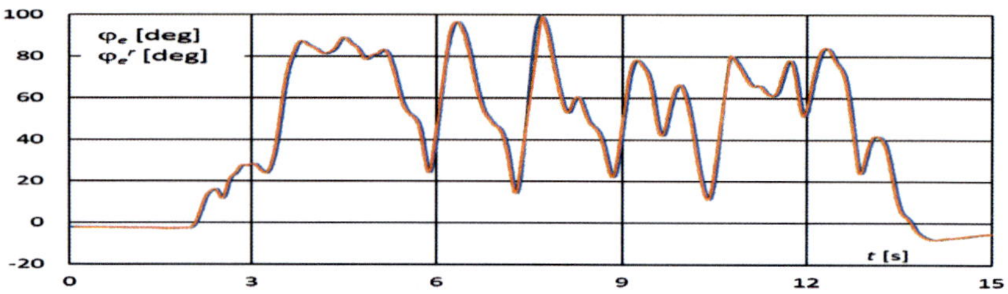

Fig. 5. The rotation angles in the elbow joint: φ_e obtained from measurements with IMU sensors, φ_e^r obtained from measurements on the test stand.

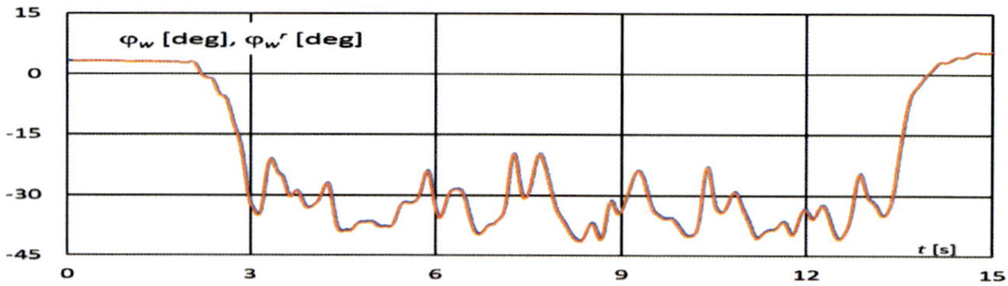

Fig. 6. The rotation angles in the wrist joint: φ_w obtained from measurements with IMU sensors, φ_w^r obtained from measurements on the test stand.

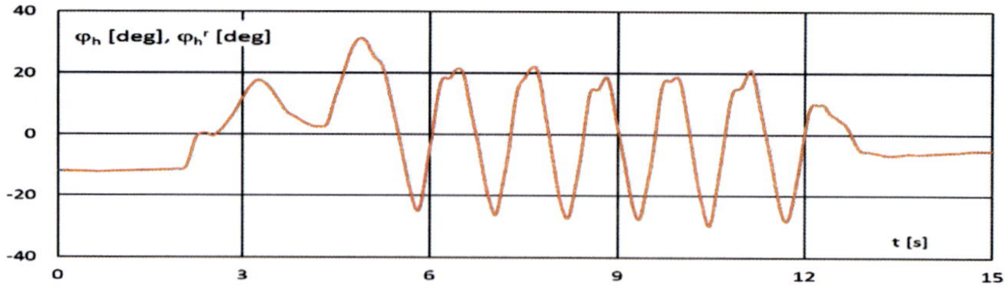

Fig. 7. The rotation angles in the hip joint: φ_h obtained from measurements with IMU sensors, φ_h^r obtained from measurements on the test stand.

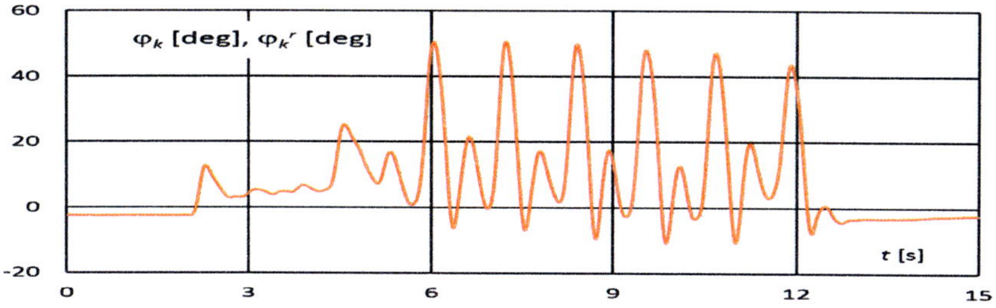

Fig. 8. The rotation angles in the knee joint: φ_k obtained from measurements with IMU sensors, φ_k^r obtained from measurements on the test stand.

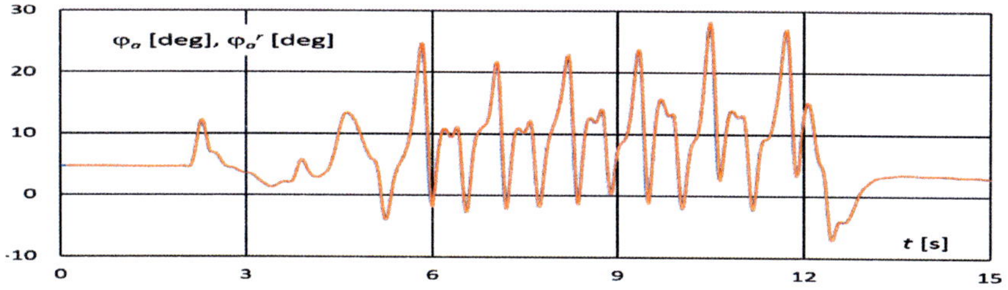

Fig. 9. The rotation angles in the ankle joint: φ_a obtained from measurements with IMU sensors, φ_a^r obtained from measurements on the test stand.

The calculated error $\Delta\varphi_i$ courses are presented in the graphs in Figs. 10, 11, 12, 13, 14 and 15.

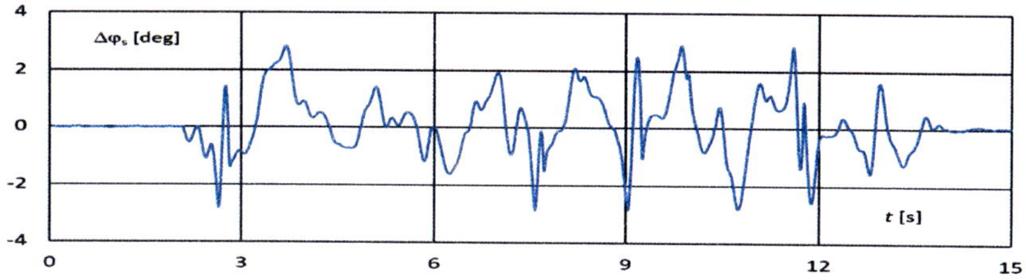

Fig. 10. Error $\Delta\varphi_s$ in the execution of the rotation in the shoulder joint φ_s on the test stand.

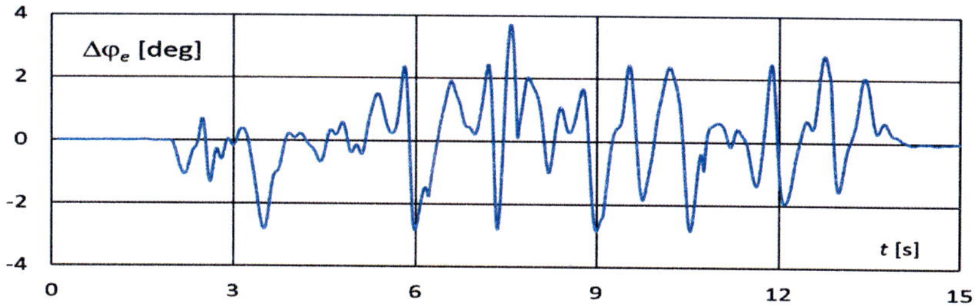

Fig. 11. Error $\Delta\varphi_e$ in the execution of the rotation in the elbow joint φ_e on the test stand.

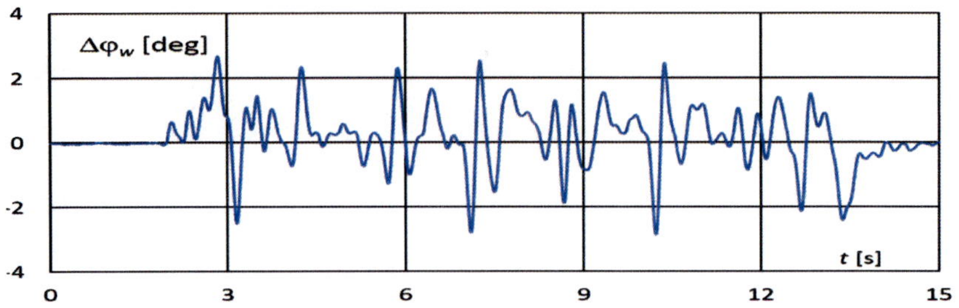

Fig. 12. Error $\Delta\varphi_w$ in the execution of the rotation in the wrist joint φ_w on the test stand.

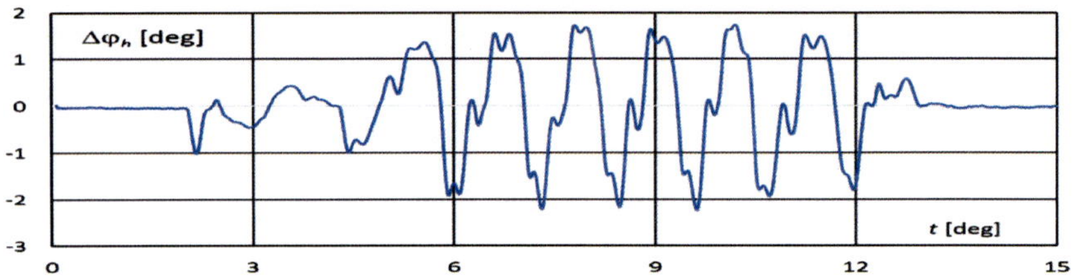

Fig. 13. Error $\Delta\varphi_h$ in the execution of the rotation in the hip joint φ_h on the test stand.

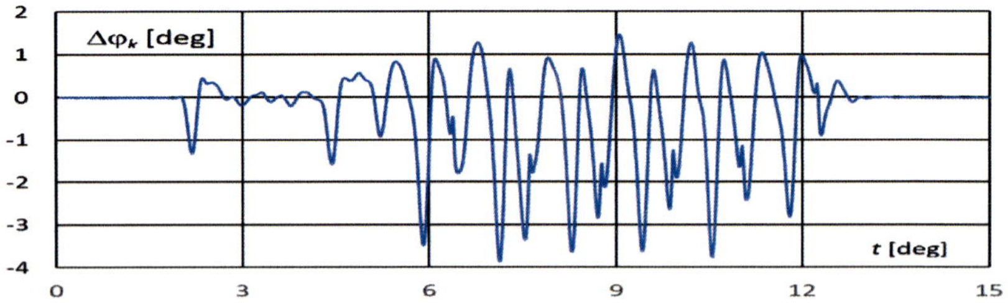

Fig. 14. Error $\Delta\varphi_k$ in the execution of the rotation in the knee joint φ_k on the test stand.

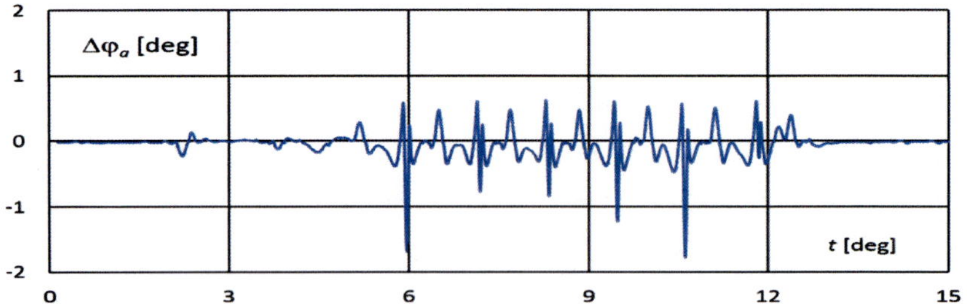

Fig. 15. Error $\Delta\varphi_a$ in the execution of the rotation in the ankle joint φ_a on the test stand.

The courses of errors in the implementation of rotation angles in the joints presented in the diagrams showed the correct operation of the drive and control systems of the developed simulation stand. Despite dynamically performed movements of the limbs during NW gait (Figs. 4, 5, 6, 7, 8 and 9), the maximum values of errors $\Delta\varphi_i$ and the simulated rotations in the joints do not exceed the values of ±4° (Figs. 10, 11, 12, 13, 14 and 15). The maximum values of $\Delta\varphi_i$ appear in the form of peaks of runs and occur in times of rapid changes in the values of rotation angles in the joint (e.g. the greatest values of errors $\Delta\varphi_k$ in the knee joint occur for t = 5.94 s, 7.05 s, 8.22 s, 9.53 s, 10.70 s (Fig. 14) which corresponds to the points of the fastest changes in the angle of rotation φ_k^r in this joint (Fig. 8).

4 Conclusions

The extensive research and the analysis of the results showed that the developed stand for simulating the movements of the upper and lower limbs during NW gait performs the given movements with sufficiently good accuracy. This means that the developed mechanical structure, the applied drive systems, controls and operation algorithms have been properly designed and made.

Therefore, the constructed simulation stand can be used for research on the mechatronic system to monitor NW poles' movement. A significant advantage of the constructed stand is that it enables comparative tests of the developed prototype systems to monitor the angle of the pole orientation. In particular, research on electronic circuits placed on NW poles and their operation algorithms.

The main goal of the design of the NW mechatronic poles is the ease of use and low cost. The use of a commercial universal pole angle measurement system would be so expensive that such a solution would not be widely available to patients.

Acknowledgments. This research is supported by the National Science Centre (NCN) under the project Opus 12 no 2016/23/B/NZ7/03310.

References

1. Eijkeren, F., Reijmers, R., Kalda, R., Tahepold, H., Maaroos, H.: Nordic walking improves mobility in Parkinson's disease. Mov. Disord. **23**, 2239–2243 (2009)
2. Fritz, T., Caidahl, K., Osler, M., Ostenson, C.G., Zierath, J.R., Wändell, P.: Effects of Nordic walking on health-related quality of life in overweight individuals with type 2 diabetes mellitus, impaired or normal glucose tolerance. Diabet. Med. **28**(11), 1362–1372 (2011)
3. Keast, M., et al.: Randomized trial of Nordic walking in patients with moderate to severe heart failure. Can. J. Cardiol. **29**(11), 1470–1476 (2015)
4. Kocur, P., Deskur-Smielecka, E., Wilk, M., Dylewicz, P.: Effects of Nordic walking training on exercise capacity and fitness in men participating in early, short-term inpatient cardiac rehabilitation after an acute coronary syndrome - a controlled trial. Clin. Rehabil. **23**(11), 995–1004 (2009)
5. Oakley, C., Zwierska, I., Tew, G., Beard, J.D., Saxton, J.M.: Nordic poles immediately improve walking distance in patients with intermittent claudication. Eur. J. Vasc. Endovasc. Surg. **36**(6), 689–694 (2008)
6. Wilson, J., Torry, M., Deckert, M., Kerzonek, T., Steadman, J.: Effects of walking poles on lower extremity gait mechanics. Med. Sci. Sports Exerc. **33**, 142–147 (2001)

7. Morss, G., Church, T., Earnest, C., Jordan, A.: Field test comparing the metabolic coast of normal walking versus Nordic walking. Med. Sci. Sport Exerc. **33**, 23–30 (2001)
8. Pellegrini, B., Bortolan, L., Zoppirolli, C., Fumene Feruglio, P., Schena, F.: Application of principal component analysis for the assessment of ground reaction forces in walking and Nordic Walking. J. Sport Med. PhysFit. **51**(3) (2011). suppl1:22
9. Winter, D.: Biomechanics and Motor Control of Human Movement. Wiley, Hoboken (2009)
10. Szrek, J., et al.: Force measurement module for mechatronic Nordic Walking poles. In: Rusiński, E., Pietrusiak, D. (eds.) CAE 2018. LNME, pp. 790–794. Springer, Cham (2018). https://doi.org/10.1007/978-3-030-04975-1_91 ISSN 2195-4356
11. Wudarczyk, S., et al.: Research on the mechatronic gait monitoring system with Nordic Walking poles. In: 15th International Conference Mechatronic Systems and Materials, MSM 2020, Bialystok, Poland, 1–3 July 2020, pp. 1–6. IEEE, cop., Piscataway (2020)
12. Wudarczyk, S., et al.: Experimental research on mechatronic Nordic Walking poles. In: 15th International Conference Mechatronic Systems and Materials, MSM 2020, Bialystok, 1–3 July 2020, pp. 1–6. IEEE, cop., Piscataway (2020)
13. https://www.dspace.com/en/pub/home/products/products.cfm. Accessed 23 May 2021
14. https://www.stt-systems.com/motion-analysis/inertial-motion-capture. Accessed 23 May 2021

Vibration of a Rigid Vertical Rotor Supported by a Shear Radial Magnetic Bearing

Jaroslav Zapomёl[1,2(✉)], Petr Ferfecki[2], Jan Kozánek[1], Jan Košina[1], and Jan Cibulka[1]

[1] Institute of Thermomechanics, Prague, Czech Republic
{zapomel,kozanek,kosina,cibulka}@it.cas.cz
[2] VŠB - Technical University of Ostrava, Ostrava, Czech Republic
{jaroslav.zapomel,petr.ferfecki}@vsb.cz

Abstract. Vertical rotors are often used as components of energy machines or flywheels for energy storage. Application of the contactless bearings makes it possible to minimize resistance against their rotation, wear, and energy losses. The advanced technological solution consists in utilization of permanent magnetic bearings or magnetic rings lifting the vertical rotors mounted in rolling element bearings to reduces their axial loading. The carried out research was focused on investigation of a rigid rotor supported at its upper end by a rolling element bearing inserted in a squeeze film damper and coupled with the stationary part by a shear magnetic bearing at its lower end. The bearaing is composed of two permanent magnets, one is attached to the stationary part and one to the rotor. The magnets are mutually attracted, which returns the rotor always to the equilibrium position during its vibration. In the developed computational procedure the rotor is represented by an absolutely rigid body and the permanent magnets by electric coils powered by equivalent current. The frequency response shows that application of the proposed magnetic bearing reduces amplitude of the rotor vibration in the range of lower velocities.

Keywords: Vertical rotors · Permanent magnets · Shear magnetic bearing · Mathematical model

1 Introduction

Energy machines or flywheels for energy storage are often designed as devices with vertical rotors. Their lateral and axial loading leads to resistance against rotation and to wear and energy losses in the support elements. The advanced technological solution making it possible to reduce these undesirable effects consists in application of contactless bearings based on utilization of magnetic forces.

The basic information on the design and properties of permanent magnetic bearings in the field of rotordynamics can be found in [1]. Some historical review and formulas for designing permanent magnetic bearings are reported in [2]. The approach to determination of stiffness of axial bearings using permanent magnets can be found in [3]. The advanced technological solution based on reducing axial loading of rolling element bearings supporting vertical rotors consists in their lifting by means of annular permanent

© The Author(s), under exclusive license to Springer Nature Switzerland AG 2022
J. Beran et al. (Eds.): TMM 2020, MMS 85, pp. 183–190, 2022.
https://doi.org/10.1007/978-3-030-83594-1_19

magnets. The details on application of this technology are reported in [4]. Efficiency of lifting the rotors was analysed in [5].

This paper deals with a shear concept of a contactless passive magnetic bearing intended for supporting vertical rotors and with its modelling. The bearing consists of two permanent magnets, which are mutually attracted. The results of the simulations show that application of this bearing stabilizes the rotor vibration and reduces amplitude of its oscillation in the range of lower angular velocities.

2 Proposal and Modelling of the Magnetic Bearing

The investigated system is a rigid vertical rotor supported at both its ends (Fig. 1). The upper support is formed by a radial-thrust bearing, the outer race of which is connected with the bearing housing by a coupling element exhibiting elasticity and damping (rubber ring, oil squeeze film damper, elastomeric squeeze film damper, etc.). The lower end of the shaft is coupled with the stationary part by a magnetic bearing that consists of two concentric axially magnetized cylindrical permanent magnets. One magnet is coupled with the rotor, the other is attached to the stationary part. The magnets are oriented in such a way that they attract each other. If the rotor vibrates and its lower end is displaced in the lateral direction, the attracting magnetic force returns the rotor to the equilibrium position.

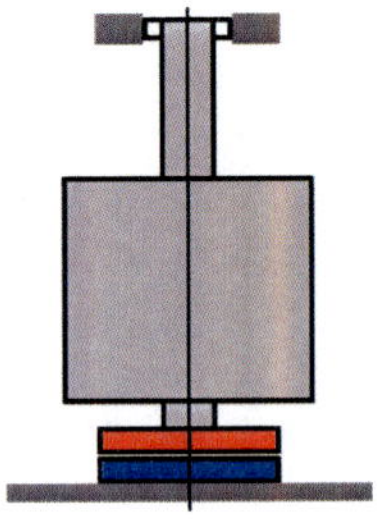

Fig. 1. Scheme of the investigated rotor system.

The magnetic force has two components, axial and lateral. The axial force increases axial loading of the upper bearing. It can be compensated either by adding a pair of permanent magnets to the rotor upper bearing (Fig. 2 left) or by application of a symmetric arrangement of the shear magnetic bearing (Fig. 2 middle and right).

The usually applied approach to modelling permanent magnets consists in their replacement by electric coils fed by equivalent current. It must hold that the magnetic moment of the permanent magnet is of the same magnitude as that of the electric coil. Then it holds

$$\pi\, r_{PM}^2 h_{PM} M = \pi\, r_{PM}^2 N I_{eqv} \tag{1}$$

r_{PM} is the radius of the permanent magnet, h_{PM} is the height of the permanent magnet, M is the permanent magnet magnetization, N is number of the coil turns, and I_{eqv} is the equivalent current. Radius of the coil is the same as of the magnet.

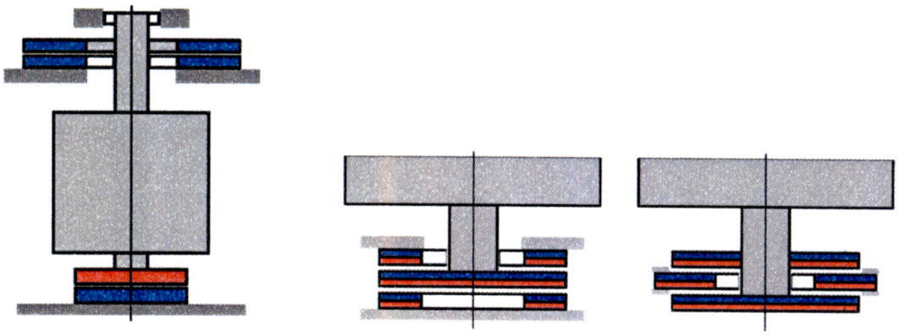

Fig. 2. Arrangement of the movable and stationary permanent magnets.

The force, by which one turn (turn Z_1) of one coil acts on one turn (turn Z_2) of another coil is expressed by application of the Biot-Savart-Laplace and Ampere laws. The force of infinitesimal magnitude acts on each oriented element of turn Z_2 (Fig. 3) finding itself in the magnetic field produced by the current passing in turn Z_1

$$\mathrm{d}\vec{F}_{mag} = -I_{Z2}\vec{B}_{Z1} \times \mathrm{d}\vec{L}_{Z2} \tag{2}$$

I_{Z2} is the current passing in turn Z_2, $\mathrm{d}L_{Z2}$ is the length of the infinitesimal element of turn Z_2 oriented in the direction of current I_{Z2}, $\mathrm{d}F_{mag}$ is the magnetic force of infinitesimal magnitude acting on the element of turn Z_2, and B_{ZI} is magnetic induction of the magnetic field produced by the current passing in turn Z_1 at location of the element of turn Z_2.

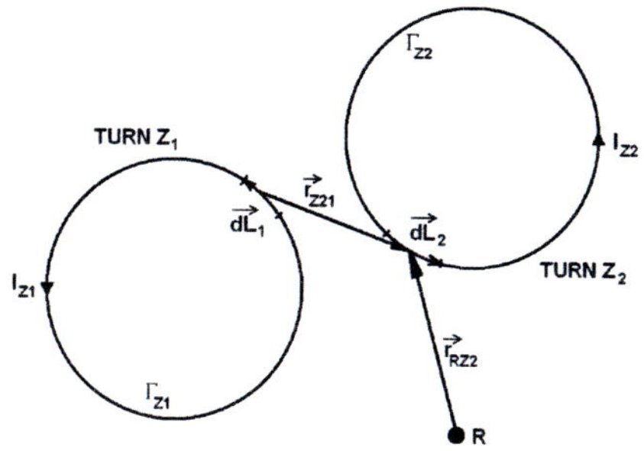

Fig. 3. Currents in two interacting coil turns.

Magnetic induction B_{ZI} of the magnetic field produced by the current passing in turn Z_1 is determined by means of the Biot-Savart law

$$\vec{B}_{Z1} = \frac{\mu_0 I_{Z1}}{2\pi} \int\limits_{\Gamma_{Z1}} \frac{d\vec{L}_{Z1} \times \vec{r}_{Z21}}{|\vec{r}_{Z21}|^3} \tag{3}$$

μ_O is the relative permeability of the vacuum, I_{ZI} is the current passing in turn Z_1, $\mathrm{d}L_{ZI}$ is the element of the infinitesimal length of turn Z_1 oriented in the direction of current I_I, r_{Z2I} is the position vector of the length element of turn Z_2 relative to the length element of turn Z_1, and Γ_{ZI} is the path of turn Z_1.

The relation for the infinitesimal moment of the elementary magnetic force relative to a reference point reads

$$d\vec{M}_{mag} = -I_{Z2}\,\vec{r}_{RZ2} \times \left(\vec{B}_{Z1} \times d\vec{L}_{Z2}\right) \tag{4}$$

r_{RZ2} is the position vector of the element of infinitesimal length dL_{Z2} of turn Z_2 relative to the reference point.

3 The Simulated Rotor System

The studied rotor is rigid. It consists of a vertical shaft and of one disc. At its upper end the rotor is supported by a ball bearing inserted in the elastomeric squeeze damper. The lower end of the shaft is coupled with the stationary part by the proposed magnetic bearing. The rotor rotates at constant angular speed, is loaded by its weight, and is excited by the imbalance.

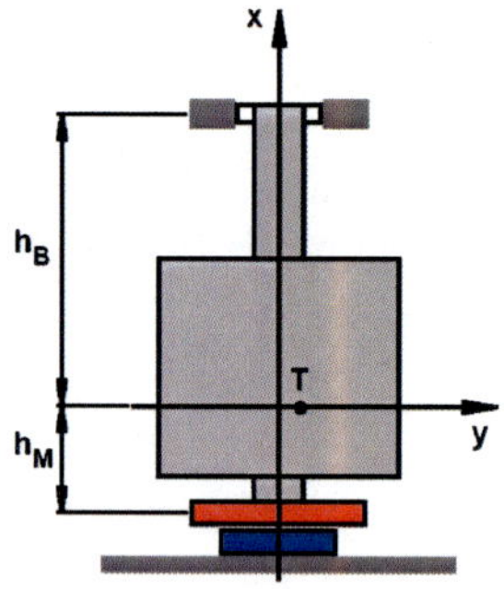

Fig. 4. Scheme of the rotor system and the introduced frame of reference.

Axis x of the introduced frame of reference is vertical and goes through the centre of the rotor upper support. The system origin lies at the intersection of the plane perpendicular to axis x passing through the rotor centre of gravity and the rotor centreline.

In the computational model the rotor is considered to be absolutely rigid and the stiffness and damping of the upper support element as linear. The ball bearing constrains displacement of the shaft in the radial direction but enables its rotation about the axes perpendicular to the shaft centre line. Damping and other resistances produced by the environment are neglected. The permanent magnets forming the lower magnetic bearing are represented by one coil turn each.

The rotor can perform two independent displacements in the lateral direction and two rotations about axes perpendicular to the shaft centre line. The point of intersection of the shaft centre line and the perpendicular plane passing through the rotor centre of gravity is referred to as the rotor centre.

The rotor centre of gravity was chosen as the reference point to set up the motion equations. In accordance with [1] the governing equations read

$$0 = F_A - mg + F_{magx}, \tag{5}$$

$$m\ddot{y}_T = -b_B\dot{y}_B - k_B y_B + F_{magy}, \tag{6}$$

$$m\ddot{z}_T = -b_B\dot{z}_B - k_B z_B + F_{magz}, \tag{7}$$

$$J_d\ddot{\phi}_y - \omega J_a\dot{\phi}_z = -h_B b_B \dot{z}_B - h_B k_B z_B + M_{Ay} + M_{magy}, \tag{8}$$

$$J_d\ddot{\phi}_z + \omega J_a\dot{\phi}_y = -h_B b_B \dot{y}_B - h_B k_B y_B + M_{Az} + M_{magz}. \tag{9}$$

m is the rotor mass, J_d is the rotor moment of inertia relative to its diametral axis going through rotor centre of gravity, J_a is the rotor moment of inertia relative to its rotation axis, k_B is lateral stiffness of the upper rotor support, b_B is the coefficient of damping of the rotor upper support, ω is angular speed of the rotor rotation, h_B is the dimension defining vertical position of the upper rotor support in the introduced coordinate system (Fig. 4), F_{magx}, F_{magy}, F_{magz} are the x, y and z components of the magnetic force acting on the permanent magnet attached to the rotor at location of the lower support, respectively, F_A is the axial constraint force acting on the rotor at location of the upper support in the x direction, M_{magy}, M_{magz} are the y and z components of the moment produced by the magnetic force acting on the permanent magnet attached to the rotor relative to the reference point, M_{Ay}, M_{Az} are the y and z components of the moment produced by force F_A relative to the reference point, g is the gravity acceleration, y_T, z_T are displacements of the rotor centre of gravity, y_B, z_B are displacements of the point of the shaft centre line related to the upper rotor support in the y and z directions, ϕ_y, ϕ_z are angles of rotation of the rotor about y and z axes, and (.), (..) denote the first and second derivatives with respect to time.

Utilizing the kinematic relationships

$$y_T = y + e_T\cos(\omega t + \psi_T), \tag{10}$$

$$z_T = z + e_T\sin(\omega t + \psi_T), \tag{11}$$

$$y_B = y + h_B\phi_z, \tag{12}$$

$$z_B = z - h_B\phi_y \tag{13}$$

the motion Eqs. (6)–(9) can be rewritten

$$m\ddot{y} + b_B\dot{y} + h_B b_B\dot{\phi}_z + k_B y + h_B k_B\phi_z = m e_T\omega^2\cos(\omega t + \psi_T) + F_{magy}, \tag{14}$$

$$m\ddot{z} + b_B\dot{z} - h_B b_B\dot{\phi}_y + k_B z - h_B k_B\phi_y = m e_T\omega^2\sin(\omega t + \psi_T) + F_{magz}, \tag{15}$$

$$J_d\ddot{\phi}_y - \omega J_a\dot{\phi}_z - h_B b_B \dot{z} + h_B^2 b_B \dot{\phi}_y - h_B k_B z + \left(h_B^2 k_B + mgh_B\right)\phi_y = M_{Ay} + M_{magy}, \tag{16}$$

$$J_d\ddot{\phi}_z + \omega J_a\dot{\phi}_y + h_B b_B \dot{y} + h_B^2 b_B \dot{\phi}_z + h_B k_B y + \left(h_B^2 k_B + mgh_B\right)\phi_z = M_{Az} + M_{magy} \tag{17}$$

y, z are the displacements of the rotor centre in the y and z directions, e_T is the rotor eccentricity, t is the time and ψ_T is the phase shift of the centre of gravity.

The constraint force F_A is expressed from the equation of equilibrium (5). y and z components M_{Ay}, M_{Az} of its moment relative to the reference point (the rotor centre of gravity) are given by the following relations

$$M_{Ay} = (z_B - z_T)\,F_A, \tag{18}$$

$$M_{Az} = -(y_B - y_T)\,F_A \tag{19}$$

Components of the magnetic force and its moment relative to the reference point (the rotor centre of gravity) are obtained by integration of the elementary magnetic force and its moment relative to the reference point along the length of the turn corresponding to the magnet attached to the rotor utilizing relations (2) and (4).

The motion Eqs. (14)–(17) are nonlinear because their right hand side is a nonlinear function of the rotor generalized coordinates. The trigonometric collocation method was applied to obtain their steady state solution.

4 Results of the Simulations

The main technical parametres of the investigated system are: 27.7 kg the rotor mass, 0.37 kgm^2 the rotor diametral moment of inertia, 0.13 kgm^2 the rotor axial moment of inertia, 500 kN/m lateral stiffness of the upper rotor support, 100 Ns/m damping coefficient of the upper rotor support, 300, 600 mm design dimensions (Fig. 4) h_B, h_M, 50 μm eccentricity of the rotor centre of gravity, 120 mm diametre of the stationary permanent magnet, 160 mm diametre of the movable permanent magnet, 20 mm thickness of the permanent magnets, 5 mm vertical distance between the permanent magnets, 795.8 kA/m magnetization of the permanent magnets. Each permanent magnet was represented by one coil turn. The equivalent currents are 15.9 kA.

The task was to study effect of the pair of permanent magnets on the rotor lateral vibration.

Dependence of the axial and radial magnetic force of the radial displacement of the movable permanent magnet can be seen in Fig. 5.

Figure 6 shows the frequency response characteristic related to the rotor centre displacement in the y direction for the design case when no magnetic bearings are applied. The results show that the rotor system has two critical speeds, about 6 and 460 rad/s, in the interval of examined running velocities.

The frequency response of the rotor for the case when the rotor is constrained by the permanent magnets is depicted in Fig. 7. It is evident that the magnetic coupling suppresses the lower critical speed and reduces amplitude of the rotor vibration in the interval of low angular velocities.

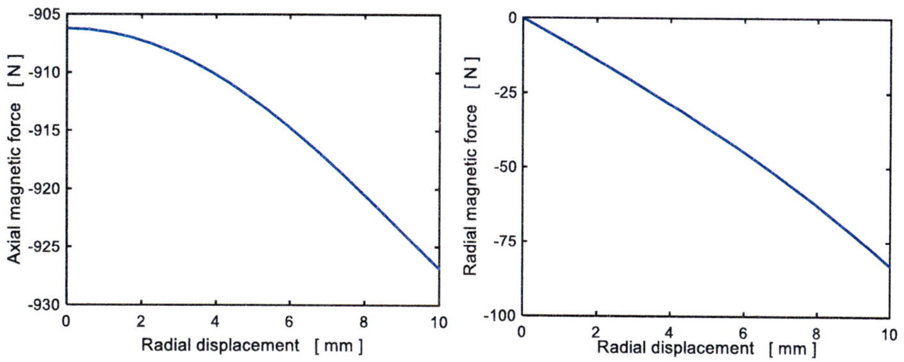

Fig. 5. Dependence of the axial and radial magnetic force of the magnet radial displacement.

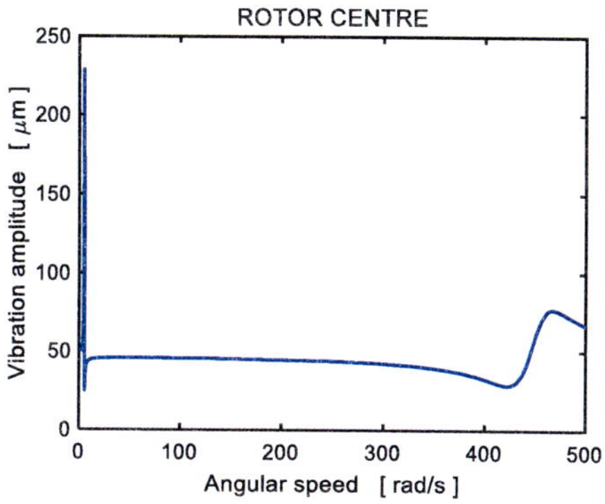

Fig. 6. Frequency response (design without the magnetic bearing).

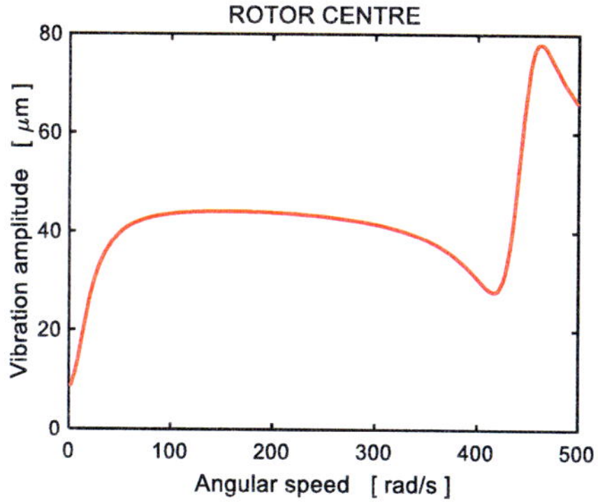

Fig. 7. Frequency response (design with the magnetic bearing).

5 Conclusions

The main objective of the performed research was to propose and to study a shear variant of a permanent magnetic bearing and its application for supporting vertical rotors. The bearing consists of two permanent magnets. The magnetic force is attractive, because of which it always pulls the rotor to the equilibrium position. The analysis was focused on modelling of the shear magnetic bearing and on implementation of the developed mathematical model in the computational procedures for determination of the steady state vibration of vertical rotors. Evaluation of the frequency responses obtained by computational simulations shows that mounting the rotor in the investigated bearing reduces the rotor vibration amplitude in the range of lower angular velocities. The performed research contributed to the development of the mathematical modelling and enabled to get to know more on influence of magnetic bearings on oscillation of vertical rotors.

Acknowledgements. This work was supported by the Czech Science Foundation (grant project 19-06666S) and by the Ministry of Education, Youth and Sports within the National Programme of Sustainability (NPU II, project LQ1602 - IT4Innovations excellence in science).

References

1. Gasch, R., Nordmann, R., Pfützner, H.: Rotordynamik, 2nd edn. Springer-Verlag, Berlin, Heidelberg (2002)
2. Paden, B., Groom, N., Antaki, J.: Design formulas for permanent-magnetic bearings. Trans. ASME – J. Mech. Des. **125**, 734–738 (2003)
3. Ravaud, R., Lemarquand, G., Lemarquand, V.: Force and stiffness of passive magnetic bearings using permanent magnets. Part 1: axial magnetization. IEEE Trans. Magn. **45**(7), 2996–3002 (2009). https://doi.org/10.1109/TMAG.2009.2016088
4. Jiang, S., Wang, H., Wen, S.: Flywheel energy storage system with permanent magnet bearing and a pair of hybrid ceramic ball bearing. J. Mech. Sci. Technol. **28**(12), 5043–5053 (2014)
5. Zapoměl, J., Ferfecki, P., Kozánek, J.: Application of permanent magnetic rings to reducing energy losses in the supports of vertical flywheels. In: Proceedings of the 30th Nordic Seminar on Computational Mechanics, pp. 210–213, Lyngby (2017)

Robots and Manipulators

Determination of the Workspace of the System Based on the 3-PRRR Mechanism for the Lower Limb Rehabilitation

Maksat Ahmetzhanov[1], Larisa Rybak[2], Dmitry Malyshev[2](✉), and Santhakumar Mohan[3]

[1] Institute of Information and Computational Technologies, 125, Pushkina, Almaty 050010, Kazakhstan
[2] BSTU named after V.G. Shukhov, 46, Kostyukova, Belgorod 308012, Russia
[3] Indian Institute of Technology Palakkad, Palakkad, Kozhippara, Kerala 678557, India

Abstract. The article proposes a lower limb rehabilitation system based on a passive orthosis and an active parallel robot 3-PRRR. Numerical algorithms have been developed to determine the workspace of the parallel mechanism, taking into account the interference of the links. The optimization of the geometrical parameters of the mechanism has been carried out, and the configuration of the robot has been selected based on the criterion of maximizing the workspace. The analysis of the workspace and the influence of various geometric parameters on it is carried out. The simulation results of the workspaces, as well as the positions of the robot at which interference occur, are rendered using the transformation of a set of 3D boxes to an STL file.

Keywords: Approximation set · Parallel robot · Workspace · Non-uniform covering · Optimization · Link interference

1 Introduction

The methods of robotic mechanotherapy are currently widely used in rehabilitation medicine to restore the motor functions of the limbs of patients with various disorders of the musculoskeletal system. Upper limb treatment focuses on restoring the patient's nervous system, muscle capacity and arm strength, while lower limb treatment focuses on the various articular movements of the legs and their synchronization. So, in work [1] it is shown that there is a significant progress in the recovery of patients with the help of robotic physiotherapy. A promising direction is the use of parallel robot architectures in medical applications such as rehabilitation and kinesitherapy. Currently, there are many devices for lower limb rehabilitation. The Gait Trainer is a wheeled device that helps a person who cannot walk on their own. Rehabilitation complex WalkTrainer [2] has a mobile frame design, which includes a system for unloading the weight, due to which it is possible to regulate the dynamic load on the patient's lower limbs. The functionality of such devices does not allow for the rehabilitation of the lower limbs of patients in the early stages of rehabilitation. Thus, these devices are not suitable for patients who are

© The Author(s), under exclusive license to Springer Nature Switzerland AG 2022
J. Beran et al. (Eds.): TMM 2020, MMS 85, pp. 193–203, 2022.
https://doi.org/10.1007/978-3-030-83594-1_20

only able to be in a sitting and lying position, whose motor functions of the lower limbs are impaired.

Among the rehabilitation systems, which is possible in the early stages of rehabilitation, one can single out the LokoHelp movement therapy station, an electromechanical gait trainer with a weight-unloading system. However, this system is not applicable to patients who cannot be suspended. A gait rehabilitation device based on a 3DOF parallel arm [3], which generates the required gait pattern by moving the patient's foot while the body weight is supported by a seat belt system. The KARR rehabilitation system [4] allows for the rehabilitation of patients in a sitting position. Also worth noting is the Lambda robot rehabilitation system [5], which involves the rehabilitation of bedridden patients by mobilizing the ankle joint. An upright table with an integrated orthopedic device and synchronized functional electrical stimulation Erigo Pro allows for intensive cyclic movement therapy in the form of passive dynamic movements of the lower limbs of recumbent patients. The listed devices as physiotherapeutic movements provide flexion-extension in the knee joint. In contrast, such a movement as adduction-abduction in the hip joint is impossible, which is a significant drawback.

2 Construction of a Mathematical Model of the Robotic System

Let us consider the design of RS for lower limb rehabilitation, a 3D model of which is shown in Fig. 1a. The RS includes an active parallel 3-PRRR mechanism and a passive orthosis based on the RRRR mechanism to support the patient's leg. A diagram of an active parallel mechanism is shown in Fig. 1b.

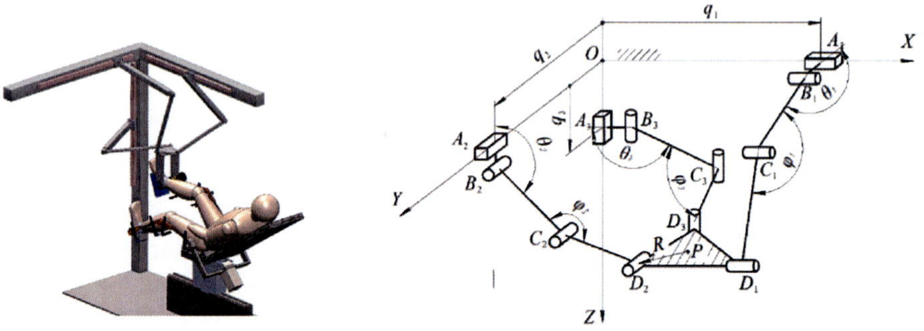

a b

Fig. 1. Conceptual design of RS for lower limb rehabilitation: a) 3D model, b) 3-PRRR design scheme

Due to the active 3-PRRR mechanism, the proposed robotic rehabilitation system provides the required physiotherapeutic movements, and the passive part based on the RRRR orthosis provides the necessary angles, and, accordingly, the correctness of the exercise. This 3-DOF mechanism with translational motions along the axes was proposed by Kong and Gosselin in [6]. The mechanism consists of three kinematic chains

$A_iB_iC_iD_i$. The position of the movable platform on which the patient's foot is placed is an equilateral triangle $D_1D_2D_3$ centered at point P and the radius of the circumscribed circle R. The movement of the platform is determined by linear displacements $\boldsymbol{q} = (q_1q_2q_3)$. Let's denote a_i - the distance between points A_i and B_i, b_i- between B_i and C_i, c_i - between C_i and D_i, d_i - between B_i and D_i.

Let us further consider the problem of determining the workspace of the parallel mechanism to ensure the required movements of the passive orthosis under clinical data for gait training. We will apply the non-uniform coverings method to determine the workspace. It is described in detail in [7–9], and the methods of interval analysis [10]. We will take into account the following design constraints on the parameters of the mechanism:

- on the drive coordinates q: $q_i \in [q_{min}; q_{max}]$.
- the distance d between the joints by the centers of the joints $B_i$and D_i: $d_i \in [0; d_{max}]$, where $d_{max} = b + c$.

Determine the drive coordinates and angles of rotation in the joints $B_i$and C_i for certain coordinates of the center point P of the platform $D_1D_2D_3$. To do this, we write down the intervals that describe the ranges of changes in the coordinates of the point P:

$$\boldsymbol{X_P} := [\underline{X_P}, \overline{X_P}] = \{\underline{X_P} \le x_P \le \overline{X_P}\},\ \boldsymbol{Y_P} := [\underline{Y_P}, \overline{Y_P}] = \{\underline{Y_P} \le y_P \le \overline{Y_P}\}$$

$$\boldsymbol{Z_P} := [\underline{Z_P}, \overline{Z_P}] = \{\underline{Z_P} \le z_P \le \overline{Z_P}\} \tag{1}$$

The intervals describing the ranges of change of the drive coordinates have the form

$$\boldsymbol{Q_i} := [\underline{Q_i}, \overline{Q_i}] = \{\underline{Q_i} \le q_i \le Q_i\} \tag{2}$$

The drive coordinates q_i are determined according to the expressions

$$q_1 = x_{D1} = x_P + \frac{\sqrt{3}}{2}R,\ q_2 = y_{D2} = y_P + \frac{R}{2},\ q_3 = z_{D3} = z_P. \tag{3}$$

The intervals describing the ranges of the distance d_i are of the form

$$\boldsymbol{D_i} := [\underline{D_i}, \overline{D_i}] = \{\underline{D_i} \le d_i \le D_i\}$$

The distances d_i are defined as the distance between B_i and D_i:

$$d_1 = \sqrt{\left(y_P + \frac{R}{2} - a\right)^2 + (z_P)^2} \tag{4}$$

$$d_2 = \sqrt{\left(x_P - \frac{\sqrt{3}}{2}R\right)^2 + (z_P - a)^2} \tag{5}$$

$$d_3 = \sqrt{(x_P - a)^2 + (y_P - R)^2} \tag{6}$$

The obtained expressions (1–6) allow us to determine the values of the geometric parameters of the mechanism that determine the boundaries of its workspace.

3 Determination the Workspace of the Mechanism

Various numerical methods of interval analysis and grid approximation are used to determine the workspace [11]. However, their use is associated with significant computational difficulties, since the task of determining the workspace of parallel robots has a large dimension. In [7], the application of the method of non-uniform coverings to approximate the set of solutions to a system of nonlinear inequalities is considered, and in [8, 9], the application of this method to determine the workspace of some types of planar robots. Let us consider the application of this approach to constructing and analyzing the workspace of a robotic system for lower limb rehabilitation.

The algorithm is synthesized using formulas (1)–(6). It works with two lists of three-dimensional boxes $\mathbb{P}$ and $\mathbb{P}_I$, each of the dimensions of which corresponds to the coordinates x_p, y_p, z_p of the center of the movable platform P. At the first step of the algorithm, a box is specified that is guaranteed to include the workspace and approximation accuracy δ. The box is listed in $\mathbb{P}$. For the intervals $\boldsymbol{X_P}$, $\boldsymbol{Y_P}$ and $\boldsymbol{Z_P}$ of the box, the ranges $\boldsymbol{Q_i}$ and $\boldsymbol{D_i}$ are computed and constrained. If at least one of the calculated ranges is invalid, the box is excluded, in other cases it is divided in half into two equal boxes and the procedure is repeated. The algorithm ends its work when the size of the boxes becomes less than δ. Such boxes are listed in the $\mathbb{P}_I$ list.

The algorithm is implemented in the C++ programming language using the Snowgoose interval analysis library. The visualization of the simulation results was carried out by converting the list of boxes $\mathbb{P}_I$ into a universal format of 3D models - stl-file. The computational experiment was carried out for the following geometric parameters of the active mechanism: $a = 50$ mm, $b = c = 500$ mm, $R = 50$ mm $q_{min} = 0$ mm, $q_{max} = 1000$ mm. The simulation results are shown in Fig. 2. The computation time for the approximation accuracy $\delta = 2$ mm on a personal computer was 141 s.

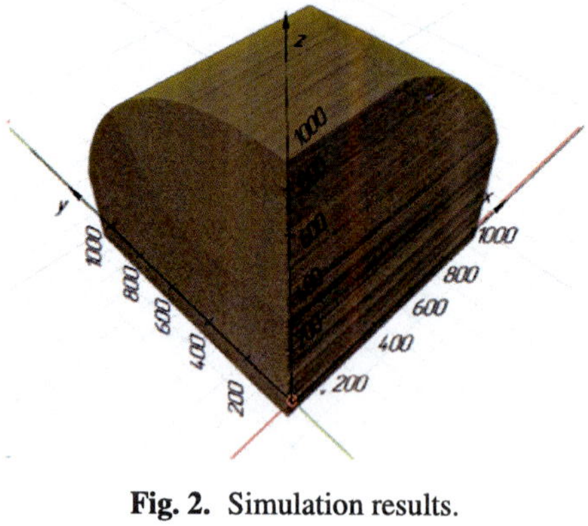

Fig. 2. Simulation results.

4 Determination of Interference of Links of the Mechanism

The interference of the links of the mechanism can be divided into two groups:

- interference at small angles between links connected by rotary joints.
- the interference of links that are not connected to each other.

The first group can be determined, taking into account the restrictions on the angles of rotation in the joints B_i, C_i and D_i:

$$\left\{ \begin{array}{l} \theta_i \in [\theta_{min}; \theta_{max}] \\ \varphi_i \in [\varphi_{min}; \varphi_{max}] \\ \psi_i \in [\psi_{min}; \psi_{max}] \end{array} \right\}. \tag{7}$$

We define the angles φ_i:

$$\varphi_i = \cos^{-1}\left(\frac{b^2 + c^2 - d_i^2}{2bc}\right), \ \varphi_i \in [0; \pi]. \tag{8}$$

Considering that the points C_i are the interference points of circles with radii b and c and cents at points B and D, respectively, the angles ψ_i can be determined using the formula of cosines between vectors:

$$\psi_1 = \frac{(x_{C2} - x_{D2})(x_{D1} - x_{D2}) + (y_{C2} - y_{D2})(y_{D1} - y_{D2}) + (z_{C2} - z_{D2})(z_{D1} - z_{D2})}{\sqrt{3}cR}. \tag{9}$$

$$\psi_2 = \frac{(x_{C3} - x_{D3})(x_{D1} - x_{D3}) + (y_{C3} - y_{D3})(y_{D1} - y_{D3}) + (z_{C3} - z_{D3})(z_{D1} - z_{D3})}{\sqrt{3}cR} \tag{10}$$

$$\psi_2 = \frac{(x_{C3} - x_{D3})(x_{D2} - x_{D3}) + (y_{C3} - y_{D3})(y_{D2} - y_{D3}) + (z_{C3} - z_{D3})(z_{D2} - z_{D3})}{\sqrt{3}cR} \tag{11}$$

where $z_{Di} = z_{C3} = z_P$, $x_{D1} = x_P + \frac{\sqrt{3}}{2}R$, $y_{D1} = y_{D2} = y_{C2} = y_P + \frac{R}{2}$, $x_{D3} = x_P$, $x_{D2} = x_P - \frac{\sqrt{3}}{2}R$, $x_{C2} = \frac{(s_2 * x_{D2}) \pm \sqrt{b^2 - s_2^2} * (z_P - a)}{d_2}$, $z_{C2} = \frac{ad_2 + s_2(z_P - a) \mp \sqrt{b^2 - s_2^2} * (x_{D2})}{d_2}$, $x_{C3} = \frac{ad_3 + s_3(x_{D3} - a) \mp \sqrt{b^2 - s_3^2} * (y_{D3})}{d_3}$, $y_{C3} = \frac{(s_3 y_{D3}) \pm \sqrt{b^2 - s_3^2} * (x_{D3} - a)}{d_3}$, $y_{D3} = y_P - R$, $s_i = \frac{b^2 - c^2 + d_i^2}{2d_i}$.

Note that the presence of the $\pm$ or $\mp$ signs is due to the ambiguity of the problem of determining the coordinates of the points C_i. An example of two possible positions of the kinematic chain $A_1B_1C_1D_1$ is shown in Fig. 3.

The first position in $\pm$ and $\mp$ corresponds to the upper sign, the second position to the lower one. The cosines of the angles θ_i are defined as:

$$\theta_1 = \cos^{-1}\left(\frac{\mp z_P\sqrt{b^2 - s_1^2} - s_1(y_{D1} - a)}{d_1 b}\right), \ \theta_1 \in [0; 2\pi]. \tag{12}$$

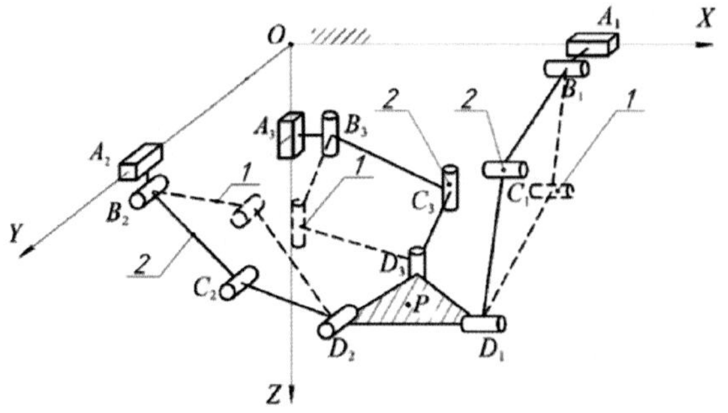

Fig. 3. Ambiguity of the problem of determining the coordinates of points C_i

$$\theta_2 = \cos^{-1}\left(\frac{\mp x_{D2}\sqrt{b^2 - s_2^2} - s_2(z_P - a)}{d_2 b}\right),\ \theta_2 \in [0; 2\pi]. \tag{13}$$

$$\theta_3 = \cos^{-1}\left(\frac{\mp y_{D3}\sqrt{b^2 - s_3^2} - s_3(x_{D3} - a)}{d_3 b}\right),\ \theta_3 \in [0; 2\pi]. \tag{14}$$

We define the second group of interferences using an approach based on determining the minimum distance between the segments drawn between the centers of the joints of each of the links. In [12], a similar condition is used, but the approach has drawbacks. In particular, the authors propose to determine on the auxiliary plane the interferences of the segments, and not the distance between the nearest points. This does not allow identifying such interference of links in which there is no interference of the axes. The approach proposed in the current work is as follows. To determine the interferences of the links of the mechanism, we construct an auxiliary plane, which is parallel to the axis of one of the links and to which the axis of the other link belongs. In this case, the condition for the absence of interferences of the links will take the form:

$$\sqrt{u_1^2 + u_2^2} > \mathrm{D}_{link}, \tag{15}$$

where u_1 is the distance between the axis of the link that does not belong to the plane and the auxiliary plane,u_2 is the distance between the nearest points of the segments connecting the centers of the joints of each of the links when projecting a segment that does not belong to the auxiliary plane onto this plane, D_{link} is the diameter of the links. It is worth noting the special case when the links are parallel to each other and the construction of an auxiliary plane is not required. The authors described the method in detail in [13].

5 Analysis of the Workspace Taking into Account the Interference of the Links

To determine the positions of the output link inside the workspace, in which the interference of links occurs, the list of boxes $\mathbb{P}_I$ obtained at the stage of determination the workspace is used. For the coordinate of the center of each of the boxes, conditions (7) and (15) are checked. If all the conditions are met, the box is excluded from $\mathbb{P}_I$ and added to the new $\mathbb{P}_A$ list. As a result, the $\mathbb{P}_I$ list will contain only boxes corresponding to the interference areas of the links, and $\mathbb{P}_A$ - only boxes corresponding to the workspace without interference.

Let's analyze the change in the volume of the workspace for various configurations of the 3-PRRR mechanism. The correspondence of the configurations and the position of the links, and, accordingly, the choice of the sign in Eqs. (9)–(14), is given in Table 1.

Table 1. Configurations of 3-PRRR mechanism.

Kinematic chain	Link positions for configurations							
	1	2	3	4	5	6	7	8
1	1	1	1	1	2	2	2	2
2	1	1	2	2	1	1	2	2
3	1	2	1	2	1	2	1	2

Workspace for configuration 8 after excluding areas of interference, as well as areas of interference for $b = c = 500$ mm, $a = 50$ mm, $R = 50$ mm $q_{min} = 0$ MM, $q_{max} = 1000$ mm, $\varphi_{min} = \theta_{min} = \psi_{min} = 10°$ mm, $\varphi_{max} = \theta_{max} = \psi_{max} = 170°$, $D_{link} = 20$ mm are shown in Fig. 4. The computation time for the approximation accuracy $\delta = 2$ mm on a personal computer was 4 min 59 s, of which the time for determining the workspace was 2 min 17 s, the time for determining the interference of the workspace was 2 min 42 s.

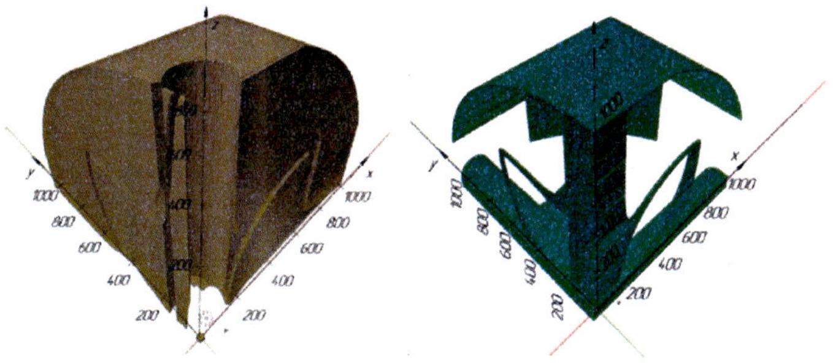

a b

Fig. 4. 3-PRRR mechanism workspace: a) without interference, b) areas of interference.

To verify the results, the position of the links was visualized at which they intersect. In Fig. 5 shows some of the link interference that occur.

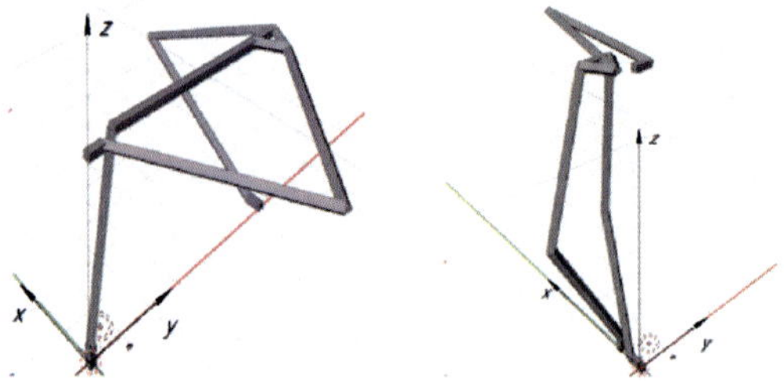

a b

Fig. 5. Examples of identified interference of links: a) link B_1C_1 with link B_3C_3 b) link C_3D_3 with link D_1D_2.

Comparison of the volume of the workspaces for the choice of the configuration was carried out for the following geometric parameters of the active mechanism: $a = 50$ mm, $R = 50$ mm $q_{min} = 0$ mm, $q_{max} = 1000$ mm, $\varphi_{min} = \theta_{min} = \psi_{min} = 10°$, $\varphi_{max} = \theta_{max} = \psi_{max} = 170°$, $D_{link} = 20$ mm. The results are shown in Table 2.

Table 2. Scope of the workspace for configurations.

b, m	c, m	Volume of the workspace for configurations, m^3*10^{-3}							
		1	2	3	4	5	6	7	8
500	500	533.3	531.6	472.5	548.1	447.8	525.1	474	647.5
500	600	640.4	627.7	570.8	644.6	569.1	630.5	572.8	751.4
500	700	657.1	637.4	606.4	664.5	624.8	656	613.8	769.2
500	800	593.4	583.5	571.9	619.0	583.3	611.7	584	712.3
600	500	588.5	591.5	523.9	612.5	461.3	576.1	523.4	748.3
600	600	717.6	710.1	624	732.2	581.6	693.4	627.7	873.9
600	700	746.8	722.7	646	743.4	628.7	715.3	648.8	887.8
600	800	676.7	647.1	602	679.9	614.1	660.9	610.9	816.9
700	500	559.5	565.5	497.9	592.4	411.8	550.7	497	764.4
700	600	681.4	671	593.8	700.1	503.7	650.9	593.3	885.2
700	700	729.7	707.9	612.7	734.4	549.6	683.1	616.7	917.4
700	800	715.6	677.6	593.2	702.6	555.5	662.2	595	882.6
800	500	482.7	476.5	424.9	585.0	324	468.5	423.1	706.3
800	600	571.3	548.8	489.9	584.7	379.9	540.6	488.7	812.3
800	700	650.2	613.6	544.6	647.9	437	598.1	543.9	880.5
800	800	692.6	648.9	554.7	678.9	483.9	628.1	560.2	901.8

It can be seen from the table that for all sizes the maximum volume of the workspace is achieved with configuration 8. Let's take this configuration to select the optimal sizes of the links at which the maximum volume of the workspace is reached. Determination of the optimal sizes of links b, c and R was carried out in several stages. At each stage, the volume of the workspace was calculated taking into account the interference of the links for various combinations of sizes. To reduce the computational complexity, the resizing ranges and the iteration step are reduced with each step.

Numerical values are given in Table 3. The last column contains the volume of the workspace, which is a criterion for excluding or including in the next stage a certain part of the size range.

Table 3. Stages of determining the optimal link lengths.

Step	bmin	bmax	cmin	cmax	Rmin	Rmax	Step	Required volume
1	300	900	300	900	50	150	50	0,8
2	525	900	525	900	50	150	25	0,9
3	602	842	602	842	50	50	8	0,919
4	643	713	643	713	50	50	4	0,9197
5	666	689	666	689	50	50	1	-

The maximum workspace is reached at b = 676 mm, c = 678 mm, R = 50 mm. For such a ratio of dimensions, the total volume of the workspace is 1 m^3, of which 0.9198 m^3 is an area without intersecting links. It should be noted that it would be possible to achieve a larger volume of the workspace with a smaller size of the movable platform, but this is structurally unacceptable.

The dependence of the volume of the workspace on the change in the length of the links for R = 50 mm is shown in Fig. 6.

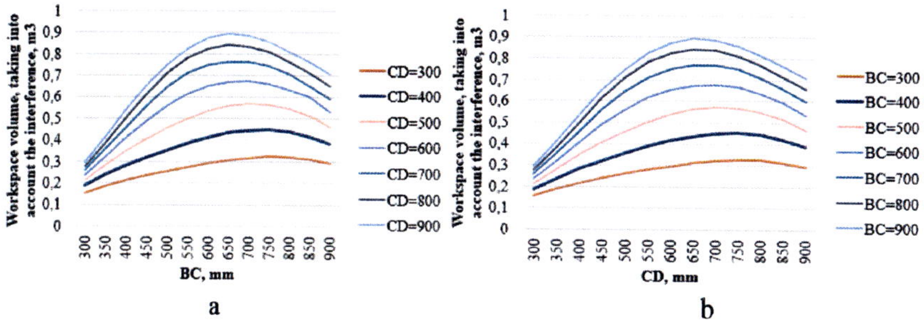

Fig. 6. Dependence of the workspace volume on the length of the links: a) BC, b) CD.

In Fig. 7 shows the dependence of the proportion of areas in which interference occur in the total volume of the workspace, depending on the lengths of the links. The graphs show that at small and large values of the link lengths, the proportion of interference areas in the workspace increases.

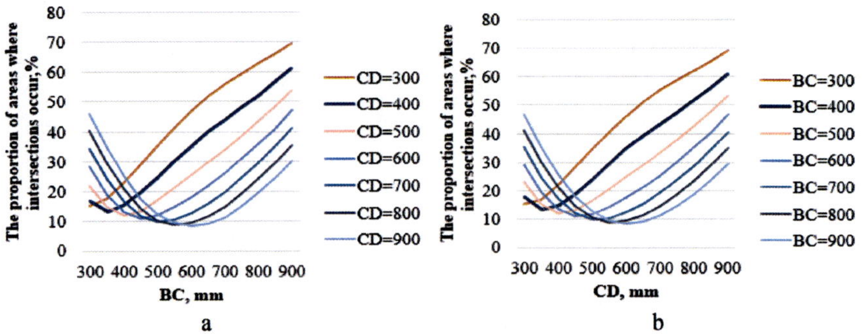

Fig. 7. Dependence of the areas of interference ratio on the length of the links: a) BC, b) CD.

6 Conclusion

For the proposed robotic system for lower limb rehabilitation, based on a passive orthosis and an active parallel robot, effective numerical methods and algorithms for determining the workspace and interference of the active robot links have been developed and tested. The configuration of the 3-PRRR robot is selected, which achieves the maximum workspace for all sizes of links. For the selected configuration, with ranges of reference coordinates of 1000 mm, the maximum workspace, taking into account the interference of the links, was achieved at $b = 676$ mm, $c = 678$ mm, $R = 50$ mm.

Acknowledgements. This work was supported by the Russian Science Foundation, the agreement number 19-19-00692, the Ministry of Education and Science, Republic of Kazakhstan, project no. AP05133190.

References

1. Vashisht, N., Puliyel, J.: Polio programme: let us declare victory and move. Indian J. Med. Ethics **9**(2), 114–117 (2012)
2. Bouri M., Stauffer Y., Schmitt C., Allemand Y., Gnemmi S., Clavel R.: The Walk Trainer TM: A Robotic System for Walking Rehabilitation. In: IEEE International Conference on Robotics and Biomimetics, pp. 1616–1621. IEEE, USA (2006)
3. Rios, A., Hernandez, E., Moreno, J.A., Keshtkar, S., Garza, R.: Kinematics analysis of a new 3DOF parallel manipulator as walking rehabilitation device. In: 15th International Conference on Electrical Engineering, Computing Science and Automatic Control (CCE), pp. 1–6. IEEE, USA (2018)
4. Almaghout, K., Tarvirdizadeh, B., Alipour, K., Hadi, A.: Design and control of a lower limb rehabilitation robot considering undesirable torques of the patient's limb. J. Eng. Med. **234**, 1457 (2020)
5. Bouri, M., Gall, B., Clavel, R.: A new concept of parallel robot for rehabilitation and tness: the Lambda. In: Proceedings of the 2009 IEEE International Conference on Robotics and Biomimetics (ROBIO), pp. 2503–2508. IEEE, USA (2009)
6. Kong, X., Gosselin, C.M.: Kinematics and singularity analysis of a novel type of 3-CRR 3 DOF translational parallel manipulator. Int. J. Robot. Res. **21**(9), 791–798 (2002)

7. Evtushenko, Y., Posypkin, M., Rybak, L., Turkin, A.: Approximating a solution set of nonlinear inequalities. J. Global Optim. **71**(1), 129–145 (2017). https://doi.org/10.1007/s10898-017-0576-z
8. Evtushenko, Y., Posypkin, M., Turkin, A., Rybak, L.: The non-uniform covering approach to manipulator workspace assessment. In: Proceedings of the 2017 IEEE Russia Section Young Researchers in Electrical and Electronic Engineering Conference (ElConRus 2017), pp. 386–389. IEEE, USA (2017)
9. Rybak, L.A., Posypkin, M.A., Turkin, A.V.: Method for approximating the workspace of the parallel robot. Int. J. Pharm. Technol. **8**(4), 25045–25055 (2016)
10. Jaulin, L., Kieffer, M., Didrit, O., Walter, E.: Applied Interval Analysis: With Examples in Parameter and State Estimation, Robust Control and Robotics. Springer, New York (2001). https://doi.org/10.1007/978-1-4471-0249-6
11. Merlet, J.P.: Determination of 6D workspace of gough-type parallel manipulator and comparison between different geometries. Int. J. Robot. Res **18**(9), 902–916 (1999)
12. Anvari, Z., Ataei, P., Masouleh, M.T.: The collision-free workspace of the tripteron parallel robot based on a geometrical approach. In: Computational Kinematics, pp. 357–364 (2018)
13. Behera, L., Rybak, L., Malyshev, D., Gaponenko, E.: Determination of workspaces and intersections of robot links in a multi-robotic system for trajectory planning. Appl. Sci. **11**, 4961 (2021)

Check for updates

Improvement of the Robotic Workplace to Prevent a Collision

Jaroslav Antoš(✉) and Martin Bušek

VÚTS, a.s., Liberec, Czech Republic
{jaroslav.antos,martin.busek}@vuts.cz

Abstract. In the serial production of air filters for the truck brake system, an important operation is to pick and place four emptied boxes of semi-finished products from the assembly space. This, due to the prescribed production cycle, must be done in one step. These emptied boxes cannot be locked and it is therefore not possible to define the position in advance. It is necessary to implement a specialized camera system. This system must reveal the different positions and shapes of these boxes. The current high demand for production cycle and the restriction of the robot's movement is the reason for scanning only one box from said four vision system. This can lead to dangerous collisions between the robot's gripper and the pick-up boxes. The article deals with the improvement of the robotic workplace so as to avoid the mentioned collisions. The modifications consist in supplementing the workplace with optical inspection gates, implementation of a special algorithm in the robot system, superior control system of the production line and HMI visualization. The whole update is carried out with regard to the high safety standards and requirements of the automotive industry.

Keywords: Robot · Gripper · PLC · Vision system · Safety

1 Introduction

Industrial automation is gaining more importance and receiving more attention. The demand for higher production of handling machines, lower costs, user comfort, versatility and the situation on the labor market, where there is a significant shortage of qualified human resources, contributes to this. In the production environment, it is important to develop and automate especially areas of mechatronic and robotic applications (automated handling and assembly, machine vision, quality control, measuring, 3D scanning, laser technology, etc.), which are closely related to electronics, mechanics, software engineering and many other scientific and technical disciplines [1].

A typical example of an automation task may be an application with a robotic pick and place operation in the serial production of air filters for truck brake systems. The robotic application is an integral part of the production line for the assembly of these filters and ensures the supply of filter components to the assembly space, the production belt and the removal of empty packaging, boxes from the line space. The workspace of the robotic application is divided into 4 sectors, see Fig. 1. Two sectors are intended

© The Author(s), under exclusive license to Springer Nature Switzerland AG 2022
J. Beran et al. (Eds.): TMM 2020, MMS 85, pp. 204–209, 2022.
https://doi.org/10.1007/978-3-030-83594-1_21

for bottom plate pickups and empty box pickups. The third sector is intended only as a storage place for empty boxes. The fourth sector is for bottom plates storage and consists of a manipulator and a conveyor carrying the components for assembly.

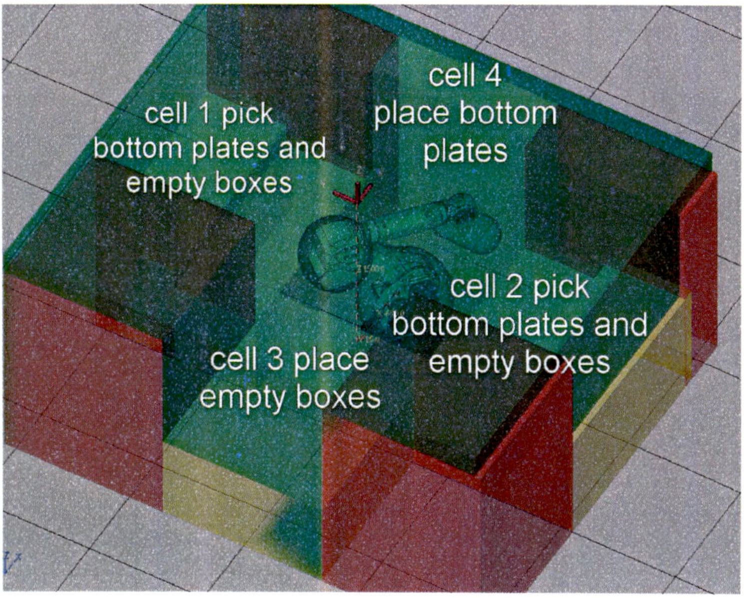

Fig. 1. The robot workplace contains 4 cells

2 The Critical Moment

Special grippers installed on the robot body ensure operations with mounted air filter components, packaging boxes, vision system and external lighting. The problematic moment of the whole workplace is the operation, which occurs at the moment of emptying all eight boxes in the given picking sector. All 8 boxes must be stored here. Due to the high cycle time of the production line, 4.1 s, it is necessary to perform this operation with respect to the stock of components on the conveyor belt of the line in the limit of 32.8 s. This time is critical because it is necessary to reliably remove 8 empty boxes. The robot must therefore enter the emptying area for empty boxes twice. Adherence to the given cycle is borderline in this operation, in the case of an error condition given by the inaccurate placement of the boxes, the production cycle is no longer observed and the entire production process stops. In addition, there is very often a collision between the gripper (Fig. 2) and the packaging box. The situation can be solved by modifying the gripper, ie by increasing its stroke, which really contains a redesign of the entire gripper. For cost and time reasons, it was necessary to detect crisis collisions of the gripper with the packaging material [3].

The potential occurrence of a collision is due to the fact that the current vision system, for time reasons, photographs only one box out of four, assuming that the remaining three are perpendicular and parallel. These boxes are not completely next to each other and there are gaps of various sizes between them due to storage and introduction into the sector. The Vision system performs a scan, the robot with a storage gripper moves over a place with an empty four boxes. Due to their poor placement, not all four boxes may be gripped correctly. Some may remain on the desktop, which is not currently signaled and is the cause of a future collision. The problem of correctly gripping empty boxes can be solved by modifying the design of the gripper or by adding a gripper pressure sensor. However, this is again financially and time consuming and complicated in terms of transmitting the alarm signal via the robotic system to the line controller.

Fig. 2. The robot multifunction gripper with two double grippers, camera and flash

The simplest way to predict a future collision of a gripper with a forgotten box is the additional installation of a security element in the form of two optical gates, the positive signal of which alerts the operator with an alarm on the HMI and no gripper collision.

3 The Optical Gate for Inspection

The optical gate used consisted of a pair of reflectors and two reflectors, which were located in the space between the pick sector and the place in the path of movement. One optical gate with a combination of movement in the vertical axis is not enough, because the first axis of the robot works in the range of $-182°$ to $182°$ and the stop is just in the control area. The beams must pass through both pairs of carried boxes, but at the same time they must not be obscured by the robot itself. Because the pair of grippers is constructed horizontally, the optical must be positioned vertically. Due to the limited minimum stroke of the gripper, it is not possible for the robot to pick up only one box, so there are only four picking options: no boxes, two boxes on the left, two boxes on the right and all boxes. For more permanent lighting conditions, the workplace area was covered with a board due to external lighting, and this cover is used to screw on the reflection points of the light gate. Because the operator goes to the cell 3 workplace, the gate had to be supplemented with a protective cover against damage by the operator. It is also necessary to treat the situation when the sensor is covered with dirt from stored boxes, dust, label paper or the robot itself covers it with a bad storage of an empty box. See Fig. 3 [4, 5].

Fig. 3. The optical gate for inspection

4 Implementation of the Algorithm into the Robot System

The process of removing empty boxes begins by checking the functionality of the optical gate, see Fig. 4. During the process, the activity of both sensors is first checked and the beam passes through the reflecting surface. If this is not the case, the robot goes into an alarm state with a request to check the optical gate and its possible covering by a foreign object, which will be performed by the operator. If both sensors are continuous, the robot can start removing empty boxes.

The robotic system has information on the number of layers, boxes with components from the PLC and increments the data when removing them. The vision system contains a subroutine for adjusting the actual height of the placed boxes. The vision system detects and corrects misaligned or non-recessed boxes in the vertical direction by making a difference between expected and actual height. Then there is the standard pick process. The Vision system takes a picture of one box from the four taken, the robot moves to the required position and the gripper removes this box and three adjacent ones. This is followed by a transfer to the storage location, which must be stopped for a moment at the optical gate checkpoint. If both gates are interrupted, no action is taken to resolve the fault condition and the boxes continue to move to the storage area. In case of incorrect gripping of any of the pick empty boxes, this condition is detected by the optical gate. This is followed by the setting of the error flag and the completion of the removal action, which ensures that the correctly gripped boxes are stored and the robot is moved to a safe home position. In the safe home position of the robot, the alarm sign is tested and in case of a negative result, the alarm of the whole line is announced and the operator is called.

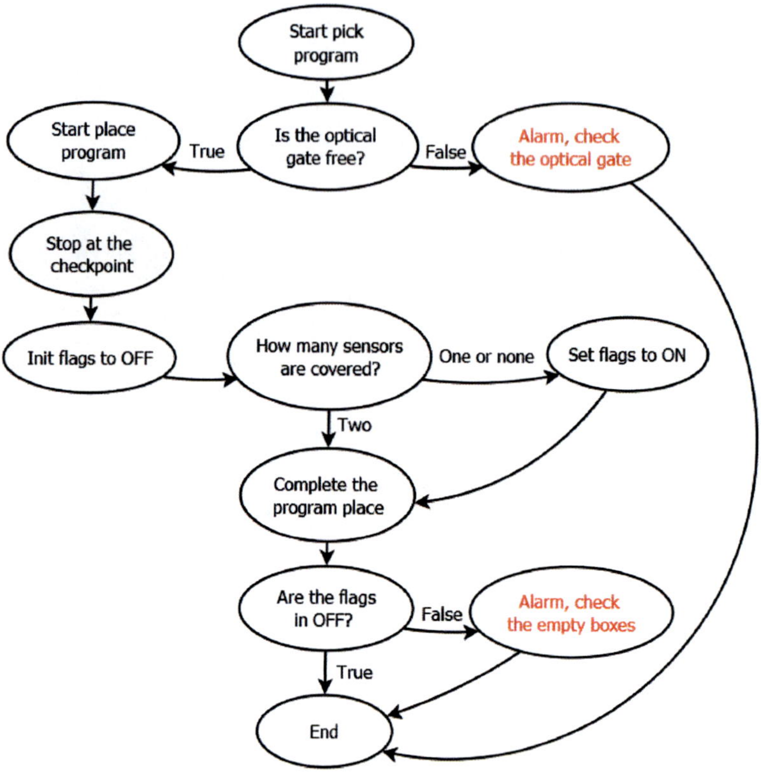

Fig. 4. The flowchart of the inspection algorithm

5 The Control System with HMI and Safety

The alarm flag is transmitted from the robot system to the PLC control system of the production line via handshake communication. The announced alarm can be of two types. The alarm type is indicated on the HMI panel. In the first case, it is an alarm caused by a fault of the optical gate inspection system, the reflectors are dirty, their location is not perpendicular or they are damaged, covered with logistics paper or the reflectors are dirty. In this case, it is up to the called operator to find out the cause and ensure remedy. In the second case, an alarm of a badly placed box is displayed on the HMI, while the robot stops in the home position and waits for the operator. The operator must manually move the badly picked up box, acknowledge the alarm, and start the manufacturing process.

Dual check safety limits the movement space and speed of the robot according to input parameters. It primarily serves to protect the operator (interaction between the machine and human) and secondarily between individual machine parts (for example, a robot versus a fence). Eight spaces have been modeled for the palletizing workstation, four are restricted zones and four are working zones. The robot system must be designed to avoid trapping and collision between the moving parts of the robot and other fixed or moving objects. The robot system must be designed to avoid trapping and collision between the moving parts of the. The operating position must satisfy the following conditions. The operator can easily operate the operator's panel or the teach pendant. The operator can easily verify the operation of the system. The operator can immediately stop the entire or partial system in the event a malfunction of the system or any dangerous condition. All robot system emergency stop devices must remain functional. When they are not functional, a measure must be taken so that the safety of personnel is ensured [2].

6 Conclusion

Improvement of the robotic workplace in the role of pick and place in the production of air filters for the truck brake system will eliminate very dangerous situations, which originally in many cases led to a large collision between the robot's gripper and the removed empty boxes for used components. Thanks to this solution, the whole operation of storing empty boxes was also accelerated, because the situation of losing boxes during the whole operation of storing was treated. The modified robot workplace is part of a production line whose production reaches up to 3,500 manufactured filters per shift. The modification ensured the stabilization of this workplace and the elimination of line downtime.

References

1. Operator's Manual: R-30+B Plus/R-30+B Mate Plus Controller. Fanuc Robot Series. https://www.fanuc.eu/cz/cs?gclid=EAIaIQobChMIiNaLo9rO6AIVlOR3Ch0CeQVvEAAYASAAEgLFnPD_BwE. Accessed 14 Apr 2021
2. Safety Handbook: B-80687EN/15. Fanuc Robot Series. https://www.fanuc.eu/cz/cs?gclid=EAIaIQobChMIiNaLo9rO6AIVlOR3Ch0CeQVvEAAYASAAEgLFnPD_BwE. Accessed 14 Apr 2021
3. Zhou, X., Xu, Z., Li, S., Wu, H., Cheng, T., Lv, X.: AI Based Robot Safe Learning and Control. Springer, Guangzhou (2020). https://doi.org/10.1007/978-981-15-5503-9 ISBN: 978-981-15-5502-2
4. Modelling and Simulation for Autonomous Systems: The Complete Reference, vol. 4. Springer, Prague (2020). ISBN: 978-3-030-20189-0. ISSN: 1860-949X
5. Lovasz, E.-C., Maniu, I., Doroftei, I., Ivanescu, M., Gruescu, C.-M. (eds.): New Advances in Mechanisms, Mechanical Transmissions and Robotics, vol. 88. Springer, Cham (2020). https://doi.org/10.1007/978-3-319-45450-4 ISBN: 978-3-030-60075-4

Prototype and Testing of LARMbot PK Arm

Axel Fort[1(✉)], Marco Ceccarelli[2], and Med Amine Laribi[1]

[1] Department of GMSC, Pprime Institute CNRS, ENSMA, University of Poitiers, UPR 3346, Poitiers, France
axel.fort@etu.univ-poitiers.fr,
med.amine.laribi@univ-poitiers.fr

[2] Laboratory of Robot Mechatronics, Department of Industrial Engineering, University of Rome Tor Vergata, 00133 Rome, Italy
marco.ceccarelli@uniroma2.it

Abstract. The design of LARMbot arm is presented with a structure of a tripod parallel manipulator, whose mobile platform moves an elbow joint for the forearm. A prototype is presented with construction using market components and 3D printed parts to achieve a design suitable for the LARMbot humanoid as ensuring proper mobility and payload capacity of the arm. The construction of the prototype is discussed to characterize the design solutions of the mechanical design, driving actuator system, sensing units, and onboard control equipment. A lab setup is assembled for testing activity whose results are reported to show the feasibility of the low-cost light prototype and to characterize the basic operation performance as arm of the LARMbot humanoid.

Keywords: Humanoid robots · Arm design · Parallel manipulators · Prototype · Testing

1 Introduction

LARMbot is a robot developed at the LARM2 laboratory of the University of Rome Tor Vergata [1, 2] as based on parallel architectures as proposed in [3]. It uses cable robot technology to enable complex movements at a low cost. Its legs have been built and validated using linear actuators to have great agility of movement while keeping a light and inexpensive structure, due to the limited number of motors required to operate them [4, 5]. The new solution that has been considered to improve the upper limbs is to take the already validated structure that was used for the legs, and add a servomotor at the elbow to set a forearm in motion [6]. Design and analysis for Forward and Reverse Kinematics have already been carried out in previous works [7, 8].

This paper presents results for design and construction of a prototype as based on previous numerical evaluations in [3, 5]. Lab testing is also discussed with results.

© The Author(s), under exclusive license to Springer Nature Switzerland AG 2022
J. Beran et al. (Eds.): TMM 2020, MMS 85, pp. 210–219, 2022.
https://doi.org/10.1007/978-3-030-83594-1_22

2 LARMbot Arm for Humanoid Robot

In its first version, the LARMbot robot had anthropomorphic upper limbs, but they could not perform complex movements with proper payload, and the hands were not able to grab objects. Figure 1(a) and (b) show an upper limb consisting of two servomotors for the shoulder, allowing it to perform only two rotational movements instead of the three normally possible for a human shoulder; a servomotor for the elbow, allowing it to perform the flexion/extension movement of the forearm; no servomotor to perform a movement equivalent to pronation/supination of the forearm to rotate the wrist.

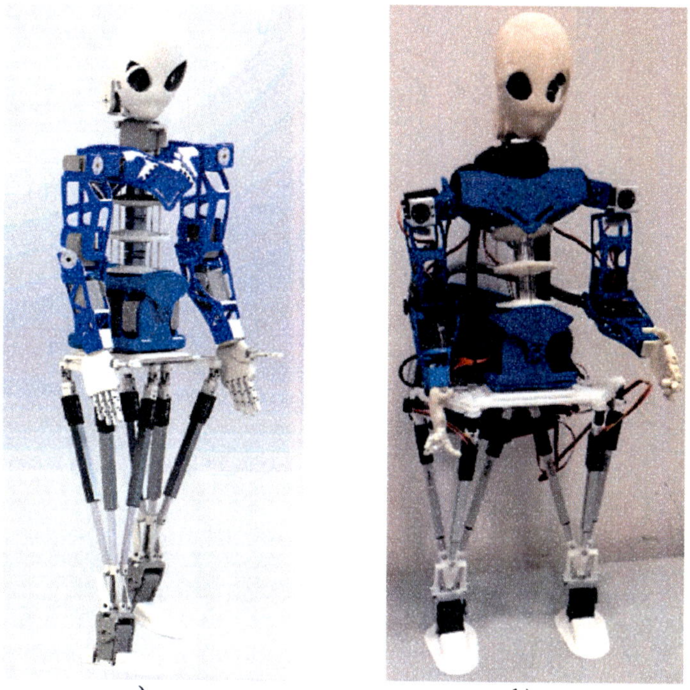

a) b)

Fig. 1. LARMbot humanoid [1, 2]: a) a CAD design; b) a built prototype

In order to increase the working space of the robot upper limbs, a tripod system identical to the one used by its legs has been designed as reported in [6]. The idea is to use the tripod for the arm, and to fix at the end of the platform a servomotor connected to a light bar to act as a forearm. A kinematic diagram and a CAD design of this upper limb are shown in Fig. 2(a) and (b). Table 1 lists chosen dimensions to fit the upper limb with the built LARMbot humanoid, as characterized by the stroke of the linear actuators for links l_i ($i = 1, 2, 3$).

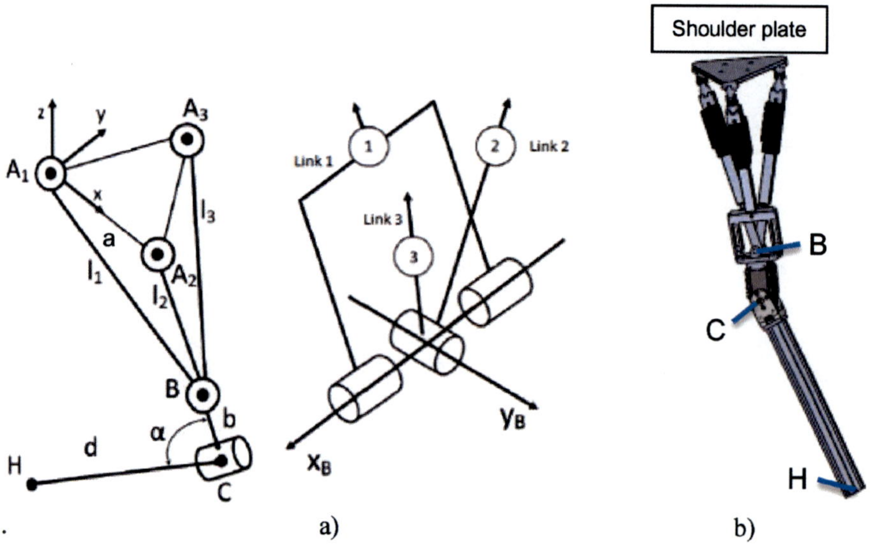

Fig. 2. Design of a new LARMbot arm [6, 8]: a) kinematic structure; b) a CAD design

Table 1. Design parameters

a	li	b	d
90 mm	(230–330) mm	30 mm	320 mm

3 Prototype Design and Construction

The prototype is designed to give the arm composed of three linear actuators from market. This structure allow to have a light structure and a limited number of motors. On the platform corresponding to the elbow, a market rotative servomotor is used. This servomotor is used to achieve the flexion/extension movement of the forearm, which completes the structure of the upper limb. The tripod platform, which is an equilateral triangle, is designed to be fixed on the trunk of the robot, and the three linear actuators meet on the elbow pad, Fig. 2.

Figure 3 shows the conceptual design of the prototype with main units using market components. In this diagram, the $M_i - L_i$ set ($i = 1, 2, 3$) correspond to linear actuators, while RS1 is a rotating servomotor for elbow motion. IMUs are used to monitor acceleration and rotation data on the extremities of each limb of the arm.

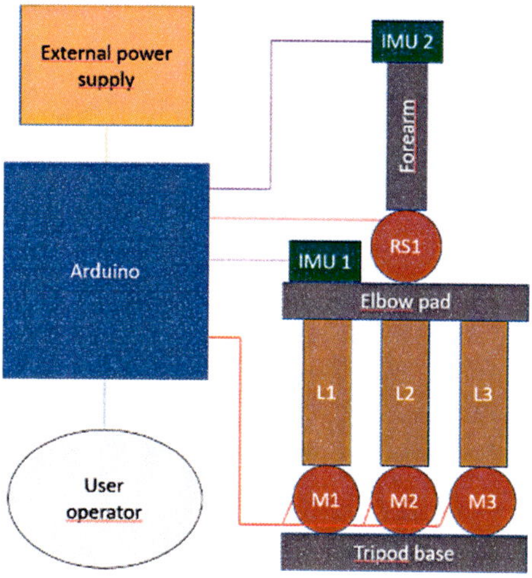

Fig. 3. A conceptual diagram for a lab prototype

The chosen linear actuator (Actuonix L16-100-63-12-P) consists of a DC motor with a potentiometer inside that gives feedback on the stroke, so that a control over the stroke of the linear actuator must be done in a program. The test shown in Fig. 4(a) was worked out to learn how to control the operation of these linear actuators, with satisfactory results in Fig. 4(b).

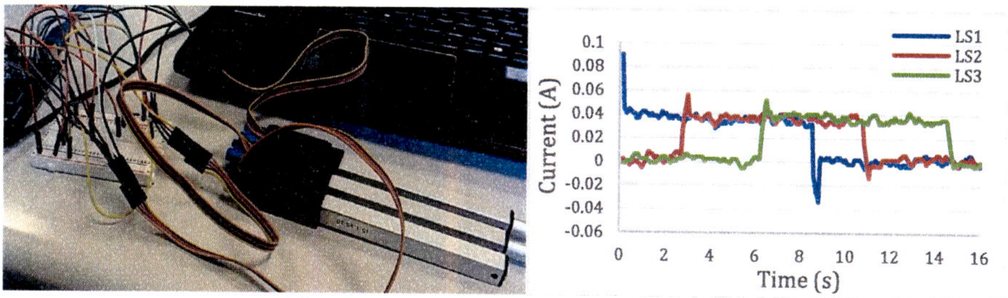

(a) (b)

Fig. 4. The three linear actuators: a) a first setting up; b) a test results

The tests are carried out with an Arduino board together with an L293D motor driver shield. The Fig. 5 shows how a setup design for linear actuators using the lessons learned from the test that was performed as in Fig. 4.

According to Fig. 4(b), the maximum value of the current which is consumed by a linear actuator at maximum speed is approximately 0.4 A.

Then, the LARMbot upper limb has been assembled according to the conceptual diagrams in Figs. 3 and 4, as shown in Fig. 6(b), whereas the device without forearm is shown in Fig. 6(a). The forearmless upper limb was useful for early testing, as it provides results that are not influenced by the weight or the torque that the forearm can produce on the device.

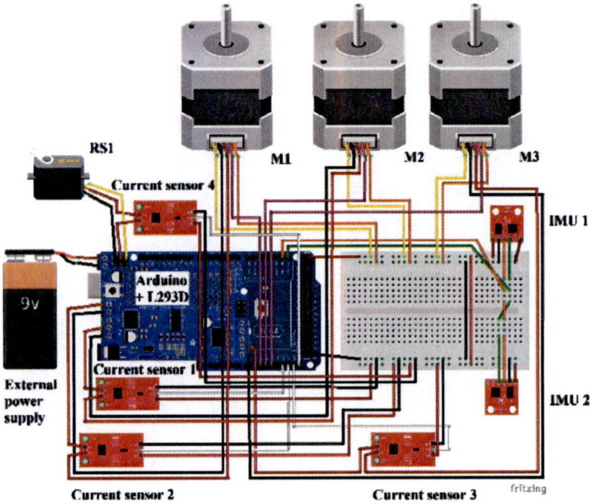

Fig. 5. The circuit design for a lab prototype

The mechanical parameters of the fully assembled prototype of LARMbot upper limb are listed in Table 1 and actuator data are listed in Table 2.

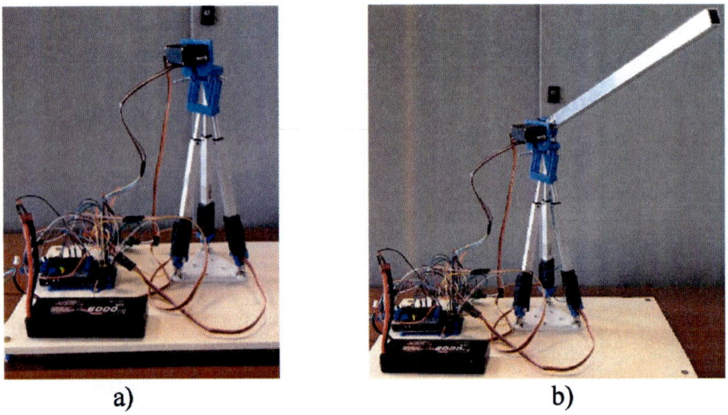

Fig. 6. The built prototype of LARMbot upper limb at LARM2 in Rome: a) arm setup; b) full setup

Table 2. Actuators parameters for the built prototype in Fig. 6

Motor type	Max force/torque	Mass	Voltage	Speed
Servomotor	150 N-cm	55 g	4.8–7.2 V	461.5 deg/s
Linear actuator	100 N	74 g	12 V	20 mm/s

4 Testing Layout and Results

Testing has been carried out to check operation performance by running three characteristic modes with results which are summarized in Table 3 and Fig. 7.

Table 3. Characteristics of reported test modes, Figs. 2

Test number	Initial position	Final position	W_{max}	a_B max	a_H max
Test 1 (Tripod only)	$(0;\ 0.08;\ 0.24)_B$	$(0.12;\ 0;\ 0.28)_B$	0.62 W	1.04 g	–
Test 2 (Full upper limb without payload)	$(0.045;\ -0.1;\ 0.24)_B$ $\alpha = 0°$	$(0.045;\ 0.2;\ 0.24)_B$ $\alpha = 90°$	1.30 W	1.52 g	2.12 g
Test 3 (Full upper limb with payload)	$(0.045;\ -0.1;\ 0.24)_B$ $\alpha = 0°$	$(0.045;\ 0.2;\ 0.24)_B$ $\alpha = 90°$	2.93 W	1.07 g	1.21 g

First, we carry out a tripod movement test alone. In agreement with Table 3, a simple movement is achieved involving movement in the three directions of space. Figure 7 shows a snapshot of the test. Thanks to sensors IMU 1 the movement can be characterized with useful data such as the acceleration of the elbow pad shown in Fig. 8(a). Current sensors also make possible to obtain in Fig. 8(b) the power consumed by each motor, which is very useful for sizing the robot's battery and estimating arm autonomy.

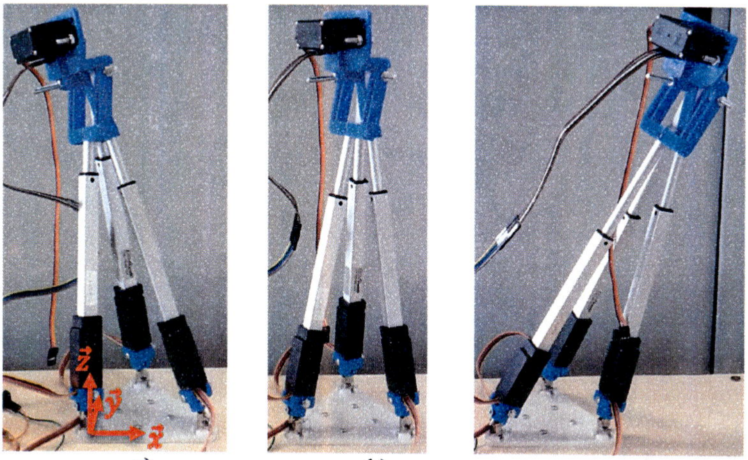

a) b) c)

Fig. 7. A snapshot of test 1, Test 3: a) beginning; b) middle; c) final

Figure 8(a) shows smooth acceleration for the point B (located at the elbow pad as shown on Fig. 2) within a range of 1.0–1.04 g (g gravity unit). Figure 8(b) shows an average of the power consumption of 0.2 W during the movement for the linear actuators, with a total average power of 0.5 W.

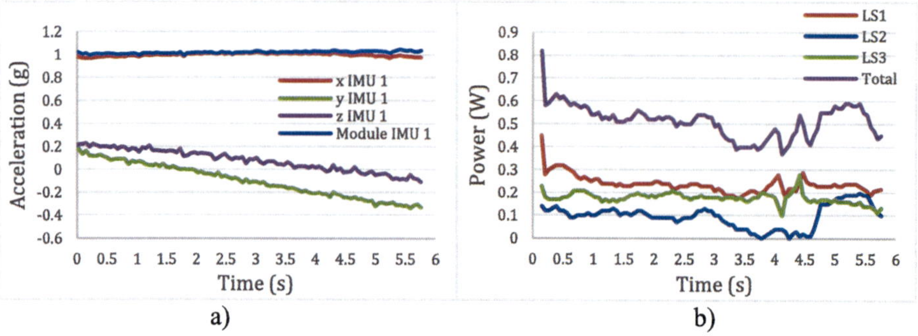

Fig. 8. Acquired results from test 1, Table 3, in terms of: a) acceleration of point B (IMU 1); b) power of the linear actuators

The second test performs a complete movement of the full upper limb, without payload, to characterize the movement. Figure 9, according to Table 3, shows the movement data for this test 2. In addition to the previous data, the data of the IMU 2 and the current sensor of the servomotor that activates the elbow are acquired.

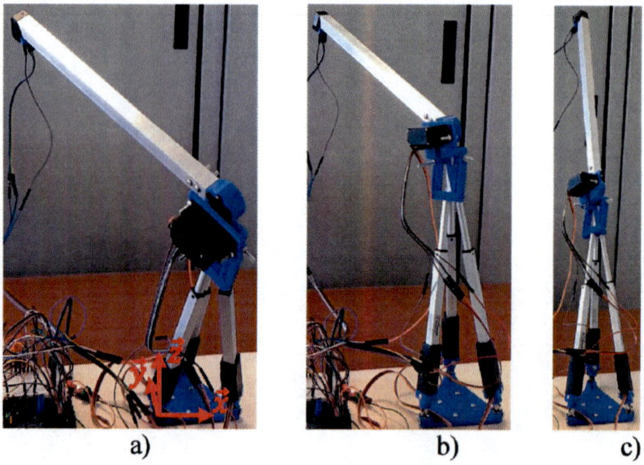

Fig. 9. A snapshot of test 2, Table 3: a) beginning; b) middle; c) final

Figure 10(a) shows like the Fig. 8(a) a smooth acceleration for the point B within a range of 0.99–1.06 g, the same for point H within a range of 0.93 g–1.03 (g gravity unit), while one can note the peak at 1.52 g and 2.12 g, for IMU 1 and 2 respectively, at the start that is due to the sudden acceleration when all motors come into operation. Figure 10(b) still shows an average of the power consumption of 0.2 W during the movement for the linear actuators, while the power consumed by the elbow servomotor increases towards the end of the movement. There is no brake on this servomotor, and to keep an angle, it is forced to consume current to balance the effect of gravity which is very important at the elbow when the arm reaches its final position. The total average power consumed by all the motors during this movement is 0.7 W.

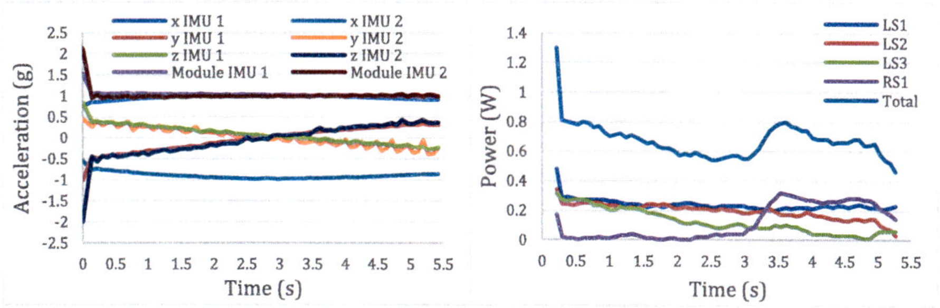

a) b)

Fig. 10. Acquired results from test 2, Table 3, in terms of: a) acceleration of point B (IMU 1) and H (IMU 2); b) power of the linear actuators and the servomotor

For test 3, the same movement as for test 2 was carried out in order to be able to better compare the differences. A load of 100 g was placed at the end effector of the arm, at point H. Figure 11 shows a snapshot of this movement.

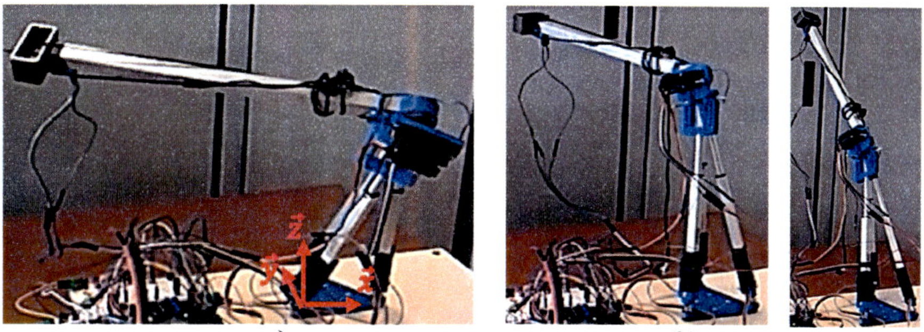

a) b) c)

Fig. 11. A snapshot of test 3, Table 3: a) beginning; b) middle; c) final

Figure 12(a) shows like the Fig. 8(a) and 10(a) a smooth acceleration for the point B within a range of 0.95–1.07 g (g gravity unit). Point H shows more noisy acceleration, because the load is a little bit heavy and therefore the rotation of the servomotor is not smooth. Nevertheless, it operates successfully in a range within 0.92–1.21 g. Figure 12(b) still shows an average of the power consumption of 0.2 W during the movement for the linear actuators, while the power consumed by the elbow servomotor increases during the movement. The explanation is the same as for the previous test, but the power consumed is even greater because of the 100 g load. The total average power consumed by all the motors during this movement is 1.8 W.

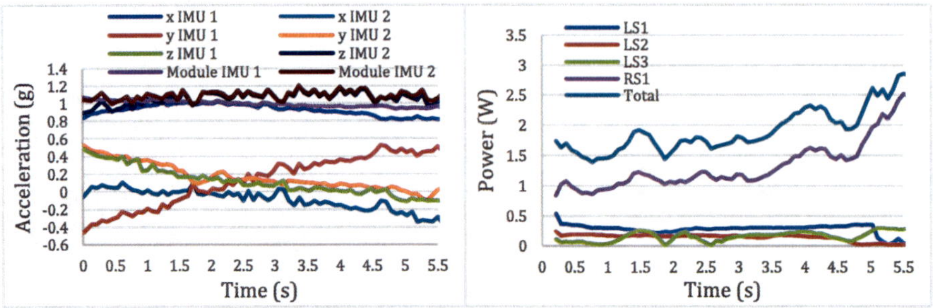

a) b)

Fig. 12. Acquired results from test 3, Table 3, in terms of: a) acceleration of point B (IMU 1) and H (IMU 2); b) power of the linear actuators and the servomotor

The main results from the tests are summarized in Fig. 13. Figure 13(a) shows that point B remains fairly stable despite the increased load, while point H quickly shows noisy accelerations. The power consumed by the arm in Fig. 13(b) is also very striking because it increases sharply when a load is added to the forearm extremity. Referring to the results in Fig. 8(b), 10(b) and 12(b), this increase in power is only due to the elbow servomotor.

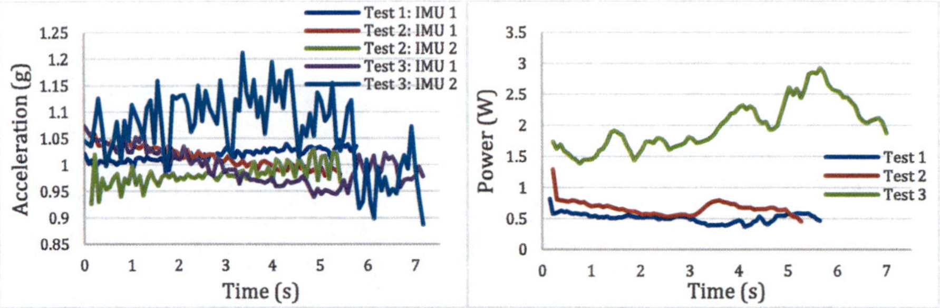

a) b)

Fig. 13. Summary of the results acquired for tests 1, 2 and 3 in terms of: a) module acceleration of point B and H; b) total power consumption

5 Conclusion

A new upper limb of the LARMbot is presented with a prototype whose design is based on a tripod structure for the arm with one elbow pad for the forearm link. The prototype is constructed by using market components and 3D printing manufactured parts. Successful tests give results characterizing the large mobility range and feasibility for high payload in manipulation, provided that the servomotor used for the elbow is changed or improved.

Acknowledgements. The first author wishes to gratefully acknowledge the Erasmsus program for permitting his period of study at LARM2 of University of Tor Vergata in the A.Y. 2020-21.

References

1. Ceccarelli, M., Cafolla, D., Russo, M., Carbone, G.: LARMbot humanoid design towards a prototype. MOJ Int. J. Appl. Bionics Biomech. **1**(2), 48–49 (2017). https://doi.org/10.15406/mojabb.2017.01.00008
2. Russo, M., Ceccarelli, M.: Design and Simulation of a parallel-serial LARMbot arm. In: Doroftei, I., Oprisan, C., Pisla, D., Lovasz, E.C. (eds.) New advances in Mechanism and Machine Science. MMS, vol. 57, pp. 379–386. Springer, Cham (2018). https://doi.org/10.1007/978-3-319-79111-1_38
3. Ceccarelli, M., Russo, M., Morales-Cruz, C.: Parallel architectures for humanoid robots. Robotics **9**, 75 (2020). https://doi.org/10.3390/robotics9040075
4. Russo, M., Cafolla, D., Ceccarelli, M.: Design and Experiments of a Novel Humanoid Robot with Parallel Architectures. Robotics **7**(4), 79 (2018). (Best Paper Award MEDER 2018). https://doi.org/10.3390/robotics7040079
5. Russo, M., Ceccarelli, M.: Dynamics of a humanoid robot with parallel architectures. In: Uhl, T. (ed.) Advances in Mechanism and Machine Science. IFToMM WC 2019, vol. 73, pp. 1799–1808. Springer, Cham (2019). https://doi.org/10.1007/978-3-030-20131-9_178
6. Ceccarelli, M., Russo, M.: Meccanismo di braccio per robot umanoide (Arm mechanism for humanoid robot), n. 102017000021920, Italy. 27 February 2017. Patent no. IT. 102017000021920, 18 June 2019
7. Russo, M., Herrero, S., Altuzarra, O., Ceccarelli, M.: Kinematic analysis and multi-objective optimization of a 3-UPR parallel mechanism for a robotic leg. Mech. Mach. Theory **120**, 192–202 (2018). https://doi.org/10.1016/j.mechmachtheory.2017.10.004. ISSN: 0094-114X
8. Russo, M., Ceccarelli, M., Takeda, Y.: Force transmission and constraint analysis of a 3-SPR parallel manipulator. Proc. Inst. Mech. Eng. Part C J. Mech. Eng. Sci. **232**(23), 4399–4409 (2018). https://doi.org/10.1177/0954406217750190

Object Picking by a Method Combined Cascade Classifier and Deep Reinforcement Learning

Koji Makino(✉), Kazuyoshi Ishida, Takahiro Tanaka, and Hidetsugu Terada

University of Yamanashi, Kofu, Japan
kohjim@yamanashi.ac.jp

Abstract. It is important problem to pick up the objects that is arranged at random automatically. And it is desirable to adapt the problem such that the mobile robot gathers spread objects in the field in the real world. In this paper, the method that consists of an approaching process using the cascade classifier and a picking process using the deep reinforcement learning is proposed to solve the problem. First, in the approaching process, the method to detect the object and to estimate the position using two camera images is proposed. Second, in the picking process, the reward matrix to learn it using the deep reinforcement learning is introduced newly. The effectiveness of each process is investigated by the simulations built by three-dimensional physical simulation using PyBullet. Final, we confirm the realization of the method that consists of the approaching process and the picking process by the experiments using the real robot at real-time.

Keywords: Picking task · Deep reinforcement learning · Cascade classifier

1 Introduction

It is an important problem that the objects that are arranged at random are picked up by the robot arm, automatically [1]. This problem is known as a picking task. The robot has to detect the position and the orientation of the object exactly, and to select the suitable point to pick it up. It is difficult to solve the picking task, since the detection and the pick-up motion must be dealt with at the same time.

A lot of method to solve the picking task was studied and developed [2]. For example, Zhen et al. used the Support Vector Machine (SMV) [3]. The methods using the deep reinforcement learning (DRL) [4] are addressed [5], recently. In many cases, there are the objects in a cage [6], on a shelf [7]. On the other hand, it is also important that the robot gathers the objects spreaded in the field [8]. We think that an agricultural chemical become unnecessary if the noxious insect

© The Author(s), under exclusive license to Springer Nature Switzerland AG 2022
J. Beran et al. (Eds.): TMM 2020, MMS 85, pp. 220–228, 2022.
https://doi.org/10.1007/978-3-030-83594-1_23

can be picked by the robot, for example. This problem is more difficult than the picking task, since the robot must move to search the object. This problem is named as "collection picking problem" in this paper. The detection and pick-up motion of the picking task including the collection picking problem is performed using the same algorithm, generally. In this case, the high performance computers with graphic boards are necessary, since the calculation cost of the learning using the deep learning is increased.

In this paper, the method that consists of an approaching process and a picking process is proposed to solve the collection picking problem. And the processes of the proposed method are employed by different algorithms; the cascade classifier and the DRL. First, the system for the collection picking problem is described, in Sect. 2. And the approach process and the pick process are proposed in Sect. 3 and 4, respectively. Next, the proposed methods are realized by using the real robot arm, and the effectivenesses are confirmed in Sect. 5. Section 6 is conclusion.

2 System for a Collection Picking Problem

The proposed system for the collection picking problem consists of two process; an approaching process and a picking process. The abstract of the relation of the two processes is shown in Fig. 1.

(a) Approaching process (b) Picking process

Fig. 1. The abstract of the relation of the approaching process and the picking process.

The advantage points using two processes are described. The robot must detect the object and move to it roughly if the distance between the robot and the object is far. And, the robot picks it up after the robot approaches it. The motion planning using the exact position and orientation is necessary. Therefore, the different two processes are effective to solve the collection picking problem.

Next, the robot and the object are described. The robot is an arm type robot (C4-A601 of EPSON) that is six degrees of freedom as shown in Fig. 2(a). The wide field angle camera (C922 of Logicool) is attached to the tip of the robot arm. And a hand that has two fingers to pick up the object is set. We assume

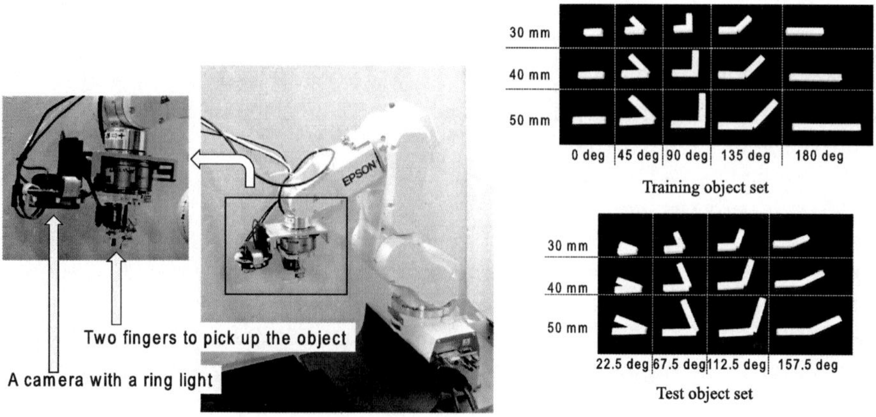

(a) Robot arm with a camera and fingers (b) Object data set

Fig. 2. The robot arm with the camera and the fingers to pick up the object, and the 15 and 12 kinds of object for training and test.

that the approaching process is executed if there is the object on peripheral region of the image, and the picking process is performed if there is the object in the closer to center of the image to simulate the robot moves. The object consists of two bars that are the same length, and the tips of the two bars are bonded as shown in Fig. 2(b). The kinds of angles of the bars are five (0, 45, 90, 135, and 180°), and the kinds of the lengths of the bars are three (30, 40, and 50 mm). Therefore, the 15 kinds of the object are used in training phase in the simulations. And the 12 kinds of the object that is different from the angle (22.5, 67.5, 112.5, and 157.5°) and is the same length are adapted in the test phase in the simulations and the experiments using the real robot to investigate the robustness. These objects are named as a "training object set" and a "test object set", respectively. And the following simulations and experiments are performed by the PC (CPU: Intel(R) Core(TM) i9-790X @3.3 GHz, RAM: 32 GB) with graphics board (NVIDIA GeForec CTX 1080Ti).

3 Approaching Process Using Cascade Classifier

In this section, the method for approaching the object that is detected using the cascade classifier to the center of the image is proposed. And the effectiveness of the proposed method is confirmed by the simulations. In the approaching process, the position of the object in the image obtained by the camera attached on the robot is detected using the cascade classifier of the OpenCV, though various methods, for example YOLO [9] based on the deep learning, SVM [10], Decision tree [11] and so on [12] are developed and studied. The time cost for training and the number of the training data using the cascade classifier is less than the deep learning, and the accuracy is higher than the other detection method except the deep learning, generally.

The proposed method for approaching the object is described. In the preliminary experiments, it is not enough accuracy to estimate the position of the object because of the detection error of the cascade classifier using one image. Therefore, the new method for decreasing position error by averaging the positions obtained by two images is proposed. For the following explanation, we assume that there is the object on the right upper in the image as shown in Fig. 3 at initial position. The position of the object in the image at first position is detected, and the center position (c_{x1}, c_{y1}) in camera coordinate system is calculated. Here, the origin of the coordinate system is the center of the image. The center of the second position (c_{x2}, c_{y2}), after the robot arm with the camera move to a diagonal quadrant, is calculated by the same manner. The length of the movement of the robot arm is expressed as L_x and L_y in real world. The initial position (p_x, p_y) in world coordinate system is calculated as follows;

$$p_x = \frac{L_x}{c_{x1} - c_{x2}} c_{x1}, p_y = \frac{L_y}{c_{y1} - c_{y2}} c_{y1} \tag{1}$$

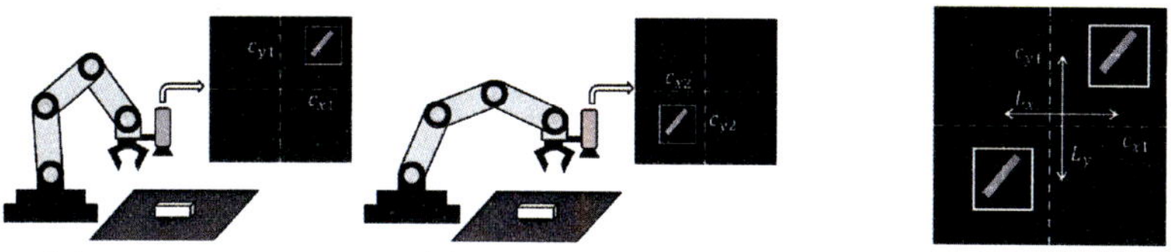

(a) First position (b) Second position (c) Relation between two images

Fig. 3. Procedure to estimate the position of the object using two images.

To investigate the effectiveness of the proposed method using two images, three-dimensional physics simulations built into PyBullet are performed. First, the training of the object detection using the cascade classifier is described. An object is selected among the training object set. The first and second position are stored. The length L_x and L_y are decided to 798 mm by preliminary trials in the simulator. The one object is arranged on either 100 lattice points. And the orientation of the object is selected among 8 poses (45° interval). The next object of the different shape is selected, and the same simulation is performed. Second, the results of the test using the test object set are described. 9600 (the 12 kinds of objects *times* 100 different initial position *times* 8 poses) results are calculated. The 9590 images among the 9600 images (99.90%) can be detected in the simulations. The difference between the initial position and the estimated position obtained by Eq. (1) are shown in Fig. 4. It is clear that the maximum error is ±9.6 mm, and the error of 80.9% images is less than ±4 mm, and the distribution of the errors is shown in Fig. 4(a). To compare the method of the original cascade classifier with the proposed method, the estimation errors are investigated under the same simulation condition. The maximum error is 55.2 mm, and the distribution is shown in Fig. 4(b). As a result, we confirm that the accuracy of the position obtained by the proposed method is higher than one using only cascade classifier.

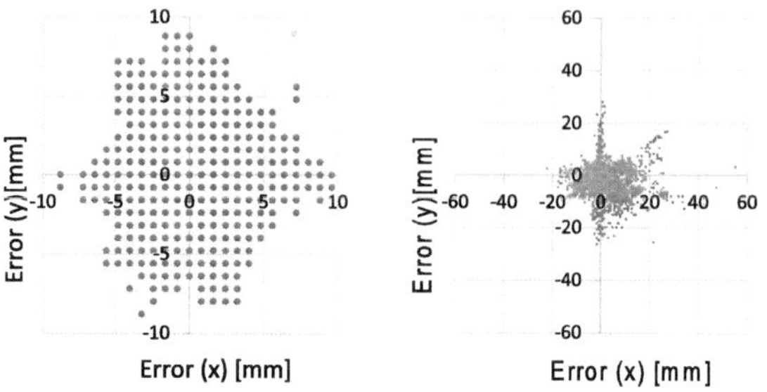

(a) Propsed method (b) Original cascade classifier

Fig. 4. Estimation errors using the proposed method and original cascade classifier.

4 Picking Process Using Deep Reinforcement Learning

In this section, the method to pick up the object using the DRL is described after the robot is close to the object. It is important to define the reward of the DRL, therefore the reward that is suitable for picking process is proposed. And the effectiveness of the method using the proposed reward is confirmed by the simulations.

The feature of the DRL is described, briefly. The DRL is composed of the reinforcement learning (RL) and the deep learning (DL). In the RL, the desirable sequential action is acquired, if the goal that means the good state is defined.

It is difficult to integrate the RL and the DL, since the RL and the DL are semi-supervised learning and supervised learning respectively. Therefore, the frameworks, for example TensorFlow and PyTorch and so on, are developed, recently. There are many methods to realize the RL, for example, Q-learning, Sarsa, Monte Carlo method and so on. Especially, the DRL using the Q-learning is called as the Deep Q-network (DQN). In this paper, the DQN using TensorFlow is employed.

First, the DL included into the DQN is described. The network of the DL is shown in Fig. 5. The image is converted using three layers of the Convolution Neural Network using Max Pooling. After the flatten, the four Dense layers are used. The data is processed by two different path with three layers of the Dense layer to employ the Dueling Network, and the two data paths are incorporated by the Concatenate layer. Here, the output is 27, since the robot selects the action among 27 ($=3^3$) actions that are three kinds (movement of x direction, movement of y direction, and rotation of yaw axis) and each action is three values.

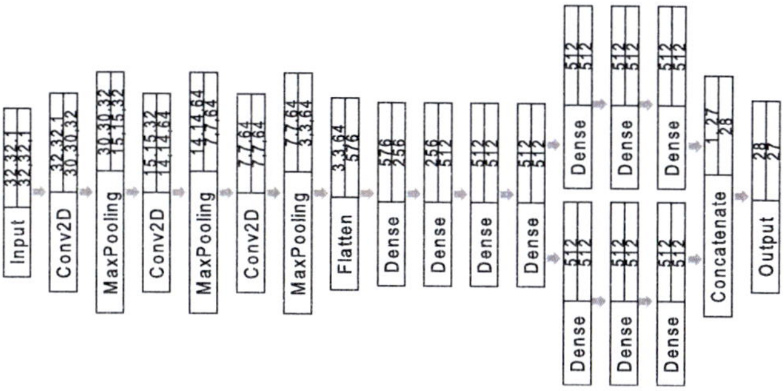

Fig. 5. Structure of the network.

Second, the Q-learning of the DQN is described. In the most case of the picking task, the reward is given if the picking is successful, or it is given according with the distance between the hand of the robot and the object. In this paper, the reward matrix is defined as shown in Fig. 6(a), newly. Assume that the image as shown in Fig. 6(b) is obtained. The example of calculation of the reward is shown in Fig. 6(c). The meaning of the reward matrix is described. In the problem in the paper, it is important that there is the object on the center of the image. Therefore, the 2×2 cells of the center of the reward matrix are 10 that is the highest value among the values in the reward matrix. And, it is desirable that the orientation of the bar of the object becomes vertical on the image, since the motion of the fingers of the hand is limited to side direction on the image. Therefore, the values of the vertical cells on the reward matrix are 1. The other cells are negative values that are different according with the distance from the center cells, since it is desirable if the object becomes near to the center. The 32×32 pixels at the center of the camera image that is converted into two values (black and white) are focused on, and the summation of the product of each pixel on the converted image and each cell on the reward matrix is calculated, and these values are summed up. And the summation value is divided by the count of the white pixels of white-and-black images. We define that the value is the reward.

Final, the effectiveness of the method is investigated by the simulations using the training object set and test object set. The results are shown in Fig. 7. Almost objects can be picked up over 95%, however the successful rate is lower in the case of the object of 30°. The main reason of low successful rate is that the connecting point of the object is picked up. The probability of the continuous failed motion is low, if the robot picks up the dropped object again. Therefore, it is not almost problem of the failed motion. It is clear that the picking motion can be trained using the DRL and the motion has robustness.

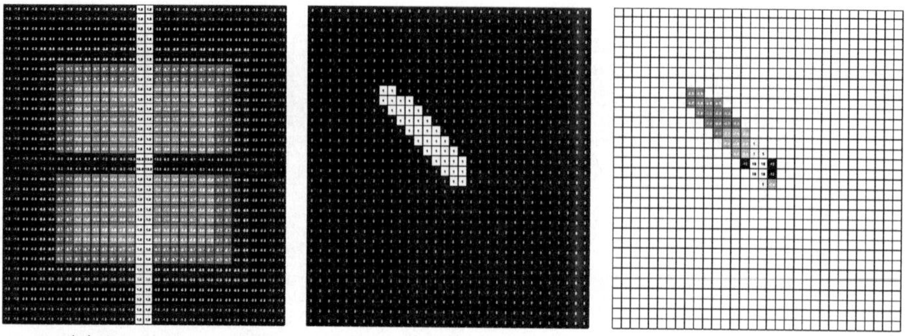

Fig. 6. Reward matrix that is composed of 32 × 32 cells, and the calculation method of the reward using the reward matrix.

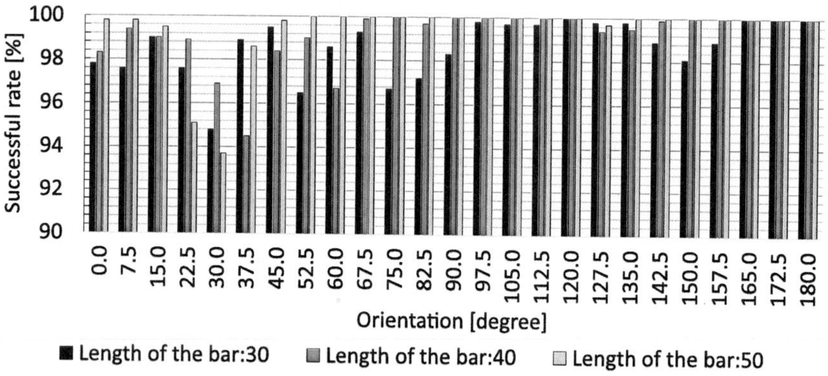

Fig. 7. The successful rate fo the picking motion

5 Experiment

The approaching process using the cascade classifier and the picking process using the DRL is combined, and the realization of the methods is confirmed by experiments using the real robot and real image.

First, the setting of the experiments is described. The both trained model of the cascade classifier and the DRL obtained by the simulation are used in the experiment. The 30 times of the trials are performed at each object that is set to random position and random direction.

Second, the experimental result is illustrated. The sequential motions are shown in Fig. 8. The robot arm moves in approaching process using the cascade classifier (Fig. 8(a)). And, the process is changed if the object becomes the center of the image (Fig. 8(b)). The object can be picked by the hand in the picking process using the DRL.

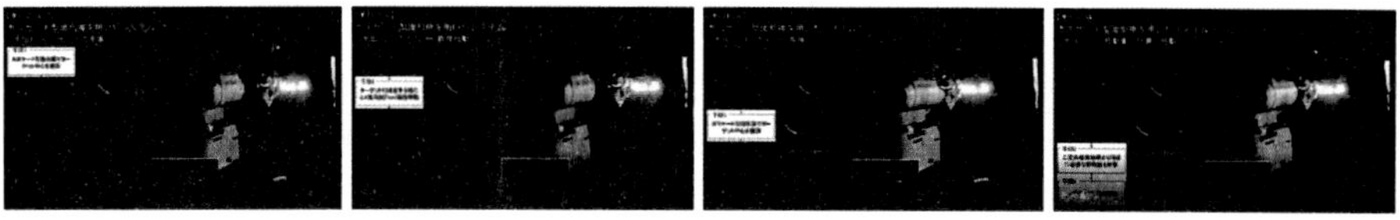

(a) Approaching process (cascade classifier)

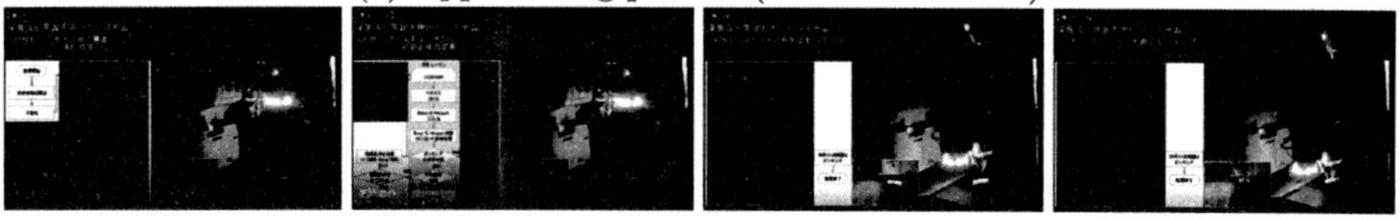

(b) Picking process (DRL:deep reinforcement learning)

Fig. 8. Sequential motions of the real robot using the approaching process and the picking process.

The results of all tests are shown in Table 1. The number of successful trials is 30 among the 30 trials except an experimental condition (50 mm, 27.5°). The failed objects in the experiments are also failed in the simulation at the same reason. Therefore, the simulator can express the real motion.

Table 1. The number of successful trials of the collection picking motion among 30 trials.

Length [mm]	Angle of test object set [degree]			
	22.5	67.5	112.5	157.5
30	30	30	30	30
40	30	30	30	30
50	28	30	30	30

It is clear that the collection picking motion is realized using the cascade process and the DRL process by the real robot.

6 Conclusion

This paper proposes the method that consists of the approaching process using the cascade classifier based on the OpenCV and the picking process using the deep reinforcement learning based on the TensorFlow to solve the collection picking problem, and describes the realization by the three-dimensional physics simulations built into PyBullet and by the experiments using real robot. First, the new method using two images in the approaching process is proposed, and the effectiveness is investigated by the simulations. Second, the reward matrix for the picking process is proposed, and we confirm that the picking motion is achieved. Final, the experiments using the method that consists of the approaching process and the picking process are executed. As a result, almost trials that is different from the arrangement of initial position are successful. It is clear that the collection picking motion is realized using the cascade classifier and the DRL by the real robot.

In the future works, the robot arm will be equipped with automobile robot. And the proposed method is adapted to the practical collection picking problem.

References

1. Tang, Y., et al.: Recognition and localization methods for vision-based fruit picking robots: a review. Front. Plant Sci. **510**, 1–17 (2020)
2. Zhao, Z.Q., Zheng, P., Xu, S.T., Wu, X.: Object detection with deep learning: a review. IEEE Trans. Neural Netw. Learn. Syst. **30**(11), 3212–3232 (2019)
3. Zhang, H., Berg, A.C., Maire, M., Malik, J.: SVM-KNN: discriminative nearest neighbor classification for visual category recognition. In: 2006 IEEE Computer Society Conference on Computer Vision and Pattern Recognition (CVPR 2006), New York, USA (2006)
4. Arulkumaran, K., Deisenroth, M.P., Brundage, M., Bharath, A.A.: Deep reinforcement learning: a brief survey. IEEE Sig. Process. Mag. **34**(6), 26–38 (2017)
5. Caldera, S., Rassau, A., Chai, D.: Review of deep learning methods in robotic grasp detection. Multimodal Technol. Interact **2**(57), 1–22 (2018)
6. Hatori, J., et al.: Interactively picking real-world objects with unconstrained spoken language instructions. In: Proceedings of International Conference on Robotics and Automation (ICRA 2018) (2018)
7. Correll, N., et al.: Analysis and observations from the first Amazon picking challenge. IEEE Trans. Autom. Sci. Eng. **15**(1), 172–188 (2018)
8. Bai, J., Lian, S., Liu, Z., Wang, K., Liu, D.: Deep Learning based robot for automatically picking up garbage on the grass. IEEE Trans. Consum. Electron. **64**(3), 382–389 (2018)
9. Redmon, J., Divvala, S., Girshick, R., Farhadi, A.: You only look once: unified, real-time object detection. In: 2016 IEEE Conference on Computer Vision and Pattern Recognition (CVPR), pp. 779–788 (2016)
10. Malisiewicz, T., Gupta, A., Efros, A.A.: Ensemble of exemplar-SVMs for object detection and beyond. In: 2011 International Conference on Computer Vision, pp. 89–96 (2011)
11. Lienhart, R., Kuranov, A., Pisarevsky, V.: Empirical analysis of detection cascades of boosted classifiers for rapid object detection. In: Pattern Recognition. DAGM 2003, vol. 2781 (2003)
12. Liu, L., et al.: Deep learning for generic object detection: a survey. Int. J. Comput. Vis. **128**(2), 261–318 (2019). https://doi.org/10.1007/s11263-019-01247-4

Unconventional Method of Robotic Machining

Marie Stará(✉), Ondřej Matúšek, and Tomáš Jarkovský

Technical University of Liberec, 461 17 Liberec, Czech Republic
marie.stara@tul.cz

Abstract. Unconventional production methods represent alternative ways of solving a wide range of manufacturing processes. The present paper deals with one such way in the form of the use of an angular industrial robot in the machining process. Firstly, a specific technical solution of the machining workplace is presented with emphasis on the end effector. Subsequently, a practical example of milling a cooler diffuser mould for a student formula-style racing car is discussed in detail and the relevant issues are demonstrated. The optimal setting of the production process, including the relevant production parameters, is documented and the entire production process is discussed in detail. The conclusion then summarizes the benefits of the applied technology and the knowledge obtained.

Keywords: Robotic milling · Robotic machining · SprutCAM · Milling effector

1 Introduction

The technology of machining, in its broadest sense, is used in a wide range of manufacturing processes, from conventional chip machining [1], laser [2] or waterjet machining [3] to welding. Chip machining, which is at the core of the present article, is one of the most common technological processes in which the desired shape of the workpiece is created by removing material in the form of chips. Machining centres are mostly used for machining complex shapes [4], while the degree of complexity of the equipment is given mainly by the number of axes. Universal 5-axis machining centres are most commonly used for this purpose, as they provide a wide range of shapes and sizes of the machined parts based on the working space of the machine [5]. In addition to the three basic linear movement axes that make up a Cartesian structure, the machines are also equipped with two configurable rotary axes. Due to the specific solution of its structure, the technology provides high-quality machined surfaces with extreme accuracy. Thanks to state-of-the-art control systems, which not only control the movement of the axes relative to each other, but also offer a wide range of options to compensate for machine inaccuracies and tool dimensions, rotary axis variations, and interpolation of different machining strategies, etc. the highest accuracy and optimum surface quality are achieved even for high-volume chip machining.

Modern and advanced ways of achieving complex shapes are mainly based on 3D printing technologies [6]. These technologies are used to achieve complicated shapes and combine several types of materials [7]. The accuracy of production is also at a

© The Author(s), under exclusive license to Springer Nature Switzerland AG 2022
J. Beran et al. (Eds.): TMM 2020, MMS 85, pp. 229–238, 2022.
https://doi.org/10.1007/978-3-030-83594-1_24

very high level. However, as in the case of machining centres, due to their structure, classic 3D printers are considerably limited by the production space in relation to the installation space of the machine. Another common negative aspect of both technologies is the relatively high purchase price of these precision production machines.

An important step in expanding the production space, both in general machining and 3D printing, is the application of a production tool to the series structure of a rigid arm, most often represented by an industrial robot (IR) [8]. Nowadays, industrial robots with serial kinematics are used in a wide range of industries, which mainly include handling and assembling, but also several technological operations, including robotic machining [9]. Compared to the existing machining methods, robotic machining offers much more flexibility and, above all, a greater working range, especially when the robot is mounted on a track motion platform. Robotic machining is most often applied to the following specific manufacturing processes and materials:

Milling – classical 3D or 5D milling of materials (non-ferrous metals, wood, plastics, polystyrene).
Finishing – deburring of castings (aluminium alloys, cast iron).
Edging, polishing, grinding – cutting of complicated plastic or composite materials.
Machining – automobile models at a scale of 1:1.
Machining – large models made of wood or polystyrene.

Moreover, robots may also work as automatic tool changers and change tools during the machining. There are several aspects that favour the use of robots. The first is cost, 5-axis machining is more affordable than computer numerically controlled (CNC) machining. Other advantages include the machining of large parts, their considerable variability in the use of machining technologies (cutter, laser, water jet, plasma, …) and last but not least the off-line programming of robot arm paths [10]. According to the International Federation of Robotics (IFR), installations of industrial robots more than tripled within ten years (2010–2019) [11]. Furthermore, according to the available IFR data, 1.4% of IR was used in 2016 for unconventional production processes [12]. There is no and has never been any effort to draw a thick line between a 5-axis machining centre and a machining robot. Similar to machine tools, machining robots are able to perform multiple tasks through an end-effector. Machining robots must be controlled in a similar way [13], whereby the given productivity and toolpath accuracy depend on a range of factors.

The most common kinematic structure of a robot, which is mainly used in industrial machining processes, is an angular structure with six degrees of freedom. If the robot is mounted on a track motion platform, the structure has up to seven degrees of freedom. A key parameter that significantly affects machining accuracy is the stiffness of the robot, due to the limited static and dynamic stiffness of the rotary joints of the robot. As described in [14], the stiffness of a typical machining centre is up to 50 times higher than the static Cartesian stiffness of a robot, which varies depending on the kinematic arrangement of the robotic structure throughout the workspace [15]. The effect of robot stiffness on instant tool-workpiece interaction during milling with a KUKA robot was simulated in [16, 17]. There are several articles on stiffness and the selection of a suitable spindle for robotic milling of a wide range of different materials [18–20]. Especially in milling processes, the dynamics of the robot are important, as they affect the quality and accuracy of the machining [8]. The accuracy of machining robots is in the order of tenths of millimetres. In general, their application is mainly in low-power operations such as trimming, drilling, polishing, deburring or grinding etc. In the case of significantly large transformations, it is necessary to select machining speeds and depth of engagement in relation to the type of material to be machined.

2 Experiments

2.1 Laboratory Workplace

A robotic laboratory workplace was built for the experimental purposes of robotic milling (Fig. 3a). The central part of the workplace is a KUKA KR 90 angular robot, model R2700, which the department uses for various experimental activities. The robot is mounted on a solid steel base, which has modified the vertical position of the working space. The basic parameters of the robot are summarized in Table 1. Rotational movement of the cutting tool is ensured by a high frequency Jägger spindle, Chopper 2300-40 HSK S5 with taper change and pneumatic cone clamping. Based on the manufacturer's instructions, the spindle is in primarily designed for high-speed grinding, drilling, and engraving [21]. Its basic technical parameters are shown in Table 2. The spindle was attached to the end flange of the robot, which is part of the sixth axis, by means of a special intermediate flange. The intermediate flange was manufactured for this purpose using a special tab and slot technology [22], which resulted in a considerably solid assembly with relatively low weight.

Table 1. Summary of basic parameters of the KUKA KR90, R2700 robot.

Parameter	Value
Load bearing capacity	90 kg
Working area	2696 mm
Number of axes	6
Repeated positioning accuracy	±0.06 mm
Control system	KR C4

The milling process takes place in the operating space above the workbench, which forms another crucial part of the workplace. The centre of the workbench is equipped with a frame structure to which a clamping plate with T-slots is attached for clamping the workpiece through the fixing elements. The cooling fluid is drained by means of a plastic drip pan. The whole of the kinematic structure of the robot was covered, due to the risk of the joints clogging by the fine dust generated during the milling.

Table 2. Summary of the parameters of the Jägger, Chopper 2300-40 HSK S5 spindle.

Parameter	Value
Rated power	3.5 kW
Rated speed	40000 rpm
Acceleration/braking value	10000 rpm
Clamping range up to	10 mm
Tool cooling	Cooling fluid

2.2 Machined Blank

The machined blank is represented by artificial wood. Specifically, it is Ebaboard 0700, which was selected due to its several positive properties in relation to the applied machining technology. The material has very good machinability, a fine texture, a very dense surface, high strength parameters, good edge stability (i.e. no chipping) and low dust levels during processing [23]. The initial blank was supplied in the form of glued sheet sections. For this purpose, a 285 × 280 × 50 mm section was used taking into account the machining allowance. The final workpiece is a mould for manufacturing a carbon composite radiator diffuser for a formula-style racing car of the student team SF TUL Racing, see Fig. 1.

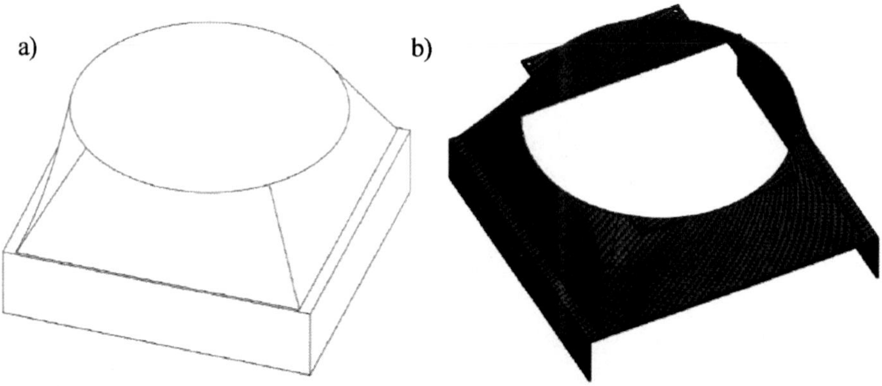

Fig. 1. a) 3D model of the machined mould, b) 3D model of the resulting carbon composite part.

2.3 Programming and Calibration of the Machining Process

The initial part of the experiments focused on the creation of the workplace. Subsequent parts of the experiment consisted of the preparation of machining program and calibration of the workplace.

Computer-aided manufacturing (CAM) software was used to program the machining centres. Based on the knowledge of the specific machine and machining technology, this software is able to design toolpaths or other aspects of production, ideally so that production is as efficient as possible, whereby using the least amount of energy and material input and producing the least amount of industrial waste at the highest possible rate of production.

SprutCAM 11 software, which has a translator for the language of the KUKA robot, was used in this work. It is necessary to create a kinematic diagram of the used KUKA KR90 R2700 robot with a Jägger Chopper 2300-40 HSK S5 spindle in the software environment (Fig. 2). Subsequently, it is necessary to import the desired final model and other peripherals (clamping table, tools) into the environment. The individual technological operations of the machining process are set after selecting the blank.

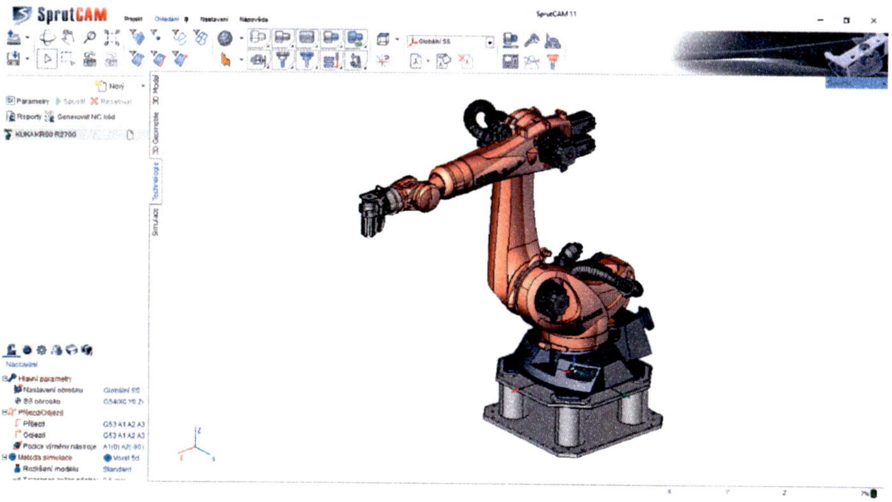

Fig. 2. Software environment with appropriate robot and spindle implemented.

Precise robotic machining is conditioned by the adequate alignment of the workpiece and definition of the tool centre point (TCP). The position of the workpiece in the real space was primarily chosen with regard to the elimination of robot singularities during the machining process, while taking into account the variable stiffness of the robot depending on the position of the robot [24, 25].

The TCP coordinates may be read directly from the CAD model of the robot's end effector assembly. However, this procedure does not include mounting inaccuracies; therefore, a manual tip-on-tip calibration has to be performed, which is accomplished by repeatedly moving the tip onto the tip from four different directions. In order to attain the maximum level of accuracy, a third collet chuck was used. The tip was clamped into the chuck so that it reaches the same length as the milling cutters used.

After successfully defining the TCP, it was possible to create and define the coordinate system of the workpiece. The measured transformation relations were transferred to the CAM software, whereby creating a unified virtual and real coordinate system.

3 Results and Discussion

Two different end mills, which are generally recommended for shape machining, were used for the milling of the mould. The tools were attached to the high-frequency spindle using a collet chuck. Each milling cutter had its own collet chuck corresponding to the diameter of the cutter, as the workplace does not allow for automatic tool changing. The roughing operations were performed with a 12 mm-diameter cylindrical milling cutter, which has a relatively large depth and width of engagement. An end mill will an 8 mm-diameter ball head was used for the finishing (Fig. 3b). The selected milling cutters are primarily intended for thermoplastics, cast and wrought aluminium, unalloyed copper, which can be considered a group relatively close to artificial wood. The optimal cutting conditions recommended by the tool manufacturer for the machined material and the selected milling cutters are summarized in Table 3 [26]. For safety reasons, the applied cutting conditions for both processes were chosen at approximately 60% of the recommended level.

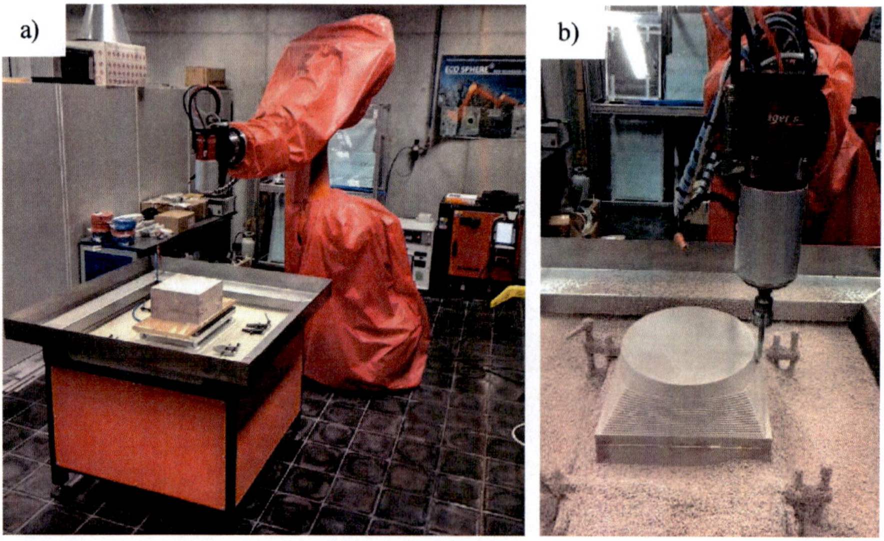

Fig. 3. a) Laboratory workplace, b) close-up of the milling cutter in the spindle during the finishing operation.

Table 3. Optimum cutting conditions during the milling process.

Parameter	End mill	Milling cutter with ball head
Feed per tooth f_z	0.160 mm	0.065 mm
Cut depth a_p	18 mm	0.22 mm
Cut width a_e	3 mm	0.22 mm
Spindle speed n	26 525 rpm	39 790 rpm
Feed speed v_f	12 730 mm/min	5 175 mm/min

Due to the fact that the laboratory workplace does not allow for automatic tool changing, the machining strategy was divided into two parts - roughing and finishing. The roughing part consisted of two operations, i.e. "face alignment" and "roughing in Z-cuts". The technological "face alignment" operation was used to align the surface of the blank to the defined surface of the workpiece. The subsequent "roughing in Z-cuts" operation removed excess blank material from around the resulting workpiece. In the roughing part, it is important to remove the maximum amount of material in the shortest possible time. In our case, however, the cutting conditions were limited due to safety in the test mode of the workplace. The roughing was followed by finishing, during which the final shape of the workpiece was finished. Finishing of the common surfaces (i.e. non-flat surfaces) was performed using the "5-axis milling" system. During this operation, the spindle speed was faster compared to the roughing, but the depth and width of the cutting (about 2% of the tool) were smaller. Comparing the overall time necessary for both processes, it was 40% roughing and 60% finishing.

The final machined mould was measured using manual gauges taking into account the possibilities of the workplace. The measured accuracy was in the order of several hundredths, with a maximum of 0.1 mm. After the final finishing, the surface of the mould was characterized by slight traces of machining, but it had a satisfactory roughness of approximately Ra 3.2. On the contrary to expectations, it was not necessary to finish the surface with an emery cloth, etc. Subsequently, the machined mould was successfully used for the production of the required part.

The result of the robotic machining was a mould (Fig. 4a) for a radiator diffuser for an open-wheel racing car of the student team from the Technical University of Liberec. The part serves to direct the air flow to the radiator. To optimize the weight, the product was designed from a carbon composite and was made by manual lamination (Fig. 4b). Before the production, it was necessary to separate the mould, i.e. coat the surface of the mould with a release agent. The individual layers of the composite were applied to the machined mould and subsequently saturated with resin. Finally, the whole system was sealed into a bag and vacuumed. After curing of the resin, a solid composite part was formed, which could be separated from the mould (Fig. 4c). In the last step, the part was trimmed and attached to the racing car.

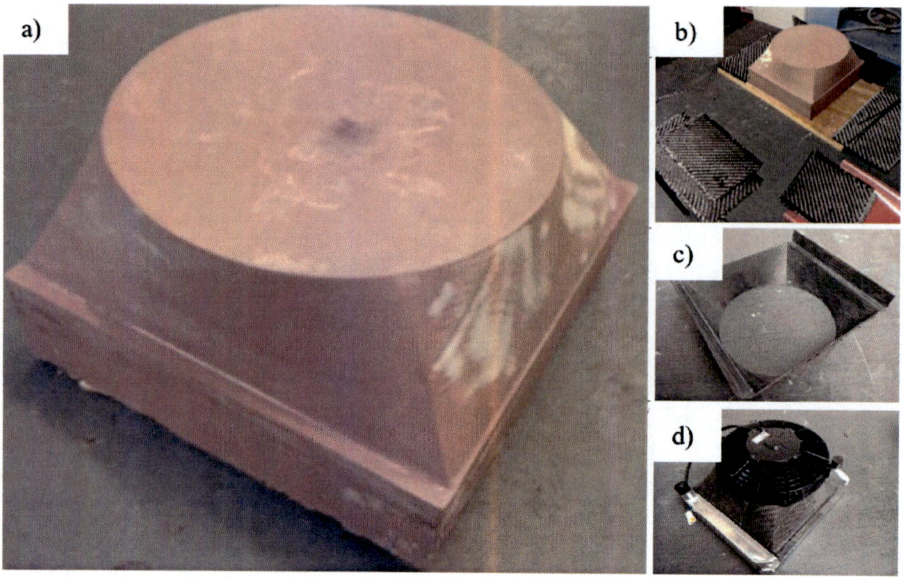

Fig. 4. a) Machined part – mould, b) preparation for lamination, c) composite part, d) radiator diffuser assembly.

4 Conclusion

This paper presents the issue of robotic machining using a practical example of the production of a mould made from artificial wood. Although robotic machining is generally suitable for products with large dimensions, the applicability of the relevant technology was also confirmed for single-piece production of relatively small components. The robot with an angular structure showed complete suitability for the given purposes at the appropriately selected cutting and production parameters. The shape deviations were achieved at the maximum level of 0.1 mm and the surface roughness ranged around Ra 3.2. These parameters were completely sufficient for demanded application, however, better results might be achieved using CNC machining. Robotic machining certainly cannot be considered as a universal replacement for large machining centres, as they cannot compete with them in terms of higher absolute achievable accuracy and stiffness. However, a robot can represent a multi-axis machining centre in terms of machining large, complex mechanical parts or from an economical point of view. Its application is potentially possible for both series and single-piece production.

Acknowledgment. This work was supported by the Student Grant Competition of the Technical University of Liberec under the project No. SGS-2019-5016.

References

1. Davin, J.P. (ed.): Machining: Fundamentals and Recent Advances. Springer, London (2008). https://doi.org/10.1007/978-1-84800-213-5

2. Marimuthu, S., Dunleavey, J., Smith, B.: Laser based machining of aluminum metal matrix composites. Procedia CIRP **85**, 243–248 (2019)
3. Selvan, C., Midhunchakkaravarthy, D., Pillai, S.R., Madara, S.R.: Investigation on abrasive waterjet machining conditions of mild steel using artificial neural network. Mater. Today **19**, 233–239 (2019)
4. Usop, Z., Sarhan, A.A.D., Mardi, N.A., Wahab, M.: Measuring of positioning, circularity and static errors of a CNC vertical machining centre for validating the machining accuracy. Measurement **61**, 39–50 (2015)
5. Lasemi, A., Xue, D., Gu, P.: Recent development in CNC machining of freeform surfaces: a state of the art review. Comput. Aided Des. **42**(7), 641–654 (2010)
6. Ackermann, M., Šafka, J., Lachman, M., Keller, P.: Properties of models produces by direct selective laser melting technology. Appl. Mech. Mater. **693**, 231–236 (2014)
7. Roach, D.J., Hamel, C.M., Dunn, C.K., Jonhson, M.V., Kuang, X., Qi, H.J.: The m^4 3D printer: a multi-material multi-method additive manufacturing platform for future 3D printed structures. Addit. Manuf. **29**, 100819 (2019)
8. Verl, A., Valente, A., Melkote, S., Brecher, C., Ozturk, E., Tunc, L.T.: Robots in machining. Manufacturing technology. CIRP Ann. **68**(2), 799–822 (2019)
9. Huynh, H.N., Assadi, H., Rivière-Lorphèvre, E., Verlinden, O., Ahmadi, K.: Modelling the dynamics of industrial robots for milling operations. Robot. Comput.-Integr. Manuf. **61**, 1–16 (2020)
10. Komák, M., Králik, M., Jerz, V.: The generation of robot effector trajectory avoiding obstacles. MM Sci. J. **2018**(02), 2367–2372 (2018)
11. IFR Press Releases. https://ifr.org/ifr-press-releases/news/top-5-robot-trends-2021. Accessed 25 Mar 2021
12. IFR Press Conference Presentation. https://ifr.org/downloads/press2018/IFR%20World%20Robotics%20Presentation%20-%2018%20Sept%202019.pdf. Accessed 15 Jan 2021
13. Altintas, Y.: Manufacturing automation: metal cutting mechanics, machine tool vibrations, and CNC design. Appl. Mech. Rev. **54**(5), 391–481 (2001)
14. Pan, Z., Zhang, H., Zhu, Z., Wang, J.: Chatter analysis of robotic machining process. J. Mater. Process. Technol. **173**(3), 301–309 (2006)
15. Abele, E., Weigold, M., Rothenbücher, S.: Modeling and identification of industrial. Robot for machining applications. CIRP Ann. **56**(1), 387–390 (2007)
16. Bondarenko, D., Pashkevich, A., Briot, S., Ritou, M., Furet, B.: Elasto-dynamic model of robotic milling process considering interaction between tool and workpiece. In: 11th Biennial Conference on Engineering Systems Design and Analysis, Nantes, France, pp. 217–226 (2012)
17. Cen, L., Melkote, S.N.: Effect of robot dynamics on the machining forces in robotic milling. Procedia Manuf. **10**, 486–496 (2017)
18. Cordes, M., Hintze, W., Altintas, Y.: Chatter stability in robotic milling. Robot. Comput.-Integr. Manuf. **55**(A), 11–18 (2019)
19. Cen, L., Melkote, S.N., Castle, J., Appelman, H.: A Method for Mode Coupling Chatter Detection and Suppression in Robotic Milling. J. Manuf. Sci. Eng. **140**(8), 081015 (2018)
20. Mousavi, S., Gagnol, V., Bouzgarrou, B.C., Ray, P.: Dynamic modeling and stability prediction in robotic machining. Int. J. Adv. Manuf. Technol. **88**(9–12), 3053–3065 (2016). https://doi.org/10.1007/s00170-016-8938-0
21. Alfred Jäger. https://www.alfredjaeger.de/en/chopper/11503003. Accessed 15 Mar 2021
22. Starý, M., Novotný, F., Horák, M., Stará, M., Vít, Z.: Experimental optimization of tab and slot plug welding method suitable for unique lightweight frame structures. J. Manuf. Process. **31**, 453–467 (2018)
23. Ebalta web catalogue. http://www.ebalta.cz/cz/katalog-produktu/ebaboard-umele-drevo/ebaboard-0700. Accessed 19 Feb 2021

24. Chen, S.F., Kao, I.: Conservative congruence transformation for joint and cartesian stiffness matrices of robotic hands and fingers. Int. J. Robot. Res. **19**(9), 835–847 (2000)
25. Dumas, C., Caro, S., Garnier, S., Furet, B.: Joint stiffness identification of six-revolute industrial serial robots. Robot. Comput.-Integr. Manuf. **27**(4), 881–888 (2011)
26. Fraisa. https://www.fraisa.com/en/online-tools. Accessed 29 Mar 2021

Development of a Wearable Assistive Robot for Baggage-Loading Operations

Hidetsugu Terada[1] (✉), Koji Makino[1], Kazuyoshi Matsumura[1], and Masato Karaki[2]

[1] University of Yamanashi, Kofu, Yamanashi, Japan
terada@yamanashi.ac.jp
[2] TOYO KOKEN K.K., Minami-alps, Yamanashi, Japan

Abstract. In this work, a wearable assistive robot is developed to assist baggage handlers load heavy luggage onto a cargo container. Therefore, the baggage handling procedure was first analyzed using an image processing method. An optimized assignment of pneumatic artificial muscle actuators and motion timings was then proposed. The designed robot is driven by two pneumatic artificial muscles to assist the leaning back and rotational motion of the upper body. Additionally, an assistive mechanism for the hip and knee joints assists joint extension using a flat spiral spring. Especially, the assistive mechanism of a knee joint is adapted for the roll-back motion. Preliminary result of the proposed wearable robot indicates that is reduced about 33% in the mean muscle activity during leaning backward, rotation of the spine, and flexion of the hip joint when handling a 10kgf baggage. Furthermore, it is confirmed that the mean muscle activity of the knee joint is reduced about 15% during lifting-operation, too. Thus, the proposed assistive robot should be further investigated for baggage-loading operations.

Keywords: Assistive robot · Wearable · Lumbar-twisting motion · Leaning back motion · Baggage loading · Pneumatic artificial muscle

1 Introduction

At most airports, baggage is loaded onto a cargo container or bay by hand, regardless of its weight. Power assistive robots would help reduce the load of the baggage handler. Various power assistive robots have been developed in Japan, generally using electrical or hydraulic actuators [1–3]. However, these devices are heavy and unsuitable for usage over a long duration. Furthermore, as most developed devices have assisted only the lifting motion of the upper body, not the lumbar-twisting motion, they are only used under limited operations. They are thus insufficient to assist in the majority of baggage-loading operations.

This work therefore aims to analyze to the motion range required during heavy baggage loading and then design a power assistive robot to address this motion. To do so, a motion analysis of the upper body with a twisting motion and hip and knee joints during heavy baggage loading was analyzed. This is because that the conventional analysis of human joint torque at lifting motion has been considered without the twisting motion

© The Author(s), under exclusive license to Springer Nature Switzerland AG 2022
J. Beran et al. (Eds.): TMM 2020, MMS 85, pp. 239–248, 2022.
https://doi.org/10.1007/978-3-030-83594-1_25

[4]. Then, an assistive robot was designed for usage in any environmental conditions. A prototype of the designed robot is then tested to characterize its impact on the required muscle activity on the back, hip, and knee joints during baggage-loading operations.

2 Motion Analysis of Baggage-Loading Operation

Cargo containers must have a near-uniform weight distribution to avoid disrupting the center of gravity of the airplane. Baggage handlers must therefore quickly assess the widely varying shape, size, and gravity center of each baggage item to determine where it should be placed. As such, loading baggage on a cargo container is the most difficult of the baggage-loading operations to automate. The development of a wearable assistive robot thus presents a more realistic alternative for reducing the load put on baggage handlers during this operation. The typical loading operation of baggage items weighing 10kgf to a cargo container can be classified into five steps:

1. Grabbing the baggage, involving hip joint and spinal flexion,
2. Lifting the baggage using both hands, involving the extension of the hip joint,
3. Moving forward and to the right, involving a side step and a lumbar-twisting motion to the right,
4. Placing the baggage on the container, involving leaning forward and then returning the upper body to the front, and then
5. Returning to the initial posture by leaning the upper body back.

Each step involves complex motions, including flexion or extension of the hip and knee and leaning the upper body forward or backward, all at varying durations and angles. In step 3, the lumbar-twisting motion necessary to stack heavy baggage on the cargo container generates a reaction force. To quantify this motion, LED markers were attached on each joint of a baggage handler. The motion of the baggage handler was then recorded by a multi-camera system to analyze each joint's motion and the rotation of the upper body using 3D-image analysis software [5] at three subjects (A, B, C). Three subjects are healthy, 23 year-old Japanese male who were 1.7 $\pm$ 0.05 m tall.

The varying duration, timing, and angle of each motion throughout the process at subject A is shown in Fig. 1. The motion of the hip joints did not synchronize with the bending of the upper body in step 2, and the flexion angle of the knee joint was considerably smaller than that of the hip joint. This is similar to the results for the other two subjects (subject B and C). Conventional power assistive robots have employed a leaning back mechanism around the hip joint. However, robots designed for baggage loading should instead allow for the assistive mechanism of the hip joint to operate independently of the assistive mechanism of the upper body.

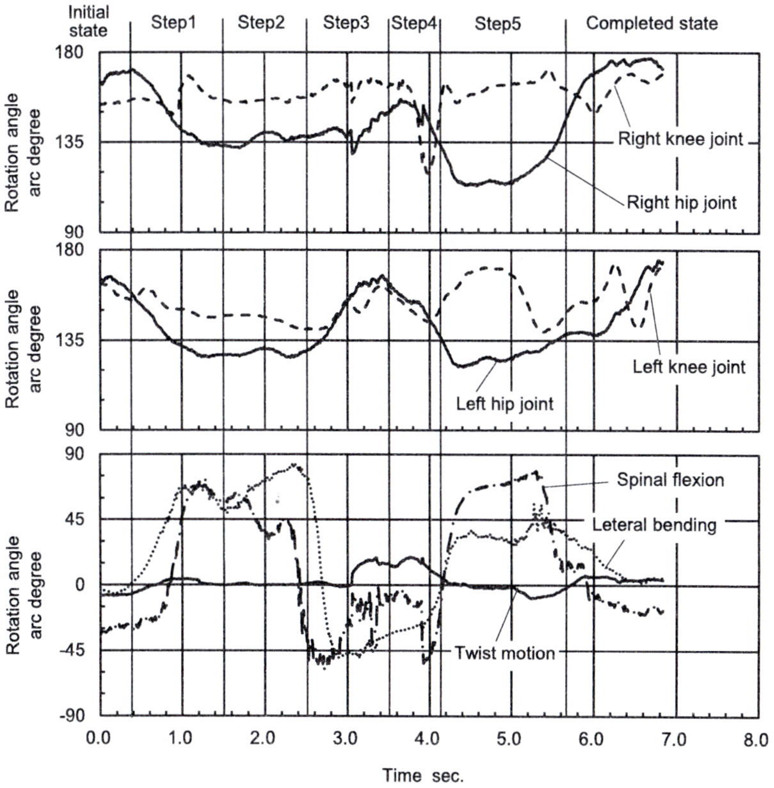

Fig. 1. Evolution of the motion of a baggage handler's spine, hip, and knee during the loading of a 10 kgf-baggage onto a cargo container at subject A

3 Design of a Wearable Assistive Robot

A wearable assistive robot was thus designed to better suit these needs. As the timing of the hip and knee flexion differed in each step and had large individual differences, the assistive mechanism for these joints were designed to be independent from the mechanism to assist the upper body. The developed robot weighing 5.85 kgf, shown in Fig. 2 [6], consisted of two sets of hip-joint assistive mechanisms and knee-joint assistive mechanisms, and an assistive mechanism for leaning and rotating the upper body. Then, considering the previous experimental analysis of a hip and knee joints motion, the assistive mechanisms for hip and knee joints were driven by flat spiral springs and were based on the structure of an assistive lifting instrument [7, 8].

The designed assistive mechanism for the hip joint contains a speed-increasing gear train (gear ratio of 1:6.25) and a flat spiral spring, as shown in Fig. 3. This mechanism accumulates the force when the hip joint is flexed, and drives the thigh when the hip joint is extended.

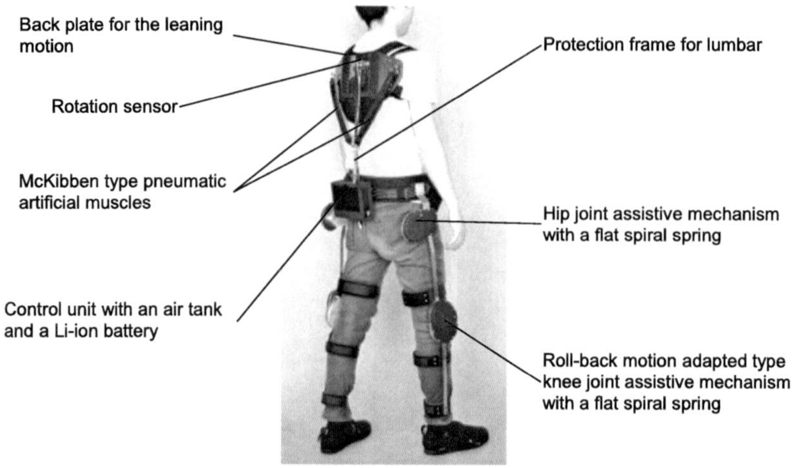

Fig. 2. Proposed wearable assistive robot providing upper-body rotation assistance for the loading

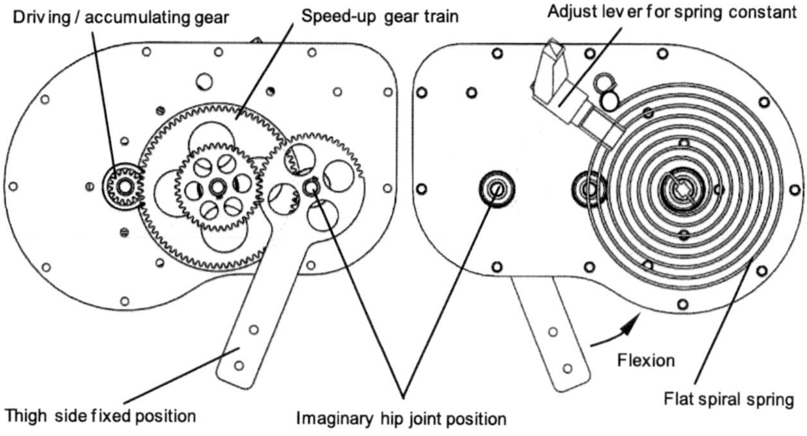

Fig. 3. (a) Inside and (b) outside view of the proposed assistive mechanism for the hip joint of a baggage handler of baggage onto a cargo container

The rigidity of the hip joint was also modified, as the motion of the hip joint differs between states. The proposed mechanism has an adjustable lever to modify the spring constant between a low- and high-stiffness modes. As shown in Fig. 4, the torque assistance provided in the high-stiffness mode is 24% to 29% greater than that in the low-rigidity mode at angles of 80°–110°, which are commonly encountered during the studied baggage-loading operation.

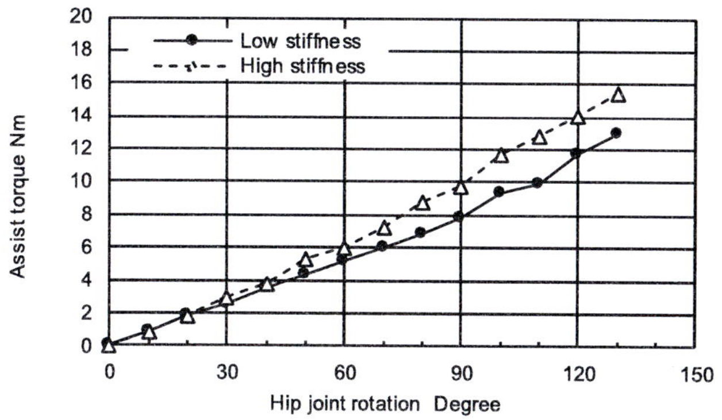

Fig. 4. Relationship between the rotation of the hip joint and the assistive torque provided by the proposed mechanism

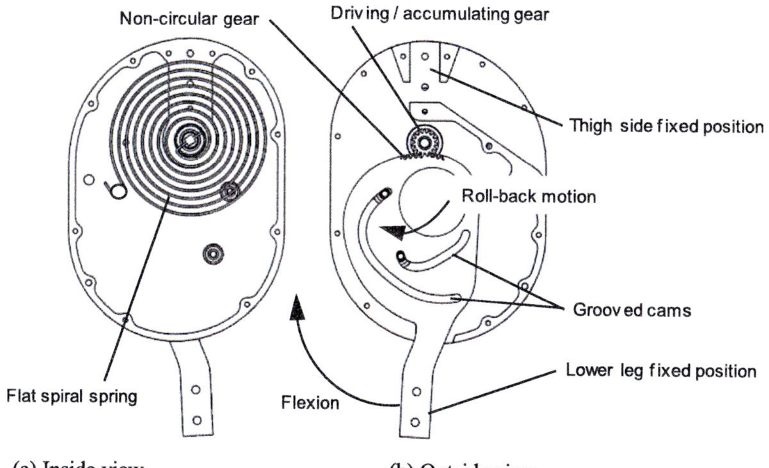

Fig. 5. (a) Inside and (b) outside view of the proposed assistive mechanism for the knee joint of a baggage handler

The proposed knee assistive mechanism uses two grooved cams and a non-circular gear to adapt to the knee flexion behavior [9], as shown in Fig. 5. Considering the physique of a mean Japanese person, the fundamental structure is the same as the conventional mechanism. The gear ratio was changed to 1:8 to assist in maintaining a knee flexion between 80° and 100°. However, in the knee-joint assistive mechanism, there is no proportional relation between them due to the shape of the non-circular gears. Additionally, there is a non-linear variation with a large jump in assistive torque provided 80°, as shown in Fig. 6.

The designed upper-body assistive mechanism consisted of two McKibben-type pneumatic artificial muscles [10], a back plate for the rotation and leaning back motion,

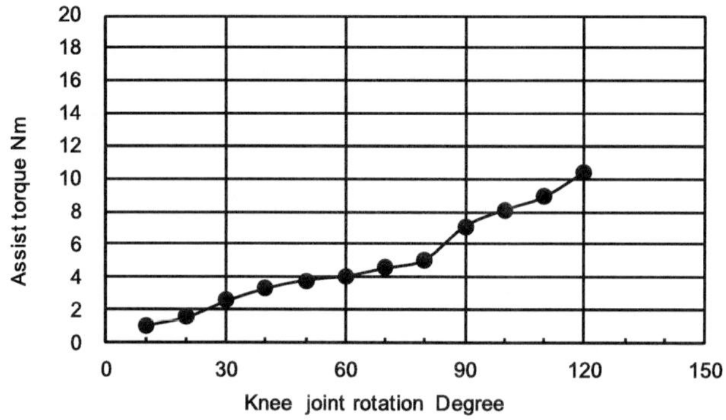

Fig. 6. Relationship between the rotation of the knee joint and the assistive torque provided by the proposed mechanism

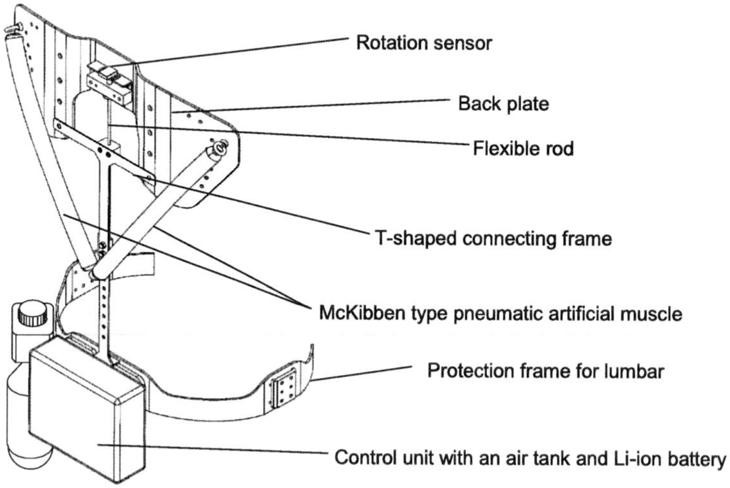

Fig. 7. Structure of the proposed upper-body assistive mechanism

a protection frame for lumbar support, a T-shaped connecting frame with an elasticity, and a rotation sensor around the spine, as shown in Fig. 7.

These artificial muscles are attached between the lumbar and each shoulder blade. When leaning the upper body back, both muscles contract simultaneously, whereas rotating requires each muscle to contract independently. The rotation sensor, attached between the back plate and the protection frame with a flexible rod, is used to detect the direction of rotation and the timing of a motion start. The motion of an upper body for baggage-loading operation is typical, and it is easy to classify the motion pattern using rotation sensor.

The upper-body assistive mechanism allowing assisted leaning back uses a control system, detailed in Fig. 8. Furthermore, the contact sensors are located on the palm and

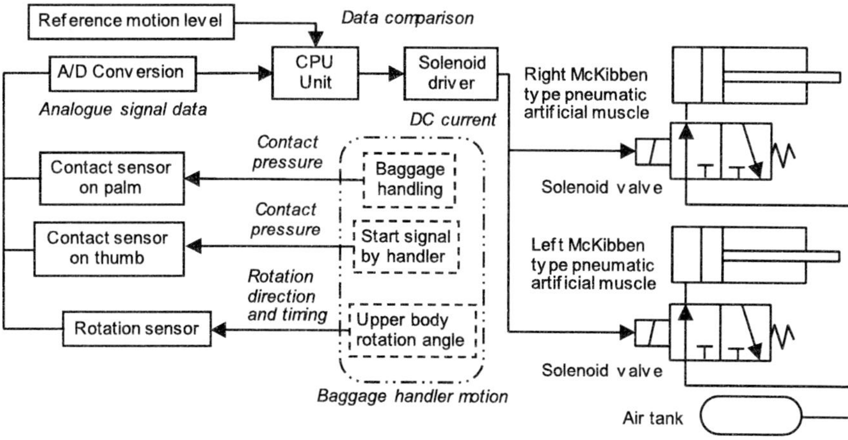

Fig. 8. Block diagram of an upper-body motion control system

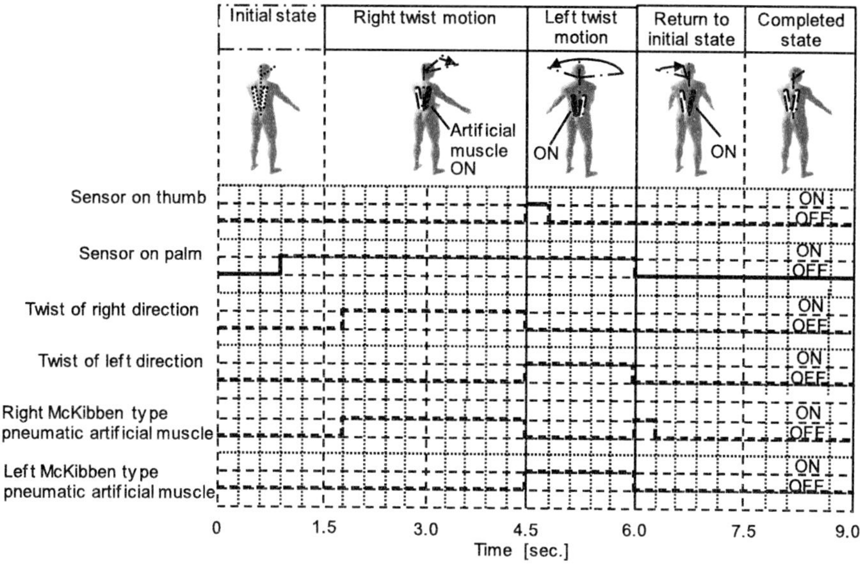

Fig. 9. Timing of the rotational motion of the baggage-loading operations

thumb of a handler's glove to prevent malfunction, and the system does not drive unless a baggage handler inputs a signal of the motion start.

A timing of the rotational motion of a baggage handler's upper body during the baggage loading onto a cargo container is shown in Fig. 9. During this motion, the rotation that generates the reaction force is considered. A contact sensor on the baggage handler's thumb controls the returning motion of upper-body to the front. To prevent any excessive counter-rotational motion, the upper body rotates to only for a short time. Both artificial muscles are driven simultaneously when leaning forward and backward.

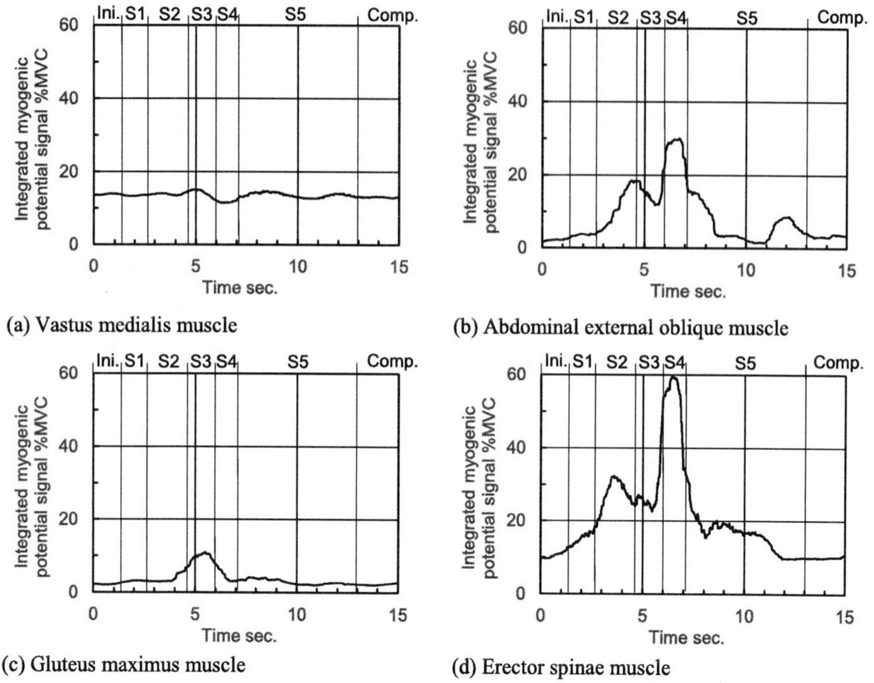

(a) Vastus medialis muscle (b) Abdominal external oblique muscle (c) Gluteus maximus muscle (d) Erector spinae muscle

Fig. 10. Muscle activity of a baggage handler equipped with the designed assistive robot during baggage loading (S1: Step 1, S2: Step 2, S3: Step 3, S4: Step 4, S5: Step 5, Ini.: Initial state, Comp.: Completed state)

Assistance leaning the upper body backward is operated using the contact sensor located on the handler's thumb; the baggage handler starts the operation by contracting the right and left artificial muscles simultaneously by the contact sensor signal of the thumb. When the contact sensor on the thumb detects again, these artificial muscles are released, ending the assistive force leaning backward.

4 Evaluation of the Prototype

The designed assistive robot for baggage-loading operations was then prototyped and tested. The muscle activity and power-assist ratio was evaluated using the integrated myogenic potential, which is the normalized maximum voluntary contraction (%MVC). The prototype was attached to the subject, a healthy, 23 year-old Japanese male who was 1.7 m tall. The handled baggage was 10 kgf (98 N) in weight, 0.53 m high, 0.38 m wide, and 0.2 m long, representative of the medium weight and size of baggage on a domestic flight in Japan. Considering the cargo container size for air-craft, the bag was then placed at a height of 1.10 m and a depth of 1.15 m and these positions show the center of gravity of the bag. The delivered assistive power to the hip and knee joints at 90° was 9.8 and 7.1 Nm, respectively; the delivered assistive power to the upper body to lean back was 18.9 N at 0.3 MPa of air pressure. Overall, the integrated myogenic potential

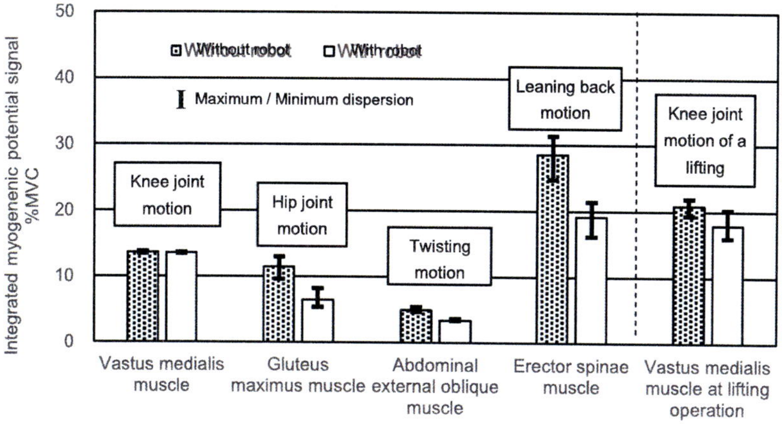

Fig. 11. Mean muscle activity of a baggage handler loading baggage onto a cargo container

Table 1. Reduction ratio of the mean muscle activity

Main acting muscle	Motion	Reduction ratio %
Vastus medialis muscle	Knee joint motion	1.5
Gluteus maximus muscle	Hip joint motion	44.3
Abdominal external oblique muscle	Twisting motion	32.0
Erector spinae muscle	Leaning back	32.6
Vastus medialis muscle at lifting operation	Lifting operation	13.9

was reduced in each of the studied pressure points, as shown in Fig. 10, where motion of the hip joint, motion of the knee joint, rotation of the upper body, and leaning back of the upper body were evaluated via the vastus medialis, the abdominal external oblique, the gluteus maximus, and the erector spi-nae, respectively. The largest reductions in integrated myogenic potential were seen in the leaning back and rotational motions of the upper body.

The mean myogenic potential of each studied point with and without the assistive robot is compared in Fig. 11. The mean muscle activity of three trials with small dispersion was reduced about 33% while leaning back, twisting, and the hip joint motion, as shown in Table 1. These preliminary result indicates that the designed assistive robot can sufficiently assist baggage-loading operations.

With the designed prototype, when the knee joint has a flexion of less than 20°, the mean muscle activity of knee joint is not reduced. However, considering the results of the previous research [7, 8], in case of the large knee flexion conditions, it is clear that the mean activity can be reduced. Therefore, the lifting operation under the condition that the initial height of the center of gravity of the baggage was changed to 100 mm from 250 mm was tested to inspect this prototype.

As the results, it was confirmed that mean muscle activity of knee joint motion at three trials was reduced about 14%. In other words, the prototype robot has a sufficient performance to assist the baggage loading operations. In addition, it is necessary to optimize the design parameters to consider the individual differences including the physique, so in the future, the method for determining design parameters should be investigated.

5 Conclusions

This work detailed the preliminary findings of a motion analysis of baggage-loading operations, which were then used to design a wearable assistive robot to reduce to force on the knees, hips, and back of baggage handlers. Preliminary result indicates that the mean myogenic potential of the hip and knee joints and erector spinae was reduced as a result of wearing the developed robot during baggage handling. Future work will aim to establish a method for determining design parameters to adapt for individual differences including the physique.

References

1. Muramatsu, Y., Tokoro, K., Kobayashi, H.: Development and assessment of muscle suit (Evaluation of the auxiliary operation using the surface EMG). Trans. JSME **83**, 847 (2017)
2. Sano, A., Suzuki, M.: Novel walking assist by non-powered passive walking, The Japan Society of Mechanical Engineers Symposium on Welfare Engineering 2012, OS1-1-7 (2012)
3. ATOUN Inc.: Powered Wear, ATOUN MODEL Y. http://atoun.co.jp/products/atoun-model-y
4. Panero, E., et al.: Influence of hinge positioning on human joint torque in industrial trunk exoskeleton. Adv. Mech. Mach. Sci. **73**, 133–142 (2019)
5. DITECT Corp: DIPP-Motion V3D. https://www.ditect.co.jp/en/software/dipp_motionv.html
6. Terada, H. et al.: Power assistive apparatus, Japanese patent publication, Tokugan 2019-095904 (2019)
7. Terada, H., et al.: Development of an assistive instrument for lifting motion using driving springs. J. Adv. Mech. Des. Syst. Manuf. **11**, 3 (2017)
8. Terada, H., et al.: Development of an assisting instrument of standing-up motion using driving springs for elderly persons. New Adv. Mech. Mech. Transm. Robot. **46**, 417–425 (2016)
9. Terada, H., et al.: Development of a wearable assist robot for walk rehabilitation after knee arthroplasty surgery. In: Beran, J., Bílek, M., Hejnova, M., Zabka, P. (eds.) Applications Advances in Mechanisms Design, Proceedings of TMM 2012, vol. 8, pp. 65–71. Springer, Dordrecht (2012). https://doi.org/10.1007/978-94-007-5125-5_9
10. Yokoyama, K., Kogiso, K.: Position control experiments of McKibben pneumatic artificial muscle using contraction ratio estimation. Trans. SICE **53**(11), 567–573 (2017)

Prototype and Testing of L-CaPaMan

Alexander Titov(✉) and Marco Ceccarelli

LARM2 Laboratory of Robot Mechatronics, University of Rome Tor Vergata, Rome, Italy
aleksandr.titov@students.uniroma2.eu,
marco.ceccarelli@uniroma2.eu

Abstract. A new version of CaPaMan design is elaborated with solutions for low-cost lightweight features. A new prototype is presented as result of design improvements by using market components and 3D printing manufacturing. The new prototype as L-CaPaMan (Light CaPaMan) is characterized with new components for a new slider solution and light-structure links. The prototype construction is discussed up to a testing layout for design validation and operation characterization. Results of testing are discussed to outline the operation performance of L-CaPaMan by using Arduino controller with basic sensors for motion and action monitoring.

Keywords: Robotics · Parallel manipulators · Prototypes · Testing · CaPaMan

1 Introduction

The first design of Cassino Parallel Manipulator (CaPaMan) was reported in 1997 [1, 2] as a new class of parallel manipulators [3, 4] with a limited number of d.o.f. This structure with its advantages, like high accuracy in positioning and stiffness with a fairly simple operation, that has been experience as an earthquake simulator [5], and in humanoid robot design [6], as examples of potential interests in many other applications as suggested for example in [7] as haptic devices, in [8] as vehicle simulators, and in [9] as telescope guiding structure.

The prototypes from 1997 [1, 2] were difficult and expensive in manufacturing because of metallic parts with large weight and, consequently, inertia, that require powerful actuators for operation. With a wide spreading of 3D printing [10], it became possible to strongly decrease weight and price of the mechanism's components, specifically for a low-cost student-oriented solution, reported in [11].

L-CaPaMan is an alternative low-cost solution of CaPaMan, keeping the original structure, using 3D-printed and available components from market.

2 CaPaMan Structures and L-CaPaMan Design

The first CaPaMan, as in Fig. 1a), is a 3-d.o.f. spatial parallel manipulator [1, 2]. The metal prototype has large weight, and it is operated by 3 industrial motors. The following

© The Author(s), under exclusive license to Springer Nature Switzerland AG 2022
J. Beran et al. (Eds.): TMM 2020, MMS 85, pp. 249–258, 2022.
https://doi.org/10.1007/978-3-030-83594-1_26

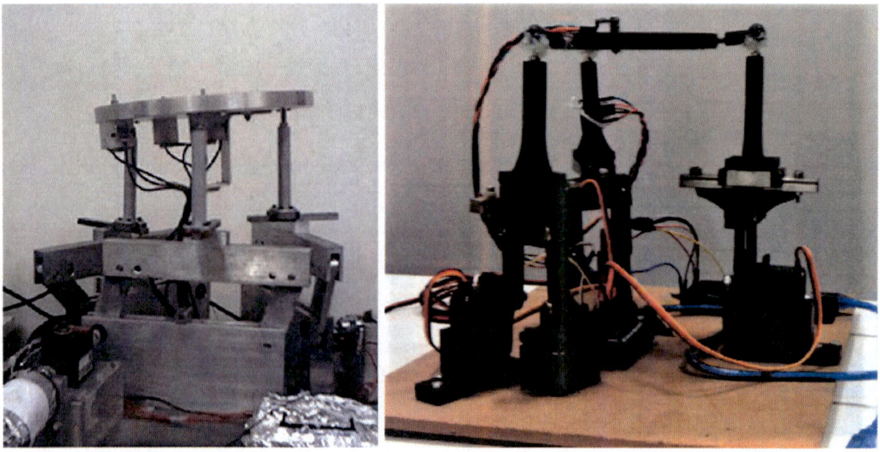

a) b)

Fig. 1. CaPaMan design structures: a) original design in 1997, [1, 2]; b) 3D printed prototype in 2017 [11]

improvements for student-oriented use, as in Fig. 1b), keep this mechanical structure in low-cost design.

The aim of a new model of CaPaMan is to reduce the costs of manufacturing and reduce weight of the mechanism. A prismatic joint in previous prototypes has a large area of contact, that can give negative effects, like high friction in sliding parts. In the new construction prismatic joint has been replaced with a wheel-based structure. In the previous work [12] a CAD simulation has been worked out and the performance has been analysed with model in CAD [13] in Fig. 2b), using data of CaPaMan prototype, as listed in Table 1.

Table 1. Design parameters of new L-CaPaMan version in Fig. 2

l1i (mm)	l_{2i} (mm)	l_{3i} (mm)	l_{4i} (mm)	l_{5i} (mm)	l_{16i} (mm)	l_{7i} (mm)
70	60	25	90	70	50	60

The calculated mass of the CAD model is equal to 0.664 kg, and overall dimensions of mechanism are defined for a movable platform equal to 200 × 200 × 210 mm.

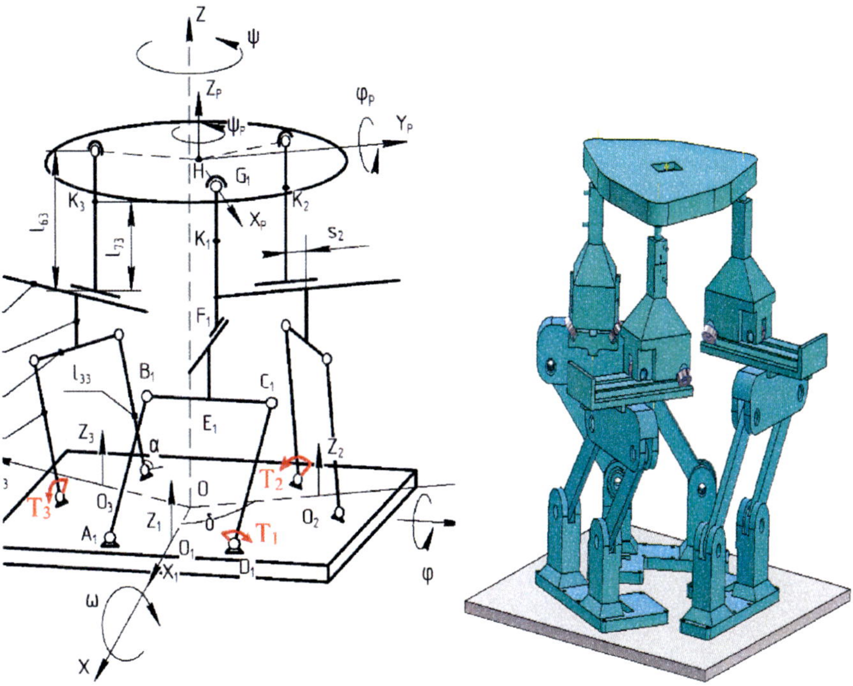

a) b)

Fig. 2. A design of L-CaPaMan version: a) kinematic diagram, b) A CAD model

3 Prototype Construction

The new L-CaPaMan prototype has been manufactured and assembled in the laboratory LARM2 of Tor Vergata university. The new design of a prismatic joint is shown in Fig. 3a) in the assembled prototype, and the wheel body with IMU set is shown in Fig. 3b). A block diagram of the design setup is in Fig. 4. Figure 5 shows the assembled prototype with connections to Arduino controller.

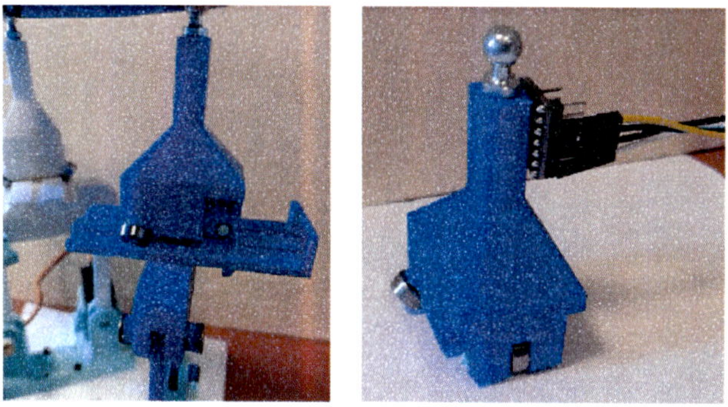

a) b)

Fig. 3. Mechanical design of the prismatic joint in L-CaPaMan prototype of Fig. 5: a) in the assembled prototype; b) with IMU sensor

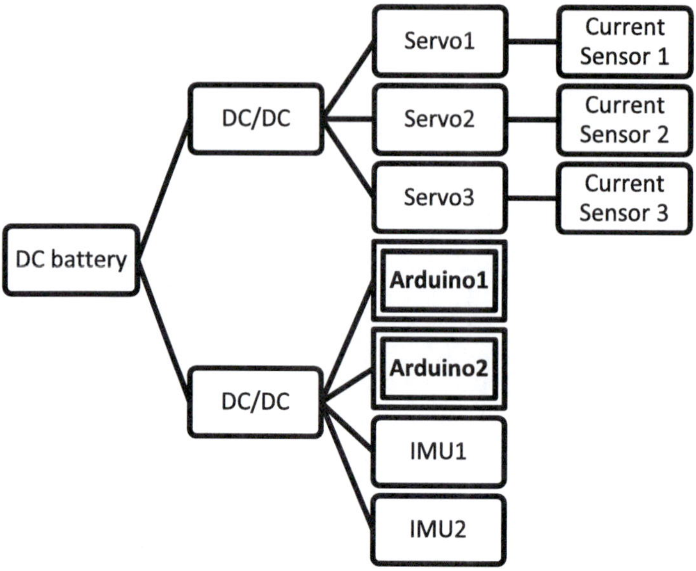

Fig. 4. A block diagram of the system design of L-CaPaMan prototype

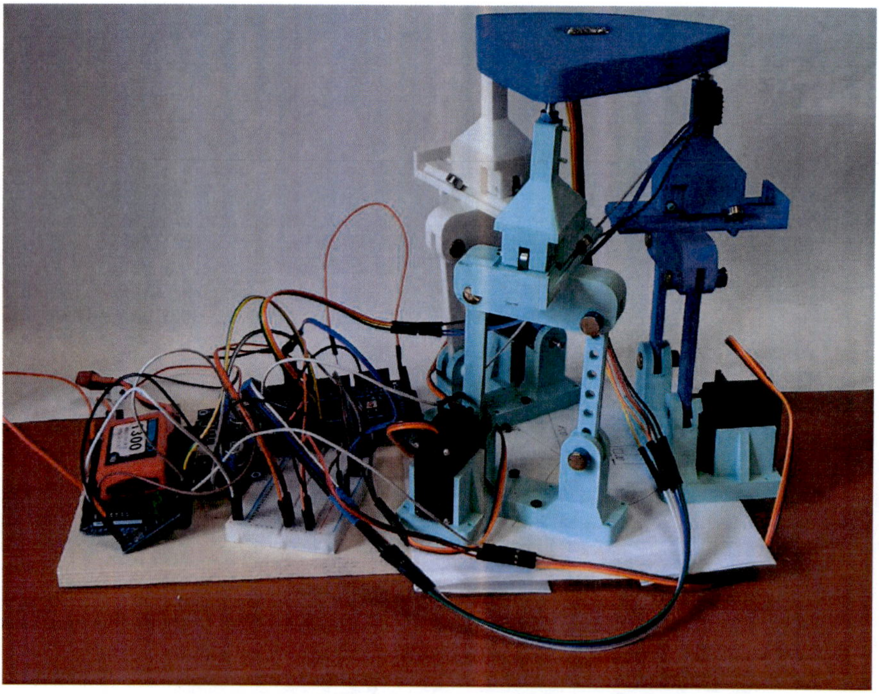

Fig. 5. A lab assembly of the new L-CaPaMan prototype in Fig. 2b)

The prototype is powered by a battery of 7.2 V, that is shown in Fig. 4. When testing the code, two Arduinos with sensors are connected to the PC by USB 5 V, and the motors have another source of current. Current sensors have been used for measuring power consumption of each motor, then torque for each motor is calculated, assuming linear dependence between the current and torque. One of the ways to increase the frequency of receiving the data is using two Arduino: the first gives commands to the motors, the second reads data from the sensors.

4 Testing and Results

In order to characterize the operation performance of the new L-CaPaMan prototype, three modes are planned, as summarized in Table 2.

Table 2. Planned testing modes for performance analysis of new L-CaPaMan in Fig. 4

Mode	Plot of input data for torques T_1, T_2, T_3
1	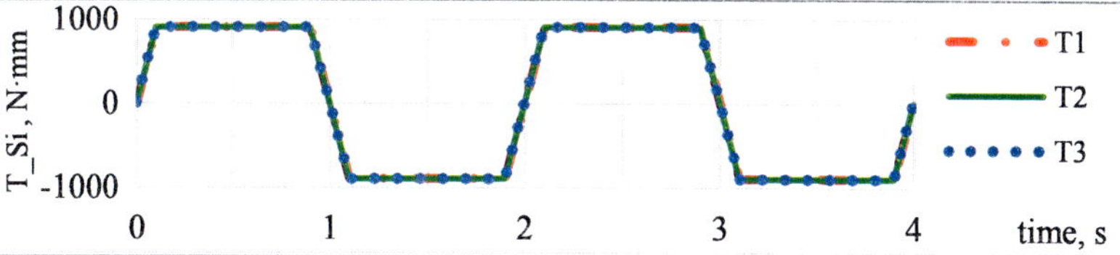
2	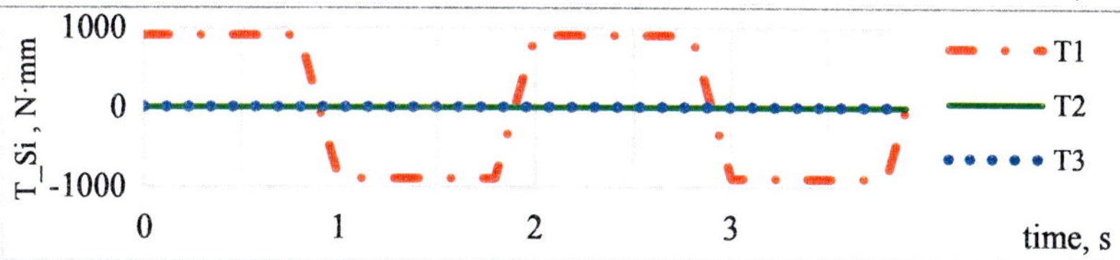
3	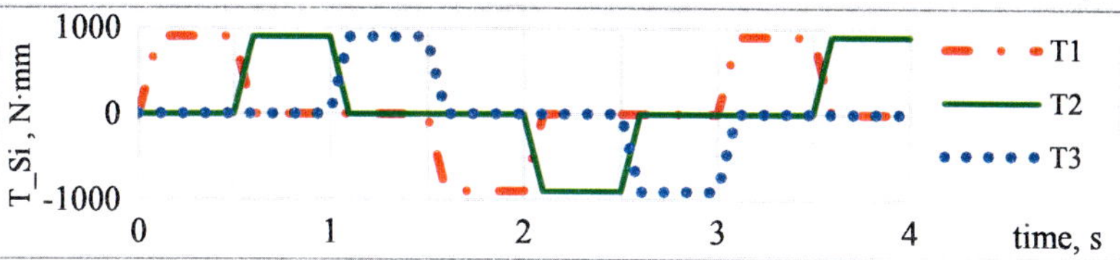

The control laws of a motor torque are represented in Table 2 as function of testing modes. T_1, T_2, T_3 are considered as torques for servomotors 1, 2, 3, respectively.

The perpendicular position of a leg to the ground is considered as 0°, counterclockwise rotation of legs is assumed as positive degree deviation and clockwise rotation is as negative degree deviation. The period of work for the Modes 1 and 2 is set on 4 s, full period of work for Mode 3 is 6 s.

The Mode 1 is worked out by rotating all the legs from −30° to +30° and vice versa for two times. The Mode 2 is worked out by rotating the first leg from −30° to +30° and vice versa for two times, while other legs are kept in position 0°. The Mode 3 is operated with legs moving in sequence as complicated movements of the platform along different axes. Illustrative results referring to a Mode 3 are reported in the Figs. 7, 8, 9, 10 and 11 from the experimental setup (Fig. 6).

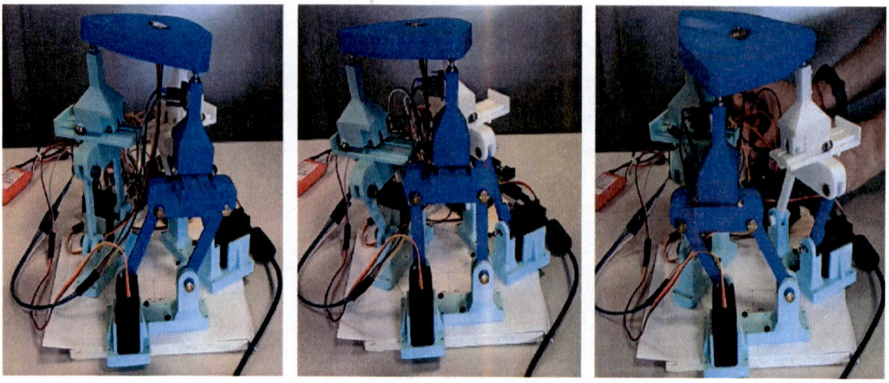

Fig. 6. A snapshot of a test mode 3

Figure 7 shows components of angular velocities of the platform. From the plot, the motions from all directions are with peaks of maximum values 165 deg/s represented for Z-component and less than 70 deg/s for another components. Due to the character of movements high peaks and strong slides are expected. The oscillations after peaks are observed because of the sharp movements of legs and rigidity of the construction.

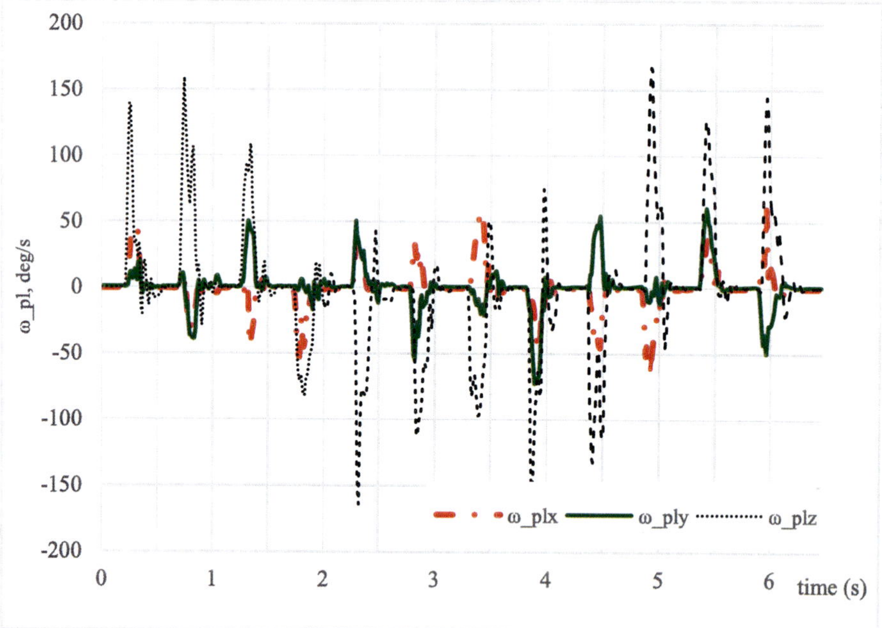

Fig. 7. Acquired results from testing of mode 3 in terms of angular velocities of the platform

Figure 8 shows the linear accelerations of point H – the central point on the platform. The highest values are acquired for X and Y components and reaches 5.68 m/s^2, while Z component of the acceleration is less than 1 m/s^2. The oscillations after passing peaks of velocities are observed, they are caused by the rigidity of the construction.

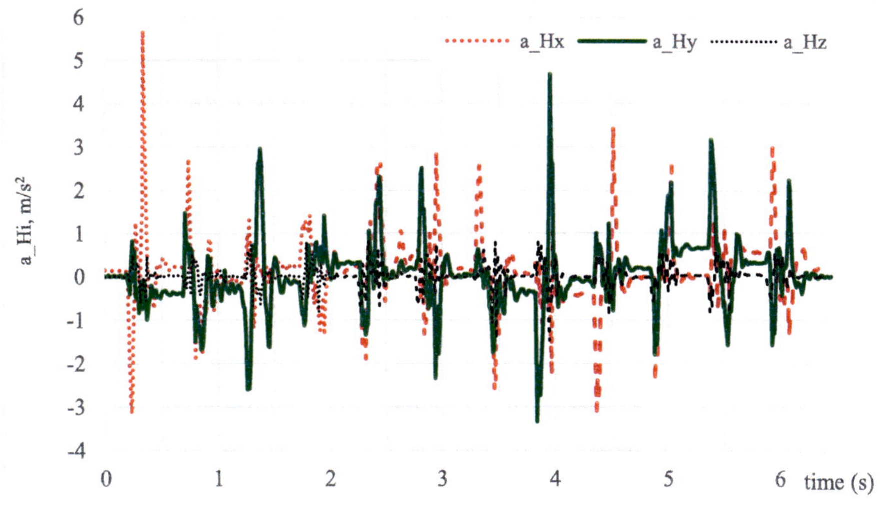

Fig. 8. Acquired results from testing of mode 3 in terms of accelerations of point H

Figure 9 shows the angular velocities of the slider. The maximum values are acquired for Y component of velocity and reach 209.7 deg/s. Angular velocities in Fig. 9 and accelerations in Fig. 10 show the effect of one leg movements on other legs of the mechanism. The measurements for these plots are acquired from the part, which is near to the moving platform, so the effect is maximal.

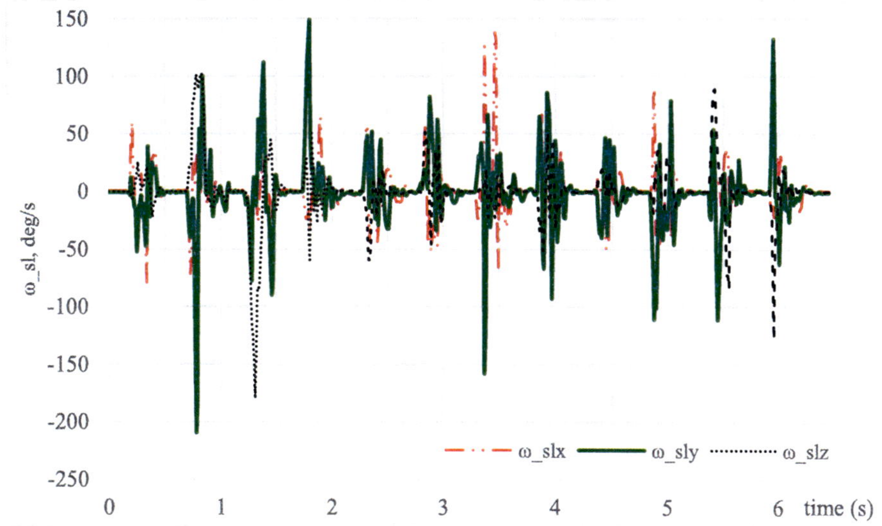

Fig. 9. Acquired results from testing of mode 3 in terms of angular velocities of the slider

Figure 10 shows the linear accelerations for the point K of the slider. The highest value is acquired for X component and equals 5.43 m/s^2, Z component of the acceleration equals 5.3 m/s^2, Y component is less than 4 m/s^2.

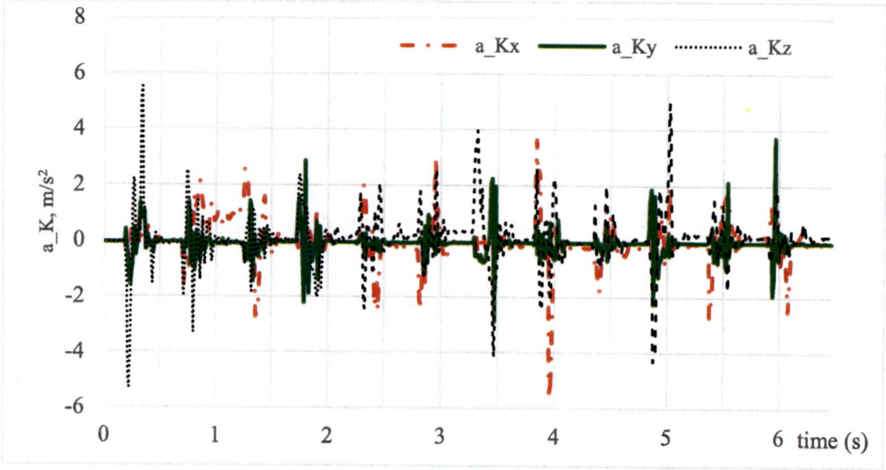

Fig. 10. Acquired results from testing of mode 3 in terms acceleration of point K of the slider

The data from current sensor is given for one motor, assuming the similarity of the motors, united source of power, and considering the motion law. The motor power is equal to 9.72 W. Despite of the increased frequency of getting data from Arduino – 0.01 s, it looks insufficiently because of high speeds of movements. This can be shown in the plots.

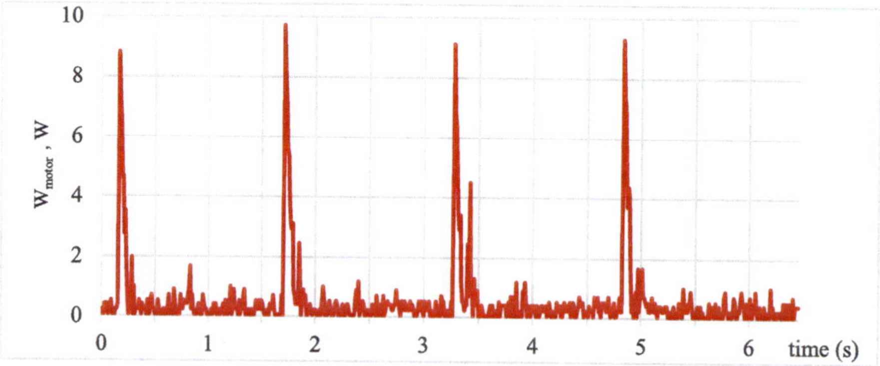

Fig. 11. Acquired results from testing of mode 3 in terms of power consumption

The results with maximum values from angular velocities of the platform (pl) and slider body (sl) and linear accelerations for the reference points H and K of the mechanism in each mode are listed in Table 3.

Table 3. Summary results of the reported test in Figs. 7, 8, 9, 10 and 11

Mode	1	2	3
$\omega_{pl\ max}$ (deg/s)	341.7	137.2	167.1
a_{Hmax} (m/s2)	2.18	4.53	5.68
$\omega_{sl\ max}$ (deg/s)	239.6	165.5	209.7
a_{Kmax} (m/s^2)	5.29	6.90	5.61

The platform shows maximum angular velocities in Mode 1, when legs move simultaneously, but linear speed of its point is less than from Modes 2 and 3. The numerical values can be satisfactory for an efficient operation of the built prototype for future testing.

5 Conclusions

The L-CaPaMan is a new version of CaPaMan, that has been designed with lightweight low-cost features. Its parts have been manufactured by 3D printer using market components. The model has been assembled and successfully tested in three modes of work. Results from sensors have been obtained and analysed to check and characterized the performance feasibility.

References

1. Ceccarelli, M.: A new 3 D.O.F. spatial parallel mechanism. Mech. Mach. Theory **32**(8), 895–902 (1997)
2. Ceccarelli, M.: Historical development of CaPaMan, Cassino parallel manipulator. In: Viadero, F., Ceccarelli, M. (eds.) New Trends in Mechanism and Machine Science, vol. 7, pp. 749–757. Springer, Dordrecht (2012). https://doi.org/10.1007/978-94-007-4902-3_78
3. Ceccarelli, M.: Fundamentals of Mechanics of Robotic Manipulation. Kluwer Academic Publishers, Dordrecht (2004)
4. Tsai, L.W.: Robot Analysis. The Mechanics of Serial and Parallel Manipulators. Wiley, New York (1999)
5. Ceccarelli, M., Pugliese, F., Lanni, C., Mendes Carvalho, J.C.: CaPaMan (Cassino Parallel Manipulator) as sensored earthquake simulator. In: IEEE/RSJ International Conference on Intelligent Robots and Systems, vol. 3, pp. 1501–1506 (1999)
6. Liang, C., Ceccarelli, M.: Design and simulation of a waist–trunk system for a humanoid robot. Mech. Mach. Theory **53**, 50–65 (2012)
7. Saracino, A., Oude-Vrielink, T.J.C., Menciassi, A., Sinibaldi, E., Mylonas, G.P.: Haptic intracorporeal palpation using a cable-driven parallel robot: a user study. IEEE Trans. Biomed. Eng. **67**(12), 3452–3463 (2020)
8. Casas, S., Coma, I., Portales, C., Fernandez, M.: Optimization of 3-DOF parallel motion devices for low-cost vehicle simulators. J. Adv. Mech. Des. Syst. Manuf. **11**(2), 17–23 (2017)
9. Jáuregui, J.C., Hernándes, E.E., Ceccarelli, M., López-Cajún, C., García, A.: Kinematic calibration of precise 6-DOF Stewart platform-type positioning systems for radio telescope applications. Front. Mech. Eng. **8**(3), 252–260 (2013)

10. Song, X., Pan, Y., Chen, Y.: Development of a low-cost parallel kinematic machine for multidirectional additive manufacturing. J. Manuf. Sci. Eng. **137**, 021005 (2015)
11. Arslan, O., Karaahmet, S. B., Selvi, Ö., Cafolla, D., Ceccarelli, M.: Redesign and construction of a low-cost CaPaMan prototype. In: Gasparetto, A., Ceccarelli, M. (eds.) MEDER 2018. MMS, vol. 66, pp. 158–165. Springer, Cham (2019). https://doi.org/10.1007/978-3-030-00365-4_19
12. Titov, A., Ceccarelli, M.: L-CaPaMan design and performance analysis. In: Machine and Industrial Design in Mechanical Engineering, KOD2021, paper 041 (in print) (2021)
13. Inventor users manual (2021). https://help.autodesk.com/view/INVNTOR/2021/ENU/. Accessed 03 Mar 2021

Mechatronics, Control and Monitoring Systems of Machines, Accuracy and Reliability of Machines and Mechanisms

Screw Thread Measurement by Camera Inspection System

Alexander Prošek(✉)

Measurement Department, VÚTS, a.S., Svárovská 619, Liberec XI – Růžodol I, 460 01 Liberec, Czech Republic
alexander.prosek@vuts.cz

Abstract. This article describes measurement of the internal screw thread by the camera system. Especially thread handedness detection is focused. For this purpose, a geometric transformation on camera image was used, followed by the FFT method to detect thread direction. Possible issues of the measurement precision are also discussed.

Keywords: Screw thread measurement · Image processing · FFT · Image moments · Camera inspection

1 Introduction

Screw thread measurement is motivated by the requirement of the manufacturing practice. According to measurement results is decided whether the inset part is the correct type or not. Main evaluated thread parameters are diameter, pitch and handedness. In our case the most difficult is to determine handedness. Two examples of tested parts are shown on the Fig. 1. Both threads are metric with diameter 39 mm, but the left-handed thread is on the left side and the right-handed thread is on the right side.

Fig. 1. Two examples of the metric screw thread M39x1.5. Left-handed thread is on the left side and right-handed thread is on the right side.

© The Author(s), under exclusive license to Springer Nature Switzerland AG 2022
J. Beran et al. (Eds.): TMM 2020, MMS 85, pp. 261–268, 2022.
https://doi.org/10.1007/978-3-030-83594-1_27

The customer's requirement was that the screw thread has to be correctly recognized with at least 99.7% reliability. Real threads have different surface quality and color, also the light conditions are variable. Due to the used method has to be sufficiently robust.

2 Measurement

2.1 Diameter Measurement

The standard image processing methods are used for the diameter measurement. The hole's ellipse is extracted from image and then the least squares method is used to find the best fit ellipse. The semi-major axis of the found ellipse corresponds with screw thread diameter.

2.2 Handedness Measurement

It is difficult to decide about handedness from the original image (see Fig. 1). Due to it is better to map the thread image back to cylinder and unfold it to the plane. This transformation will be described more precisely bellow.

The first step is to determine transformation between thread's cylinder coordinate system (denoted as T) and camera coordinate system (denoted as G).

$$\Phi_{T \to G} : \{r\}_T \mapsto \{r\}_G \tag{1}$$

Symbol $\{r\}_T$ denotes coordinates of the vector r in the coordinate system T. This transformation can be known or can be estimated by the image features (e.g. hole's ellipse orientation). Cylinder axis is identical to z-axis of the T base. This relation between camera and thread is displayed on Fig. 2.

The second transformation is projection to the focal plane. For simplicity, we assume unit focal length.

$$\begin{gathered} \Phi_p : \mathbb{R}^3 \to \mathbb{R}^2 \\ \begin{pmatrix} x \\ y \\ z \end{pmatrix} \mapsto \begin{pmatrix} x/z \\ y/z \end{pmatrix} \end{gathered} \tag{2}$$

Cylinder points can be mapped to plane by the polar transformation which is described by the equation:

$$\begin{gathered} \Phi_c : (0, 1) \times (0, 2\pi) \to \mathbb{R}^2 \\ (z, \varphi) \mapsto (Cos(\varphi), Sin(\varphi), z) \end{gathered} \tag{3}$$

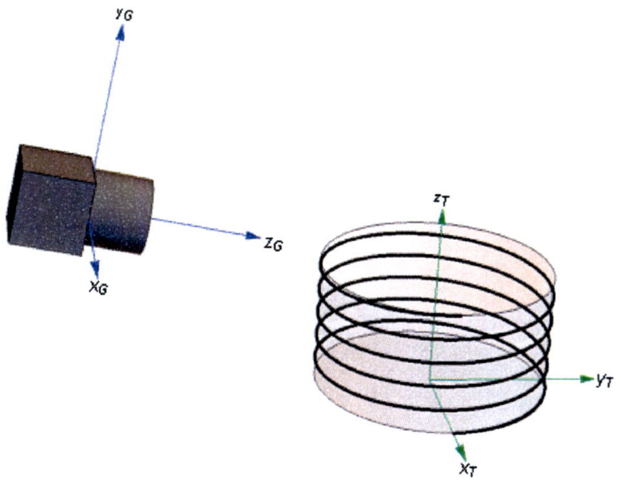

Fig. 2. Geometric arrangement visualization. Green coordinate system T corresponds with thread's cylinder and the blue coordinate system G corresponds with camera.

Composed map function $\Phi = \Phi_p \circ \Phi_{G \to T} \circ \Phi_c$ assigns image coordinates to polar coordinates of the cylindrical surface. If $I(x, y)$ is brightness function of the camera image, then function $(I \circ \Phi)(z, \varphi)$ is an image of the cylindrical surface in the polar coordinates. Example of this image transformation is shown on Fig. 3.

With real image it is also necessary to perform a transformation between physical and pixel coordinates. These details will be ignored in this article for simplicity, but the texture mapping and image warping is described more precisely in [1].

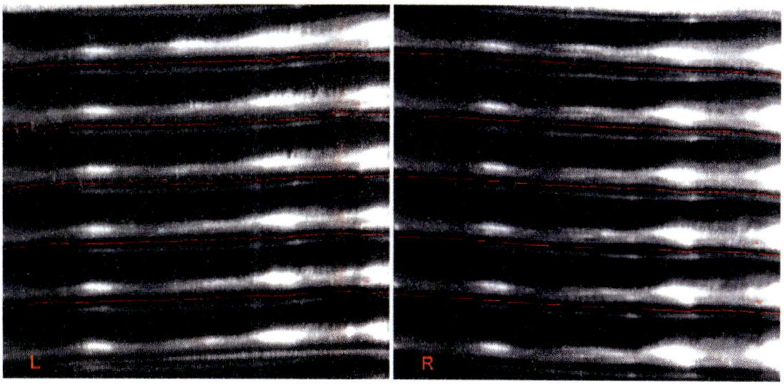

Fig. 3. Left-handed and right-handed threads mapped to detected cylindrical surface. The important edges are highlighted by red color. These edges are used to following processing.

This image can be processed by the many different approaches. Our procedure was chosen with respect to evaluation speed and robustness. To the edge detection, the image is processed by the some standard filters in the vertical direction. Then FFT is used to obtain image periodogram. 2D FFT method can be used to detect of fibres direction how is described in [2]. This idea is used to detect direction of the lines from Fig. 3. Both periodograms (left-handed and right-handed thread) are shown in Fig. 4.

Fig. 4. Image periodogram evaluated by FFT. Not all values are used to the following processing. Used points are highlighted by red color.

Image periodogram is thresholded and the selected points are processed by statistical moments. Let be $p(x, y)$ value of the thresholded and normalized image periodogram, which can be considered as probability density. Let be $S = \{(x, y) : p(x, y) > 0\}$. Then we can express the moments up to order 2 by the next equations:

$$\begin{gathered}
\mu_x = \int_S x dp, \; \mu_y = \int_S y dp \\
\mu_{xx} = \int_S (x - \mu_x)^2 dp \\
\mu_{xy} = \mu_{yx} = \int_S (x - \mu_x)(y - \mu_y) dp \\
\mu_{yy} = \int_S (y - \mu_y)^2 dp \\
M = \begin{pmatrix} \mu_{xx} & \mu_{xy} \\ \mu_{yx} & \mu_{yy} \end{pmatrix}
\end{gathered} \tag{4}$$

Eigensystem of the matrix M can be displayed as the ellipse which has axis identical with eigenvectors and axes lengths ratio is equal to ratio of the eigenvalues. These ellipses are shown in Fig. 5. The idea of image moments and moment invariants is usually used to pattern recognition. This topic is described in [3].

Note that the real situation is influenced by the variable thread surface quality, variable light conditions and metallic shiny thread surface. In these cases is not possible to find edges in the same quality which are displayed above. But evaluation by described procedure seems to be sufficiently reliable.

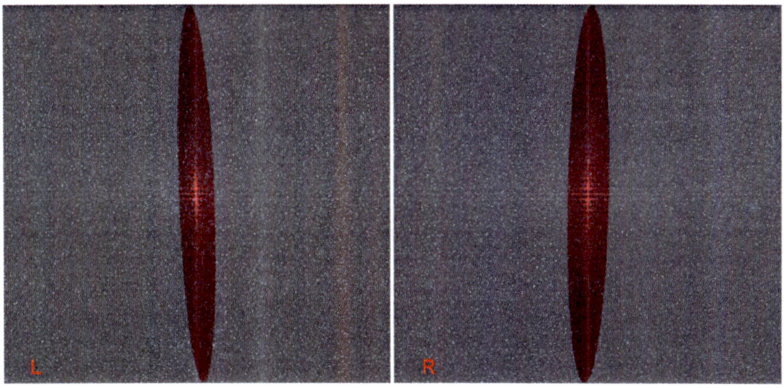

Fig. 5. Eigenvectors and eigenvalues visualized by the ellipses. Angle between ellipse semi-major axis and image vertical axis determines handedness of the thread.

2.3 Pitch Measurement

Pitch is measured from image of the cylinder map which is turned by the found angle. Image rotation makes the edge line horizontal and then is simple to measure pitch. We calculate point counts over rotated image rows. These counts as function of the row position have significant peaks. Distances between these peaks correspond to the pitch value.

3 Measurement Error

Quality of the image mapping to cylindrical surface depends on the function $\Phi_{T\to G}$. If this function is not determined exactly then the mapped image $I \circ \Phi$ is deformed. That can adversely affects the result. Some errors that can occur in our case will be discussed below. For simplicity, assume unit cylinder radius and unit focal length. Let be.

$$V = \left\{(x, y, z) \in \mathbb{R}^3 : x^2 + y^2 = 1\right\}$$
$$s(\alpha) = (Cos(\alpha), Sin(\alpha), a\alpha), \alpha \in (0, \pi), a > 0 \tag{5}$$

The point set $\{s(\alpha), \alpha \in (0, \pi)\} \subset V$ is a part of helix. Projection of this curve to the focal plane is $s_p(\alpha) = \left(\Phi_p \circ \Phi_{T\to G}\right)(s(\alpha)), \alpha \in (0, \pi)$. Example of this projection is shown on the Fig. 6.

We want to map the curve s_p to the wrongly detected cylinder V_E. Relations between the original cylinder coordinate system T and error cylinder coordinate system (denoted as E) is given by mapping $\Phi_{T\to E} : \{r\}_T \mapsto \{r\}_E$. For example, in our case, this error can be caused by detecting of the incorrect edge on the thread hole. Two cases which can occur in our task will be discussed next.

Fig. 6. Helix curve on the cylinder surface and its projection (red curve) on the focal plane (green plane). Z-axis of the thread coordinate system is represented by the arrow.

It is necessary to solve inverse problem to map the curve s_p to the error cylinder V_e. Let be P_f focal point and $\rho(t) = (1-t)\{P_f\}_E + t\{s_p(\alpha)\}_E$, $t > 0$ for some $\alpha \in (0, \pi)$. We want to find intersection $\{\rho(t) : t > 0\} \cap V_E$. This leads to quadratic equation with one unknown parameter t. Equation can be solved exactly for all t but for our case is sufficient to solve equation only for some points and use linear approximation.

3.1 Translation Error

Assume translation in x-direction. This transformation is described by function $(x, y, z) \mapsto (x + \delta, y, z)$, $d \in \mathbb{R}$. Evaluation for some discrete values of the parameter d is shown on Fig. 7.

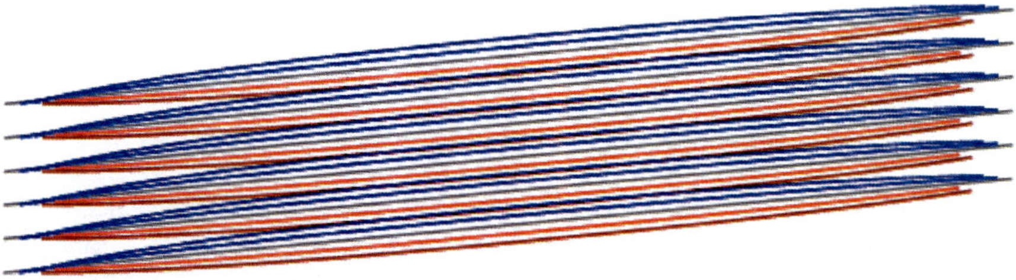

Fig.7. Helix curve mappings to shifted cylinder for shift $\delta \in \{-0.1, -0.05, 0, 0.05, 0.1\}$. Blue curves for positive shift and red curves for negative shift value.

We use image processing as it is described above. Results for the original cylinder and shifted cylinders are shown on Fig. 8.

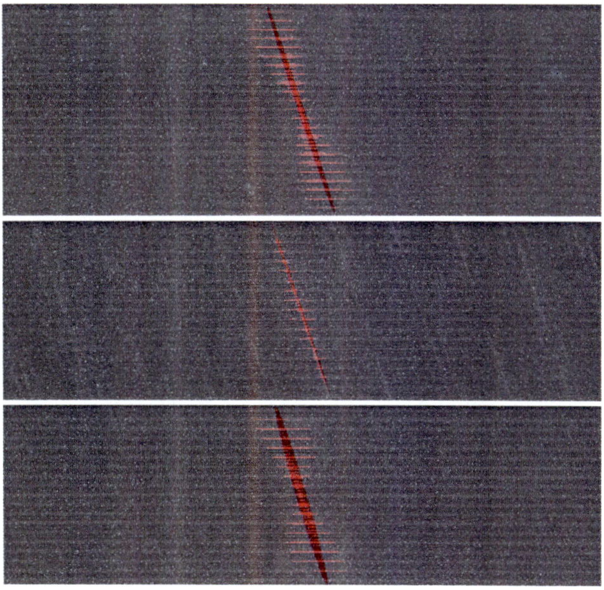

Fig. 8. Eigenvectors ellipses for three cases: $\delta = 0.2$ (top), $\delta = 0$ (middle) and $\delta = 0.2$ (bottom).

3.2 Diameter Measurement Error

In this case we can suppose that error coordinate system E is identical with the thread's cylinder coordinate system T. But the projection to cylinder surface leads to equation $\rho_x^2(t) + \rho_y^2(t) = (1+\delta)^2$, where $\rho(t) \in \mathbb{R}^3$ is parametric equation for infinite line which contains the focal point with some curve point on the focal plane. Alternatively we can suppose translation in the camera view direction. Due to projection the more distant cylinder seems smaller. It causes that the results are similar as it is mentioned for translation error in Sect. 3.1.

3.3 General Case

How it is mentioned in the previous section, the radius error can be considered as shift in the view direction. So the general case can be expressed as composition of some affine transformations. In our case the errors are caused by the detection of different edges on the screw thread (it depends how the thread is rotated around its axis). The error magnitude corresponds with the height of the thread.

Other case is when we expect the fixed position of screw thread but the real thread is shifted due to geometric tolerance. Methods mentioned above can be used to estimate maximal possible error. Alternatively, we can find conditions to ensure sufficient accuracy.

4 Conclusion

Methods mentioned in this article were used to develop camera inspection system which is used in manufacturing industry. System controls parts with internal screw thread and decides whether the inset part is the correct type. In this case some geometric conditions are known but methods can be generalized for the cases when the thread's cylinder position is unknown. Method can also be used for the external screw thread.

References

1. Heckbert, P.: Fundamentals of texture mapping and image warping. Master's thesis (1989)
2. Tunák, M., Linka, A.: Analysis of planar anisotropy of fibre systems by using 2D Fourier transform. Fibres Text. East. Eur. **15**(5–6), 86–90 (2007)
3. Hu, M.K.: Visual pattern recognition by moment invariants. IRE Trans. Inf. Theory **8**(2), 179–187 (1962)

Reliability Analysis of Rotary Table Over the Lifetime

Aleš Richter(✉) and Pavel Fišer

VÚTS, a.s., Svárovská 619, Liberec XI, 460 01 Liberec, Czech Republic
{ales.richter,pavel.fiser}@vuts.cz

Abstract. This paper deals with the evaluation of the reliability of a high-precision and fast carousel for its expected service life. The total service life is given by the weakest link of the transmission mechanism. In our case, these are the rotating moving parts of the gearbox, or their contact surfaces.

With the help of specially developed test equipment, it is possible to approximately simulate the degree of load to which the components of the carousel will be exposed under normal operating conditions. The results collected from the tests performed will allow us to modify future carousel designs to achieve the specified values of positioning cycles.

Another integral part of increasing the competitiveness of a product is its accuracy. This is given by the construction of the rotary table, production and subsequent professional assembly. The accuracy of the carousel should not change significantly during operation. It needs to be taken into account as one of the monitored factors during qualitative tests. Unfortunately, even the most accurate simulation cannot fully replace real data obtained from tables already installed in industrial production. We also have this data available and, as feedback, complete the complex optimization process of the final product.

Keywords: Rotary table · Accuracy · Reliability · Backlash · Cycles

1 Introduction

At VÚTS, we have been dealing with highly accurate and fast positioning mechanisms for many years. In terms of potential use, a carousel table with a diameter of 1000 mm is interesting, with an accuracy of 40 μm and a positioning time of 0.5 s (Fig. 1). In previous published articles, we have dealt with the mechatronic model of a carousel [1]. Then by measuring the positioning accuracy [2] and finally by improving it [3]. A prototype was created that should last several million production cycles.

The life tests are now underway. The article describes the procedures in the development and construction, which led to a gradual increase in the reliability and accuracy of carousel rotary tables. The following are the results of measuring the overall backlash of the mechanism, long-term wear tests of the contact surfaces of the rollers and the axial cam. All this is verified by the results obtained in the continuous monitoring of a single-purpose machine with an implemented carousel from the production process.

© The Author(s), under exclusive license to Springer Nature Switzerland AG 2022
J. Beran et al. (Eds.): TMM 2020, MMS 85, pp. 269–277, 2022.
https://doi.org/10.1007/978-3-030-83594-1_28

2 Rotary Table Design

2.1 Machine System Design

Figure 2 shows the carousel drive mechanism. The motor drives the axial cam by a backlash free reduction belt gear. The cam rotates the carousel plate also with a reduction gear.

During the movement of the table, the axial cam rotates, which moves a pair of rollers. The rollers encircle the cam from the outside of its collar and are attached to the table plate. One step then means turning the plate by 22.5°. The gear ratio was set to 40 by simulations.

Our design allows to reduce production and operating backlash to a minimum without fear of heating and subsequent jamming of the mechanism, which could happen in the roller-groove system. A fixed number of 16 production positions (at the same time 16 pairs of rollers) allows us to set the mechanism in the engagement of 4 rollers at one time and this significantly reduces the backlash of the carousel. At the same time, the carousel is still arbitrarily infinitely adjustable, only it is more accurate in a certain range.

Fig. 1. Rotary table assembled

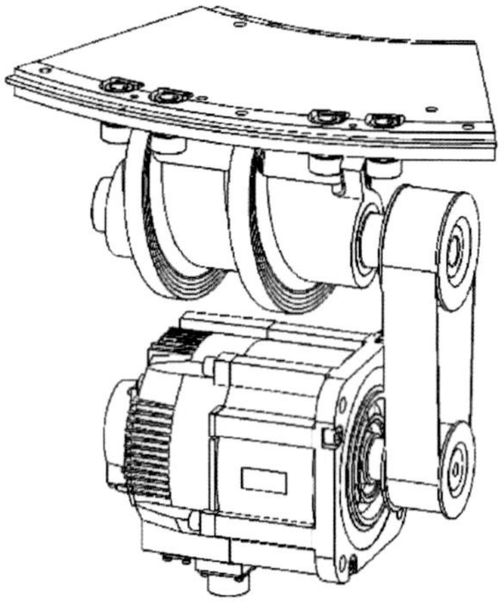

Fig. 2. Model of rotary table mechanism

A trough emerges from the gearbox, which copies the path of the rollers and pins and then enters the gearbox back. Its task is to distribute the lubricant evenly to all key parts. It is a circulating lubrication, its circulation is ensured by the same axial cam, which ensures the positioning of the table.

The external stainless steel cover provides protection against the ingress of foreign particles, keeps the mechanism clean and thus increases its service life at the cost of a slight increase in passive resistance, but the system takes this into account and does not represent a relevant load for the engine.

2.2 Stress Analysis

From the previous articles [1–3] we know what to expect the dynamic load of the plate, from which we simply determined the load of the key parts: pins, rollers and cam. These parts must be dimensioned to withstand tens of millions of load cycles. According to the Wohler curve, it is necessary to oversize the parts above the fatigue limit. The roller screw system was subjected to strength analyzes (Figs. 3 and 4). In the first phase of the design, we considered a load moment significantly higher than the measurement on the manufactured piece showed. The model assumes 70% more load than actually occurs. The maximum global stresses in the theoretical model exceed 400 MPa, they are caused by compressive stress. Table 1 shows the material values of the key components.

Table 1. Used material properties

	Axial cam	Roller	Pin
Material	18CrNiMo7–6	X38CrMoV5-1	X38CrMoV5-1
	1.7131	1.2343	1.2343
Hardness [HRc]	58	54	54
E [MPa]	212000	222000	222000
μ [–]	0,3	0,3	0,3
Re [MPa]	490	450	450
Rm [MPa]	685–980	1180–1770	1180–1770
Treatment	Cemented, hardened	Hardened, Balinit C-STAR	Hardened

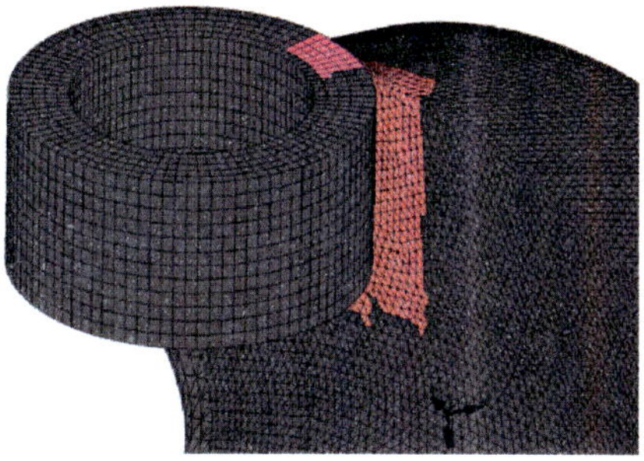

Fig. 3. Assumed contact area

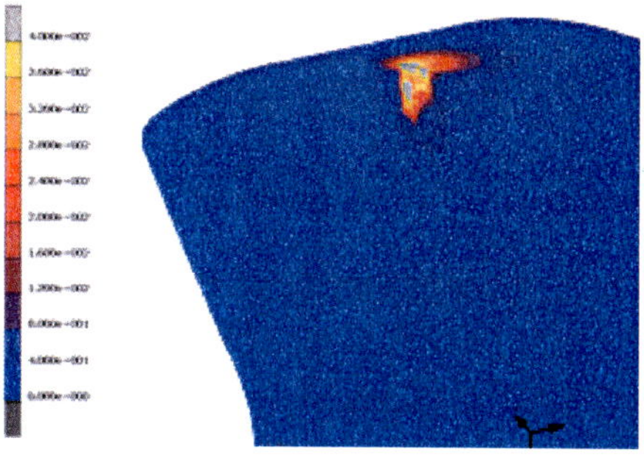

Fig. 4. Axial cam's stress analysis

2.3 Pressure Test of Roller Coating

Fatigue wear also called contact fatigue, occurs on functional surfaces that are subjected to repeated compressive stress during the rolling motion of the functional surface of a machine part. Variable elastic or elastoplastic deformations occur in the coating layers of material stressed in this way [4].

Deformations can be reduced by selecting a suitable roller coating material and thus the degree of service life and reliability can be increased as well. At the research institute, tests of semi-finished products with various surface treatments [5–7] have been carried out for many years. We use general theoretical knowledge and have specially manufactured application test equipment for this purpose.

The main input requirement will be the Hertz pressure rate set at 600–660 MPa. BALINIT C-STAR (2–3 μm) was chosen as the most suitable coating material for the test rollers (Fig. 5). The force that derives the appropriate Hertz pressure was set at 420–440 N per wall.

Fig. 5. BALINIT C-STAR coating

The roller rotates on average 7/12 when the plate is moved by one position (the roller is stressed by only 1/16 turn of the plate). The estimated life of the rotary table carousel was set at 140 million steps, which corresponds to 82 million turns of the roller and 8.75 million turns of the carousel.

In most cases, carousel tables are used as part of complex machinery and it is not possible to easily access them without dismantling the individual parts of the machine. Therefore, the first sample was tested without lubrication. This sample lasted only 20% of the planned life and then the surface layer was destroyed (Fig. 6). The Fig. 6 shows the places where the pressure disc bounced off, which resulted in unpleasant sound and shock. The machine would not certainly meet hygienic noise standards.

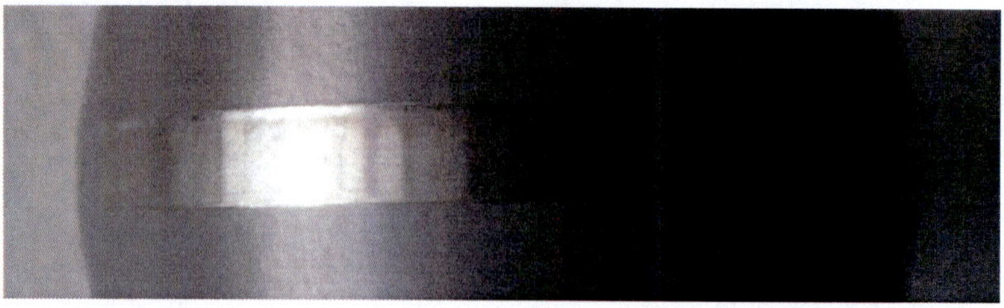

Fig. 6. Damaged sample after test without lubricant

The second test was performed with a continuously lubricated sample. It lasted the specified number of cycles. The test was continued until fatigue damage was visible on the surface of the contact area (Fig. 7) in the form of micropitting [4].

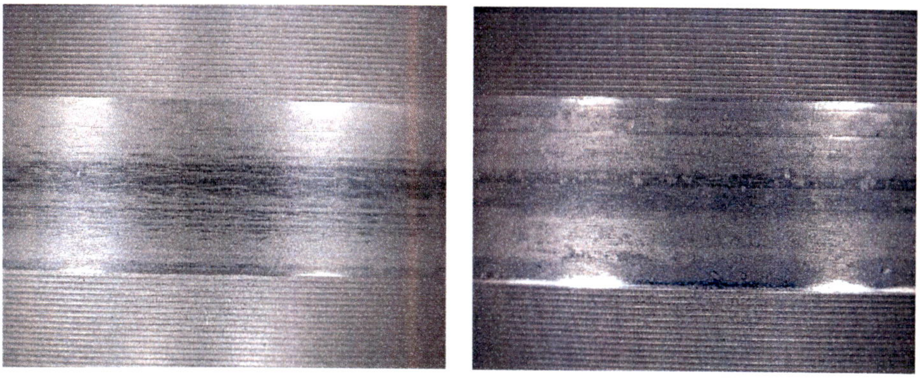

Fig. 7. Sample after 80 and 85 mil cycles test with lubrication

3 Production

A very important phase is the accurate and flawless production of the rotary table. Minor deviations can also be fine-tuned in assembly, but precise production is essential for the accuracy and long life of the carousel. The dimensions of the key parts are checked at a specialized 3D measuring station. As an example, we can use the illustrations from the axial cam measurement protocol (Fig. 8). The illustrations from the axial cam measurement protocol (Fig. 8) are stated here as an example. It was evident from the results that the width of the cam collar was relief ground by 0.01 mm and the outer diameter of the pair of rollers was then increased by this value.

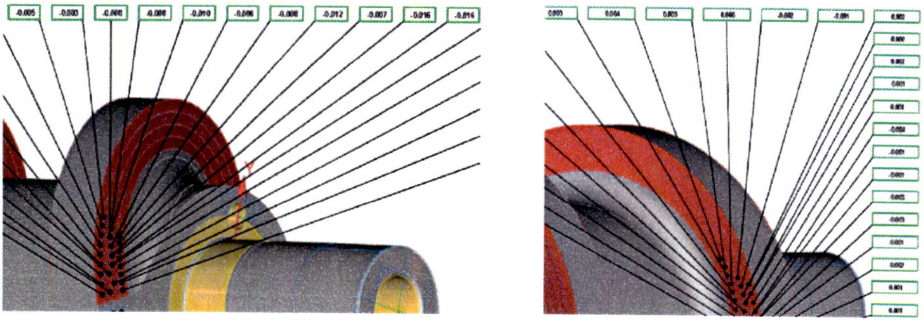

Fig. 8. Axial cam's path measurement

4 Assembly

The pairs of rollers and pins, which were measured after production, are selected so that the pin-roller-cam assembly has as little backlash as possible. Each position is marked so that any inaccuracy can be linked to a specific part.

The gearbox is precisely positioned in the carousel frame with pins so it can be removed and put back in the same place if necessary.

The belt must be properly tensioned for backlash-free and has the longest possible service life. The calculated force in the belt is converted to its natural frequency and this is checked using a Belt Tension Meter. The force in the belt must not exceed the permissible load on the motor shaft end (this would damage the motor bearings).

The quality of the assembly is verified by measuring the total backlash for each position in the motor-pulley-belt-pulley-cam-roller-pin-plate system. A digital dial indicator is connected to the outer diameter of the plate at a given location.

In the first phase, the motor performs back and forth movements of 0.4°. This corresponds to a deflection of 0.1 mm of the plate at a given radius of 568 mm. This deviation is reduced by the total backlash of the system and this is determined by measurement.

In the second phase of the measurement, we stop the engine and apply force through the lever to the carousel plate. The force is on average 600 N and by its action it is necessary to cause a deflection on the plate greater than the expected backlash in the system, i.e. 0.1 mm. Then the force stops working and we measure the deviation. In the end, we let the force work in the opposite way and the situation repeats itself. From the difference of deviations after stabilization we get the backlash in the system. Table 2 shows the backlash for each position after the specified service life has elapsed.

Table 2. Total backlash in mechanism related to gearbox distance (on plate radius 500 mm).

Position	1	2	3	4	5	6	7	8	9	10	11	12	13	14	15	16
1 [mm] * 10^{-2}	4	2	3	3	3	2	4	2	2	2	2	2	2	1	2	2
2 [mm] * 10^{-2}	3	2	2	2	1	1	0	1	1	2	3	0	1	0	1	1

The maximum total backlash was 0.04 mm and the average backlash was 0.02 mm. These values are similar to those after assembly-run-in and suggest that the accuracy of the carousel repeat will not be affected by wear.

In the middle of the test, the carousel was disassembled and inspected. The parts (Fig. 9) were measured again at specialized workplaces and deviations from the original measurement were sought. No significant damage was noted and the carousel was reassembled and reactivated.

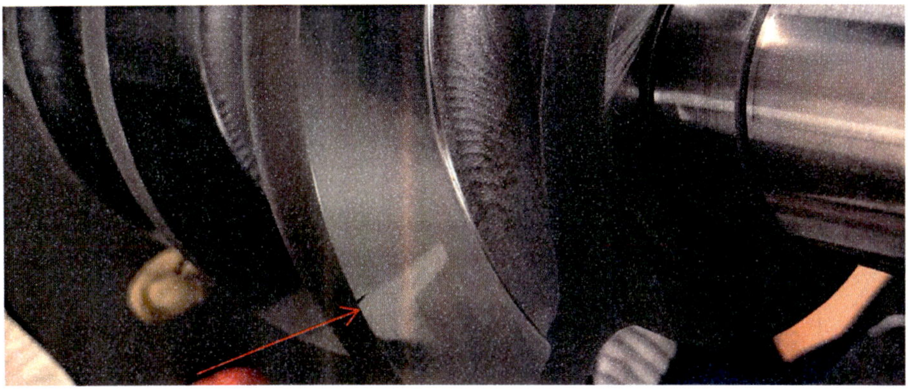

Fig. 9. Axial cam wear after lifetime expectancy tests

5 Rotary Table Application in a Manufacturing Plant

Several machines with fast and accurate carousels have been installed in many manufacturing plants. We focused only on the one that was run for the longest production time with the highest number of cycles so that mechanical components were the most worn out.

year	Trqeff [%]	steps	cycles
2017	39.80	*11 157 702*	697356.375
2018	39.90	*12 819 032*	801189.5
2019	40.53	*13 251 419*	828213.6875
2020	40.79	*12 896 699*	806043.6875
2021	40.91	*3 132 554*	195784.625
sumary		53 257 406	3 328 588

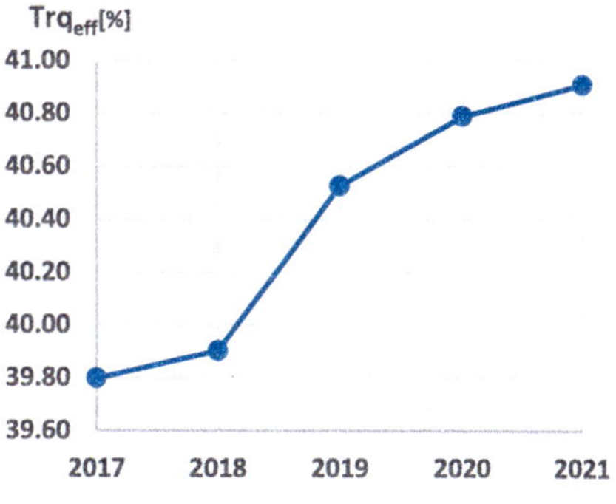

Fig. 10. Basic statistic of installed carousel

All positioning tables operate with the same step polynomial displacement diagram, which reaches 900° of motor movement in one cycle, followed by a standstill. There is 1:1 ratio between them so they have same duration. We can adjust speed up to 60 cycles per minute, this value is also preset to mentioned machine [8].

Since the displacement diagram is represented by a polynomial of the 5th degree, a continuous course of its acceleration is guaranteed, which is similar in shape to the characteristic of the output torque on the motor shaft.

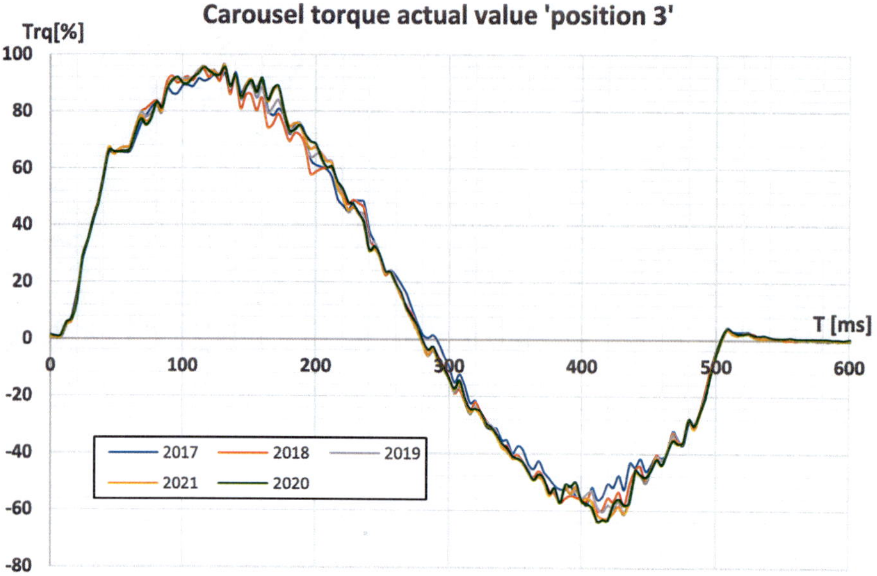

Fig. 11. Torque changes on servomotor same day different year

From several years of measuring the torques during the operation of the carousel (Fig. 11), it can be concluded that there is a slight fatigue wear of the mechanism. By calculating the average effective torque on the motor shaft over the monitored years (see graph in Fig. 10), an increase in the order of units of percent of the rated motor torque (28.4 Nm) can be seen. This corresponds to an increase of 0.315 Nm. During this time, the machine performed 53 million steps (see Fig. 10), which corresponds to 3.3 million revolutions of the rotary table. This value is equal to the number of contacts of each roller mounted on carousel plate with axial cam. Unfortunately, there is no possibility to remove the table from the machine and check the surfaces, because it is in a continuous working process. However, we can get a relatively accurate picture of their wear from a life test performed on a special test equipment. It can be assumed that if any key part of the mechanism were noticeably damaged, there would be a large significant change in characteristic of torque.

A correction table is supplied with each of the rotary tables, as it cannot be assembled with zero backlash. It is integrated to the machine control system, which adjusts each end position of the individual steps of the carousel [3]. The program is secured and inaccessible to the customer. The accuracy of the carousel cannot be adjusted other than by changing the values of corrections. During the several years of operation, there has not been a single request for their adjustment, which would be necessary to decrease the backlash in individual positions below the set limit.

6 Conclusion

In terms of practical use and commercial potential, the carousel seems to be very promising. According to the end customer, the most important feature is its service life. This is required with constant accuracy at the specified production speed in customer-set operating conditions.

It all starts with the initial design of the mechanism, continues through correctly chosen technological procedures to the subsequent assembly of machine parts. The correctness of our procedures is verified by long-term tests. On the manufactured carousel prototype, we measured the change in backlash after several million cycles. The backlash did not exceed the specified quality limit of 0.05 mm. Fatigue tests of the contact surfaces were performed on a specialized stand, where we verified that the BALINIT C-STAR roller coating chosen by us will last for the entire set life.

In addition to all tests and measurements performed, we have feedback in the form of live monitored quantities directly from the already applied rotary table. These data bring us closer to the difference between ideal laboratory conditions and real industrial conditions. It is minimal in our case. It can be said that the carousel rotary table developed and constructed by us is accurate and reliable.

References

1. Richter, A., Ondrášek, J.: Rotary table machine input parameters optimization. In: Beran, J., Bílek, M., Žabka, P. (eds.) Advances in Mechanism Design II. MMS, vol. 44, pp. 399–405. Springer, Cham (2017). https://doi.org/10.1007/978-3-319-44087-3_54
2. Richter, A., Klouček, P.: Rotary table machine accuracy verification. In: Experimental Stress Analysis 2017, p. 235 (2017). ISBN: 978-80-553-3166-9
3. Richter, A., Fišer, P.: Rotary table accuracy improvement. Mechat. Syst. Mater. (2018). https://doi.org/10.1063/1.5066527.ISBN:978-0-7354-1751-9
4. Jirásko, P., et al.: Mechatronika pohonů pracovních členů mechanismů. VÚTS, Liberec (2015). ISBN 978–80–87184–63–9
5. Hejnová M., Ondrášek J.: Life estimation of the contact surfaces. In: Beran, J., Bílek, M., Žabka, P. (eds.) Advances in Mechanism Design II. Mechanisms and Machine Science, vol. 44. Springer, Cham (2016). https://doi.org/10.1007/978-3-319-44087-3_6
6. Gromadová, M., Ondrášek, J.: Experimental method of the service life estimation of a general kinematic pair of a cam mechanism. In: EAN2019: 57th Annual Conference on Experimental Stress Analysis, Czech Society for Mechanics, Brno (2019). ISBN: 978-80-214-5753-9
7. Hejnová, M.: Service life assessment of the cam mechanisms. Procedia Eng. **96**, 157–163. https://doi.org/10.1016/j.proeng.2014.12.135
8. Fišer, P.: FDL-K 001 Wooden strip milling machine. In: Special Mechanisms and Their Drives IV, pp. 220–221. VÚTS, Liberec (2018). ISBN: 978-80-87184-87-5

Failure and Risk Analysis Based on Maintenance Reports of Machines Components in Manufacturing Industry

Aleksandra Thamm[1,2(✉)], Matthias Herz[2], Markus Wiedemann[2], Florian Thamm[1], and Andreas Maier[1]

[1] Friedrich-Alexander-University Pattern Recognition Lab, Erlangen, Germany
aleksandra.thamm@fau.de
[2] Siemens AG, Erlangen, Germany

Abstract. Manufacturing engineering is concerned with the application of engineering practices in manufacturing processes and production methods. Continuous optimization of processes, tools as well as machinery and equipment is the key to producing quality products with optimal capital investment. Hereby, maintenance plays a crucial role and allows manufacturers to optimize their processes. To enable improvements in this field it is required to analyze and evaluate the necessities for actions on component level. In this work we want to present a detailed retrospective analysis of more than 4.000 maintenance reports and analyze causes for assembly failures depending on the associated components. Furthermore, we extend the analysis and investigate the failure dependency with respect to the days elapsed from the date of maintenance request to resolving the issue. It turned out, not all cases with high failure rates are associated with subsequent higher risks considering the respective urgency related to each failure case. Furthermore, we discovered especially hydraulic and pneumatic components have a significant high to critical risk, as the labour safety and other components are threatened by their failure. Moreover we determined, that only 10% of all actions were planned maintenance and thus we propose based on the results of the risks analysis, how planned and preventive maintenance should be extended to reduce costs in this regard.

Keywords: Manufacturing · Failure analysis · Risk analysis · Spare parts · Maintenance · Preventive maintenance

1 Introduction

Maintenance describes a combination of all technical and administrative actions intended to retain or restore, e.g. machines to a state in which they can perform their required function [1]. This is closely related to "Production Management" (PM), which is concerned with the flow of information about for instance machining conditions. Thus, with the insights gained by the PM, optimization of the

© The Author(s), under exclusive license to Springer Nature Switzerland AG 2022
J. Beran et al. (Eds.): TMM 2020, MMS 85, pp. 278–286, 2022.
https://doi.org/10.1007/978-3-030-83594-1_29

maintenance becomes possible. Also the timely procurement can benefit from a good planning [2]. The reactive maintenance occurs when the components have already failed and need to be either replaced or repaired [3]. Chand et al. [4] have summarized the costs of reactive maintenance. These include: lost of production, disrupted schedules, repair, and safety costs [4]. A possibility to reduce these costs, is the introduction of a Total Productive Maintenance (TPM), which pursues a holistic plant management. TPM contains PM, a preventive maintenance and comes with improvements of maintainability [5]. Related works have shown that the use of TPM allows reduction of costs in maintenance areas, e.g. 70% reduction in loss of production or 50%–90% decrease of setups for maintenance activities [6,7]. Mobley [8] shows, that the costs for a preventive maintenance are three times lower, compared to the same repairs made once the failure occurred. The same study describes that in average 28% of the costs are incurred through maintenance actions related to the production costs in a factory [8]. To enable improvement in this regard, it is hence of importance to analyze failures especially with respect to their causes all in the context of PM. If this is known given a production line or machine, maintenance actions can be planned, following closely to the definition, to retain or restore at time points for minimal risk. Once such analysis is available TPM becomes possible.

This work presents statistics of 4.222 failure reports for several components clustered in categories from a large manufacturing factory. We analyzed the failure rates and their causes based on each component. Furthermore, we determined the average severity of the failures for each component depending on the failure causes. Combining both, the severity and the failure occurrences, it is possible to compute a quantitative risk rate. We evaluated and compared that risk rate to the failure rate of each component depending on the different failure causes and analyzed inherent patterns. Finally, we outline suggestions how planned maintenance should be extended with, such that risk is being further reduced consequently along with the cost in technician service and production loss.

2 Methods

2.1 Data

In this work we analyze more than 4.222 maintenance reports of a production line in a manufacturing factory with separating technology. These reports were recorded for two years. Each report includes the date of request, date of issue resolving, which assembly and component was concerned, as well as the cause for the respective failure. Some reports further contained free text comments regarding symptoms and remedies. We categorized the components into seven groups described in the following as categories (Cat.), which are shown as the leftmost column in Table 1. 6.1% of the reports could not be assigned to any category and 10.0% of all entries ware planned actions, such as preventive maintenance or modifications and extensions of the plants at request. Planned and uncategorized entries, together 16.1% of the reports, were excluded from the following analysis, such that 3.541 reports were left. We classified the causes manually into 7 different categories, inspired by the categories used in Pribytkova et al. [9]:

- Wear: Includes all failures caused by wear and aging which includes the case of components being outdated (like software).
- Dirt: Cases related to dirty and contaminated components.
- Operation Errors (Oper.): Failures caused by operational errors or misuse of the components.
- Construction Errors (Const.): Failures caused by false/incompatible constructions.
- False: Failures which turned out to be false alarms and were either not existent at the date of issue resolving or disappeared after a while
- Not found (N.F.): Failures which were caused by unknown reasons, e.g. which were resolved after a restart or after some time.
- Environment (Env.): Failures caused by the environment or the infrastructure. This covers cases, where components are interacting with each other. Power blackouts and subsequent damages fall as well into this group.

2.2 Analysis

The number of maintenance reports related to the respective components clustered in several categories depending on the failure causes are shown in Table 1. Each color highlighted entry of Table 1 represents the probability of each cause (Cause) given a certain component (Comp.), making them the conditional probabilities $P(\text{Cause} \mid \text{Comp.})$. The most right column represents the absolute number of reports for each component. Thus, individual absolute numbers for each entry can be calculated by taking the respective percentage of the sum absolute values. The technicians who were responsible for these production systems were obliged to eliminate the problems as quickly as possible in order to keep the downtime as short as possible. Therefore, we are able to calculate the duration of each job and use it as indicator for the acuteness of an event. To do this, we determined the median of all job durations. The duration is the difference between date of request until the date where the issue has been resolved, per component depending on each cause. Spare parts were normally available and ordered in stock, but we cannot guarantee that parts had to be ordered to solve certain cases. Therefore, the median is particularly well suited for keeping outliers out of the evaluation. Our assumption is, the faster a component needed to be fixed, the more critical its failure had to be. The results are shown in Table 2, where entries are marked as 'N/A' where no reports exist and the respective probability values count 0 in Table 1. To evaluate the risk $r(\cdot,\cdot)$ given a Cause and a component, we use a modified version of the risk priority number [10] defined as:

$$r(\text{Cause}, \text{Comp.}) = 10 \cdot P(\text{Cause} \mid \text{Comp.}) \cdot s(\text{Cause}, \text{Comp.}) \tag{1}$$

where, P represents the aforementioned conditional probability ranging from 0 to 1 and $s(\cdot,\cdot)$ the severity based on the median values of the job duration.

s reassigns the median values into a severity scoring ranging from 1 (lowest severity with median values larger than 9) to 10 (highest severity with median value of 0). s is defined as

$$s(\text{Cause}, \text{Comp}) = \begin{cases} 10 - M(\text{Cause}, \text{Comp}) & 0 \leq M(\text{Cause}, \text{Comp}) \leq 9 \\ 1 & 9 < M(\text{Cause}, \text{Comp}) \end{cases} \quad (2)$$

$M(\cdot, \cdot)$ represents the median job duration values from Table 2 between a component and cause. We have chosen 9 days as the upper bound, as 9 days is the 75 percentile of all median values. Choosing such range, we evaluate median values of e.g. 100 days equally urgent as 200 days. The left factor of Eq. 1 ($10 \cdot P(\text{Cause} \mid \text{Comp.})$) ranges from 0 to 10 and the right factor s ranges from 1 to 10, its product r ranges from 0 to 100, describing the final risk factor. These risk results are shown in Table 3.

3 Evaluation

3.1 Failure Rates

Table 1 shows, with few exceptions, the most prominent cause across all components and categories is wear. In the course of this evaluation, "wear" covers failures caused by worn/aged/outdated components. In particular, the hydraulic and pneumatic category is affected by this. Next to wear, the environment seem to be the second most appearing cause, in particular for devices integrated into the infrastructure like the PCs, controllers in general and robots. Failures caused by operational errors, seem to be dominant with measuring devices, robotics software and medium tanks. In the latter, empty tanks were not refilled in time with the corresponding medium causing errors. Tanks are furthermore prone to dirty causing failures there as well. Errors were not found and disappeared especially with the robotics software and on various devices causing false alarms. Furthermore, error occurred but were not identified only with the software on various devices. Technicians tend to have difficulties with the identification related to software issues on integrated components, as they have limited access to the low-level software procedures.

3.2 Risk Rates

Table 3 visualizes the risk rates highlighted in different colors. Entries with green color, represent the range of low risk between 0 and 25, yellow entries represent moderate risks with a range between 25 and 50, orange stands for high/significant risk with range between 50 and 75 and red values of critical risks beyond 75 to the maximal possible value of 100. Again, the wear column is dominant across many components, but not anymore for the first four categories. In general categories which affect the safety-at-work are associated with higher risks, meaning failures which occur quite often and need to fixed as soon as possible, like Safety in general and Pneumatics and Hydraulics. Speaking of the latter category, leaking

Table 1. Conditional probabilities $P(\text{Cause} \mid \text{Comp.})$ with rows as components and columns as causes.

Cat.	Comp.	Wear	Dirt	Oper.	Constr.	False	N.F.	Env.	Sum
Optics	Lasers	83	1	4	1	2	0	9	207
	Lamps/LEDs	92	1	1	1	3	0	2	95
	Others	45	10	10	0	0	0	34	29
Spindle	Spindles	68	4	4	2	2	2	20	56
	Tool changers	52	10	1	2	17	0	17	81
	Piece changers	62	8	8	8	0	0	15	13
Robots	Elec. comp.	67	10	0	7	0	0	17	30
	Gripper systems	69	7	2	4	4	1	12	81
	Robots in gen.	33	3	10	30	7	0	17	30
	Software	0	0	75	0	25	0	0	8
	Others	50	8	0	0	8	0	33	12
Devices	Balancers	67	3	6	0	9	0	15	33
	Controlers in gen.	36	4	8	3	15	1	34	165
	Elec. comp.	77	5	8	2	2	2	6	66
	Industrail PC	46	6	2	3	7	2	35	101
	Meas. devices	11	1	63	1	13	1	8	71
	Software	8	0	17	0	0	72	3	64
	Others	21	2	7	0	37	0	33	43
Safety	Cover elements	81	5	4	2	0	0	8	100
	Cabins	54	15	11	2	2	2	13	122
	Elec. comp.	32	3	29	18	0	0	18	34
	Doors	70	6	7	4	1	2	10	237
	Others	83	0	0	0	0	0	17	6
Hyd. & Pneum.	Medium tanks	36	27	24	0	3	0	10	90
	Seals s	88	0	0	12	0	0	0	72
	Filter	77	15	2	1	6	0	0	122
	Pumps	83	2	3	1	0	1	10	138
	Hoses	93	2	1	1	1	0	3	190
	Valves	92	0	3	0	1	1	3	144
	Cooling sys.	58	10	14	1	2	0	14	280
	Others	68	6	9	1	4	1	10	140
Drives	Axis	45	5	7	7	11	2	24	208
	Drives in gen.	76	5	1	1	6	2	10	246
	Measureing sys.	40	14	19	1	10	0	15	72
	Engines	72	2	6	6	6	0	9	54
	Elec. comp.	40	14	8	1	21	0	15	72
	Others	24	0	3	0	66	0	7	29

Table 2. Median of job duration per component and cause in days

Cat.	Comp.	Wear	Dirt	Oper.	Constr.	False	N.F.	Env.
Optics	Lasers	3	0	3	30	5	4	2
	Lamp/LED	13	5	22	8	24	N/A	38
	Others	10	9	16	N/A	N/A	N/A	50
Spindle	Spindles	7	133	8	92	2	1	5
	Tool changers	33	8	3	2	19	N/A	4
	Piece changers	7	98	1	0	N/A	N/A	6
Robots	Elec. comp.	39	14	N/A	9	N/A	N/A	3
	Gripper systems	4	4	133	56	3	11	2
	Robots in gen.	13	0	N/A	5	33	N/A	1
	Software	N/A	N/A	9	N/A	0	N/A	N/A
	Others	168	0	N/A	N/A	7	N/A	2
Devices	Balancers	5	0	2	N/A	13	N/A	1
	Controlers in gen.	5	3	2	16	2	0	3
	Elec. comp.	3	1	6	9	175	34	15
	Industrial PC	9	3	0	3	1	14	13
	Meas. devices	8	2	5	56	2	183	7
	Software	3	N/A	5	N/A	N/A	19	0
	Others	7	4	3	N/A	6	N/A	4
Safety	Cover elements	5	0	1	26	N/A	N/A	2
	Cabins	4	2	6	5	3	2	1
	Elec. comp.	5	16	4	251	N/A	N/A	14
	Doors	4	2	11	9	0	2	2
	Others	2	N/A	N/A	N/A	N/A	N/A	13
Hyd. & Pneum.	Medium tanks	2	2	2	N/A	1	N/A	3
	Seals	1	N/A	N/A	5	N/A	N/A	N/A
	Filters	3	6	1	71	18	N/A	N/A
	Pumps	4	0	5	103	N/A	0	3
	Hoses	1	10	15	25	25	N/A	53
	Valves	5	N/A	2	N/A	6	112	5
	Cooling sys.	4	3	4	3	1	1	4
	Others	4	4	6	220	56	3	9
Drives	Axis	3	1	13	5	4	2	1
	Drives in gen.	4	1	5	3	5	1	3
	Measureing sys.	3	1	2	2	5	N/A	2
	Engines	7	1	52	8	17	N/A	16
	Elec. comp.	3	1	2	2	2	N/A	2
	Others	5	N/A	1	N/A	149	N/A	7

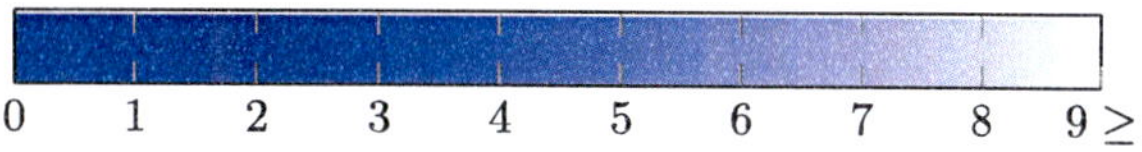

Table 3. Risk Scores ranging from 0 (most lowest risk) to 100 (most highest risk)

Cat.	Comp.	Wear	Dirt	Oper.	Constr.	False	N.F.	Env.
Optics	Lasers	58	1	2	0	1	0	7
	Lamps/LEDs	9	0	0	0	0	0	0
	Others	4	1	1	0	0	0	3
Spindle	Spindles	20	0	0	0	1	1	10
	Tool changers	5	2	0	1	1	0	10
	Piece changers	18	0	7	8	0	0	6
Robots	Elec. comp.	6	1	0	0	0	0	11
	Gripper systems	41	4	0	0	2	0	9
	Robots in gen.	3	3	11	15	0	0	15
	Software	0	0	7	0	25	0	0
	Others	5	8	0	0	2	0	26
	Balancers	33	3	4	0	0	0	13
Devices	Controler in gen.	18	2	6	0	12	1	23
	Elec. comp.	53	4	3	0	0	0	0
	Industrial PC	4	4	2	2	6	0	3
	Meas. devices	2	0	31	0	10	0	2
	Software	5	0	8	0	0	7	3
	Others	6	1	4	0	14	0	19
Safety	Cover elements	40	5	3	0	0	0	6
	Cabins	32	12	4	1	1	1	11
	Elec. comp.	16	0	17	1	0	0	1
	Doors	42	4	0	0	1	1	8
	Others	66	0	0	0	0	0	1
Hyd. & Pneum.	Medium tanks	28	21	19	0	2	0	7
	Seals	79	0	0	6	0	0	0
	Filters	53	6	1	0	0	0	0
	Pumps	49	2	1	0	0	1	7
	Hoses	83	0	0	0	0	0	0
	Valves	46	0	2	0	0	0	1
	Cooling sys.	34	7	8	0	1	0	8
	Others	40	3	3	0	0	0	1
Drives	Axis	31	4	0	3	6	1	21
	Drives in gen.	45	4	0	0	3	1	7
	Measureing sys.	28	12	15	0	5	0	12
	Engines	21	1	0	1	0	0	0
	Elec. comp.	28	12	6	0	16	0	12
	Others	12	0	2	0	6	0	2

0 *low* 25 *mod.* 50 *high* 75 *crit.* 100

liquids can cause a high risk of accidents and neighbouring components might be damaged as well, resulting in higher risks rates as they fail frequently with urgent priority. Laser components seem to be affected with higher risks in wear as well. The reason of its high risk are flash lamps installed in laser-cutting machines, which frequently failures while causing down-times and thus urgent cases. These cases can easily be avoided, if the exchange intervals of laser lamps are preventively managed and monitored such that technicians change them before failures happen. Electrical components of several devices are related to high risk with respect to worn or outdated constituents as well, because once such components fail, larger machines are subsequently affected as well at high failure rates. Operational errors on measuring devices had high failure rates, however ended up in moderate risks, as they are associated with less urgent cases (median day of 56). We claim, the risk can further be decreased if operators are better trained avoiding failures there in the first place. In contrast to operational errors with the software on robots, occurring with higher failure rates with no urgency, result in low risks.

4 Discussion

Every single report and the respective effort for the technical service cause costs. The combination of downtime and service, represented in the risk score, is related to direct costs for service but also indirectly associated with a loss in production amount. Saravanan et al. [11] showed, that the main cause of failure is the component damage by lathes and milling machines. Our evaluation has confirmed this, but we discovered that this is also true for the whole manufacturing machinery. Some categories and components result in unexpected lower risks, despite their higher failure probability as they were not associated with urgent need for resolving. In our evaluation it furthermore turned out, that only 10.0% of all reports were planned maintenance cases, meaning 90.0% were unplanned actions. We therefore recommend to reconsider the maintenance intervals. However, these numbers depend on the planned maintenance intervals for the factory at hand and might differ to other factories with different maintenance schedules. As mentioned in the introduction the costs for a preventive maintenance are three times lower compared to reactive repair actions. Our results have shown, the introduction of TPM systems especially in the high risk categories is sensible. We assume, if the planned maintenance in particular focuses more on worn and outdated components, risks can be decreased significantly especially to hydraulic and pneumatic components as they result in a lower labour safety and endanger surrounding components. In general, we recommend to follow the manufacturer's maintenance intervals, especially for hoses, valves, seals and filters (and laser flash lamps), as this will reduce costs in the long run and increase safety at work. For future work, we propose, a more comprehensive and structured maintenance report might give even more insights into the manufacturing status. For instance, reports might include the exact affected machine (by some identifier) and a fixed set of categories for the assembly and the respective cause, resulting on lower analysis effort. With such template reports, automated maintenance analysis and consequently predictive and preventive maintenance can be implemented more easily.

5 Conclusion

In this work, we have analyzed over 4000 maintenance reports of a manufacturing factory. We have categorized each report into 36 components and distinguished between 7 causes. We found out that worn, aged and outdated components result in the highest failures rated across all categories with few exceptions. However, not all cases are associated with higher risks considering the respective urgency related to each failure case. We evaluated the said risk as well using the median job duration as an indicator for the severity of each case. In the risk analysis it turned out, especially hydraulic and pneumatic components have a significant-up to critical-risk, as the labour safety and other components are threatened by their failure. In conclusion, the introduction of TPM targeting especially the categories associated with higher risks, might reduce costs in the long run and is therefore recommended.

References

1. Manzini, R., Regattieri, A., Pham, H., Ferrari, E.: Maintenance for Industrial Systems. Springer, London (2009). https://doi.org/10.1007/978-1-84882-575-8
2. Huiskonen, J.: Maintenance spare parts logistics: special characteristics and strategic choices. Int. J. Prod. Econ. **71**(1–3), 125–133 (2001)
3. Swanson, L.: Linking maintenance strategies to performance. Int. J. Prod. Econ. **70**(3), 237–244 (2001)
4. Chand, G., Shirvani, B.: Implementation of TPM in cellular manufacture. J. Mater. Process. **103**(1), 149–154 (2000)
5. Chan, F.T.S., Lau, H.C.W., Ip, R.W.L., Chan, H.K., Kong, S.: Implementation of total productive maintenance: a case study. Int. J. Prod. Econ. **95**(1), 71–94 (2005)
6. Koelsch, J.R.: A dose of TPM: downtime needn't be a bitter pill. Manuf. Eng. **110**(4), 63–66 (1993)
7. McKone, K.E., Schroeder, R.G., Cua, K.O.: The impact of total productive maintenance practices on manufacturing performance. J. Oper. Manage. **19**(1), 39–58 (2001)
8. Mobley, R.K.: An Introduction to Predictive Maintenance, 2nd edn. Elsevier, Amsterdam (2002)
9. Pribytkova M., Sahno J., Karaulova T., Shevtshenko E., Maleki M., Cruz-Machado V.: Reliability analysis module development for production route elaboration (2012)
10. Abdelgawad, M., Fayek, A.R.: Risk management in the construction industry using combined fuzzy FMEA and fuzzy AHP. J. Construct. Eng. Manage. **136**(9), 1028–1036 (2010)
11. Saravanan S., Yadava G.S., Rao P.V.: Machine tool failure data analysis for condition monitoring application. In: Proceedings of the 11th National Conference on Machines and Mechanism, Delhi, pp. 18–20 (2003)

Mechanisms of Textile Machines, Optimization of Mechanisms and Machines

Check for updates

Modal Properties of the Heald Frame of a Weaving Loom

Martin Bílek(✉), Šimon Kovář, and Josef Skřivánek

Technical University of Liberec, Liberec 46001, Czech Republic
{martin.bilek,simon.kovar,josef.skrivanek}@tul.cz

Abstract. This paper examines the problematic of determining the natural frequencies of the heald frame of a weaving loom. When calculating these frequencies, the effect is considered of the stiffness of the connection of the cross beam and side support. In solving the problematic, we performed a sensitivity analysis on the effect of the stiffness of the connection and the change of individual structural parameters on the observed values of the natural frequencies. The calculated values were verified by measuring on the actual components.

Keywords: Heald frame · Weaving loom · Natural frequency · Modal properties · Stiffness

1 Introduction

Enhancing operating frequency is an important factor in increasing the performance and production of weaving looms. It also brings with it an increase in dynamic stress of all the loom components and the condition as a whole. These high performances are limited in part by the possibilities of the weft insertion, and in part by the dynamic characteristics of the basic mechanisms. Mechanisms with the highest load are the shedding mechanism and beat-up mechanism.

The weaving loom's shedding mechanism opens the warp, thus creating a wedge-shaped space for the insertion of the weft. The basic task of the shedding mechanism is to convert the rotational motion of the main shaft into linear reciprocating motion of the heald frame. It is also the most massive part of the kinematic chain describing the shedding mechanism.

Structurally, the basic elements of the heald frame are the upper and lower beam (Fig. 1, position 1, 2) and side supports (Fig. 1, position 3). The most important properties of the heald frame are its stiffness and mass. When weaving, the deflection of the upper and lower beam and their vibration must be minimal. Excessive mass of frames not only increases the shedding mechanism load, but it also has a negative effect on the load of its own frame, and thus its service life. Vibration of the frames increases the noise level and adversely affects the vibration of the warp, and thereby the formation of the fabric.

There are precious few articles in circulation that delve into the problematic of the heald frame and its design. Selected resolved problems can be found, e.g., in [1–8].

© The Author(s), under exclusive license to Springer Nature Switzerland AG 2022
J. Beran et al. (Eds.): TMM 2020, MMS 85, pp. 289–295, 2022.
https://doi.org/10.1007/978-3-030-83594-1_30

2 Natural Frequency of the Heald Frame

The paper deals with the problematic of the basic structural node of the weaving loom: the problematic of the heald frame. The basic problematic this paper examines is an analysis of the heald frame in the case where the cross beam is connected to the side supports by a connection with a given stiffness value. Subsequent analysis of the influence of this connection stiffness on the dynamic stiffness of the heald frame, performed on an appropriately defined, simplified mathematical model, renders basic information regarding the suitability of using the given type of heald frame on the weaving loom. The size of the first two natural frequencies of the frame that occur in the plane of the frame movement will characterize for us the dynamic stiffness.

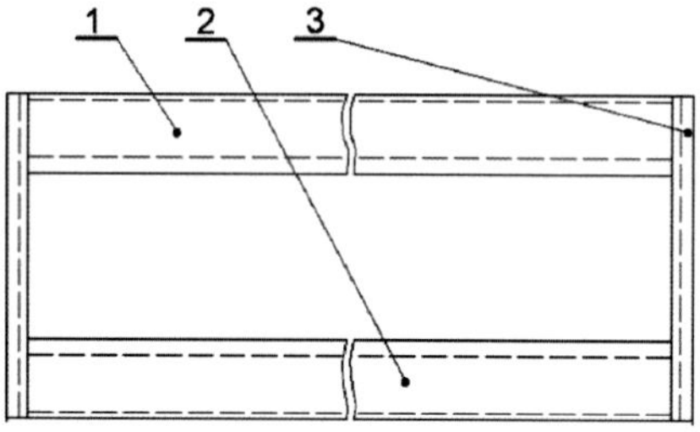

Fig. 1. Heald frame of the weaving loom.

For a description of the heald frame, we utilize a simplified mathematical model formed by longitudinal and vertical beams replacing cross beams and side supports. These are connected by a connection defining the transmission capability of a precisely determined bending moment from one beam to the other (the influence of the stiffness of the cross beam and side support connection). Properties of beams are described by their stiffness and material. The entire frame is attached to the base by means of supports enabling rotation.

In line with conclusions drawn from findings obtained by measuring the actual construction, we introduce the assumption that the behavior of the connection across the range of possible loads can be linearized. Furthermore, we assume connection stiffness as being the same for both load directions. The beams will be considered flexible bodies, i.e., as bodies that can be flexibly deformed over their entire volume and as bodies having uniform mass distribution over their entire volume. The whole heald frame can be replaced by a pair of beams attached to the base using flexible supports. If we are resolving the case of a flexible beam loaded by its own mass, we arrive at the equation of its deflection line in the form:

$$w(x) = [C_1 cosh(\lambda x) + C_2 sinh(\lambda x) + C_3 cos(\lambda x) + C_4 sin(\lambda x)] \quad (1)$$

where

$$\lambda^4 = \Omega^2 \frac{m_1}{EI} \quad (2)$$

(m_1 is the unit mass of the beam, Young's modulus E and I s moment of area). Four integration constants (C_1 to C_4) correspond to the four boundary conditions that must be determined. Using the boundary conditions, we obtain a system of linear homogeneous equations. A condition of non-trivial resolution is a zero value of the determinant. From that follows the calculation of the eigenvalues λi, respectively using (2) the calculation of natural frequencies Ωi. Assuming that the beam is secured to the base using flexible elements, we can ascertain natural frequencies by modifying the Eq. (1). If we properly combine the x-coordinates, we can define the so-called Krylov functions: S(λx), T(λx), U(λx), V(λx) [9].

$$S(\lambda x) = \frac{1}{2}[cosh(\lambda x) + cos(\lambda x)],\ T(\lambda x) = \frac{1}{2}[sinh(\lambda x) + sin(\lambda x)]$$
$$U(\lambda x) = \frac{1}{2}[cosh(\lambda x) - cos(\lambda x)],\ V(\lambda x) = \frac{1}{2}[sinh(\lambda x) - sin(\lambda x)] \tag{3}$$

Using these relations, the Eq. (1) can be rewritten into the following relation:

$$w(x) = [A.S(\lambda x) + B.T(\lambda x) + C.U(\lambda x) + D.V(\lambda x)] \tag{4}$$

Where A, B, C, D are constants dependent on the boundary conditions of each specific task. In deriving the case of support according to Fig. 2, we employ the principle described in [9], which adequately generalizes the flexible support of the beam. The free undamped vibration of the flexibly supported prismatic beam will be solved at length L. Figure 2. shows stiffnesses k_1,k_2 of the flexible support at positions 1 and 2. Stiffnesses c_1 c_2 is then the stiffness against rotation of the beam end in positions 1 and 2.

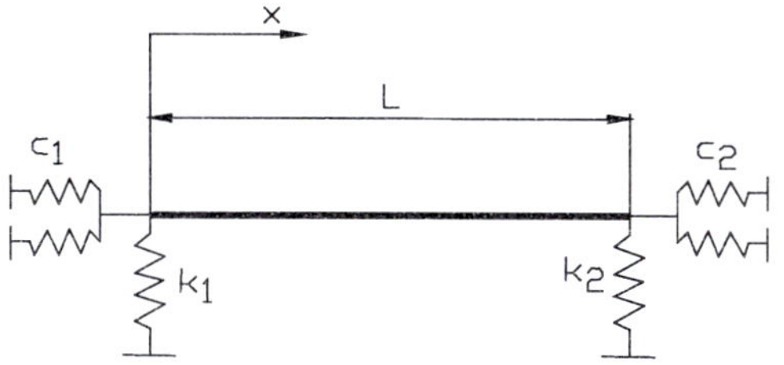

Fig. 2. Flexibly positioned supports

For the given case of the support of the beam, we can write the following boundary conditions:

$$\begin{aligned} E.I.w''(0) &= c_1.w'(0) \\ E.I.w'''(0) &= k_1.w(0) \\ E.I.w''(L) &= c_1.w'(L) \\ E.I.w'''(L) &= k_1.w(L) \end{aligned} \tag{5}$$

By differentiating the Eq. (4) to x, we progressively obtain relations for individual deformation and strength quantities of the beam. If we use these equations and using the boundary conditions (5), we obtain the Eq. (6) from whose roots it is possible to determine the natural frequencies of the beam.

$$\begin{aligned}
&\lambda^{10}\left(U^2 - TV\right) + \lambda^9(H_1 + H_2)(UV - ST) - \lambda^8 H_1 H_2\left(T^2 - V^2\right) \\
&- \lambda^7(K_1 + K_2)(TU - SV) \ + \ \lambda^6(K_1 H_2 + K_2 H_1)\left(S^2 - V^2\right) \\
&+ \lambda^6(K_1 H_1 + K_2 H_2)\left(S^2 - TV\right) \ + \ \lambda^5 H_1 H_2(K_1 + K_2)(ST - \ UV) \\
&+ \lambda^4 K_1 K_2\left(T^2 - V^2\right) + \lambda^3 K_1 K_2(H_1 + H_2)(TU - SV) \\
&- \lambda^2 K_1 K_2 H_1 H_2\left(TV - U^2\right) = 0
\end{aligned} \tag{6}$$

In the equation, the values K_1 K_2 H_1 H_2 mean the relationships:

$$K_1 = \frac{k_1 L}{EI}, \ K_2 = \frac{k_2 L}{EI}, \ H_1 = \frac{c_1 L}{EI}, \ K_1 = \frac{c_2 L}{EI},$$

Equation (6) is the general solution of the frequency equation by which we can solve any case of structural arrangement of the heald frame. Using these equations, we can solve, e.g., long frames using central reinforcements. It is only necessary to know the exact values of individual stiffnesses k_1, k_2, c_1, c_2.

In the case of the simplified frame that we are solving, we can assume stiffnesses $k_1 = k_2 = \infty$. When considering the stiffness value $c_1 = c_2 = c$, we obtain the relation for calculating natural frequencies of the beam, which replaces part of the frame (7).

$$\lambda^2\left(T^2 - V^2\right) + 2\lambda\frac{cL}{EI}(UT - SV) + \frac{c^2 L^2}{E^2 I^2}\left(U^2 - VT\right) = 0 \tag{7}$$

The theoretical analysis indicates that to calculate natural frequencies, it is important to determine the stiffness value c of the spring generating moment acting in positions of the connection with the side support against the direction of its rotation, and in doing so, simplify the calculation of the frame for calculating the flexibly positioned beam. The overall stiffness c consists of two parts: the stiffness of the side support and the cross beam connection (c_s), and the stiffness of the remaining parts of the frame (c_f). These two stiffnesses are arranged in series in a row. We select the stiffness of the connection c_s in the entire range of possible stiffnesses. That means we begin with the stiffness value of the connection of the approaching pinned support, and end with the stiffness value of the connection fully transferring the moment between the cross beam and the side support. The calculations consider stiffnesses in the interval $2e^3$ to $2e^{12}$ Nmm rad^{-1}. Stiffness c_F in the calculation is invariable; it is determined by geometric and structural parameters of the given frame.

3 Verification of the Calculation

To verify the results arising from the mathematical relations, it is necessary to compare these values with the results of the measurement of the actual components. For the given measurement, we used heald frames with connection type Standard and Dynatex. To ensure the fidelity of the results due to calculation assumptions and attaching to a shedding mechanism, it was necessary to adapt the measurement conditions.

The measured and calculated values of natural frequencies are summarized in Table 1. The measurements performed verified high concordance of the calculation results of natural frequencies with measurement. It can be stated that by using the analytical calculation, it is possible to assess with high accuracy the behavior of the actual component, with the possible subsequent optimization of the structural arrangement of the examined node of the weaving loom.

Table 1. The parameters of the heald frame.

Type of connection		Standard	Dynatex AL-AL	Dynatex AL-St
Length - L	[mm]	2112	2036	2048
Second moment of area (main beam) – I	[mm^4]	596600	596600	596600
Material of main beam		aluminum	aluminum	Aluminum
Second moment of area (side supports)	[mm^4]	8703	2349	13040
Material of side supports		aluminum	aluminum	steel
Stiffness of connection – c	[$N.m.rad^{-1}$]	1,2e7	3,0e7	3,0e7
Natural frequency determined by calculation	[Hz]	61.93	67.12	72.60
Natural frequency determined by measuring	[Hz]	62.90	66.50	73.05

4 Natural Frequency When Changing the Side Support Stiffness

Using this solution, it is possible to proceed to the sensitivity analysis of the natural frequency in relation to individual design parameters and the connection stiffnesses. In terms of a clear presentation of the results, it is possible to create three-dimensional graphs depicting the influence of the connection stiffness and change interval of the given parameter on the ascertained value of the natural frequency. An example of graph for 2 natural frequencies and a change in the stiffness of the side support is provided in Fig. 3. The graph shows us the change in the natural frequency in the stiffness interval, where one stiffness limit value approaches the pinned support connection, and a second case where a connection is used, which by its nature approaches insertion.

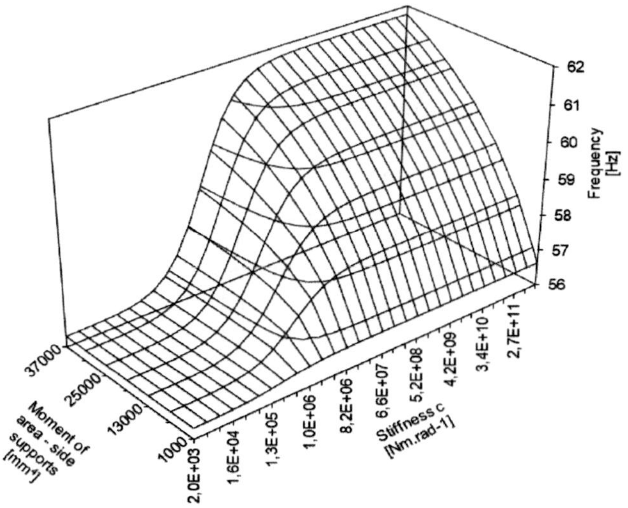

Fig. 3. First natural frequency upon a change in side support stiffness and a change in connection stiffness.

5 Conclusion

The paper shows the possibility of calculating natural frequencies of the heald frame in the case of considering the stiffness of the connection of the cross beam and side support. An example was performed of the derivation of the frequency equation for the case of calculating the natural frequency of the heald frame. The measurements performed verified concordance of the results of calculations and measurements. It can be stated that by using the analytical calculation, it is possible to assess the behavior of the actual component and then optimize the structural arrangement of the examined node of the weaving loom.

Acknowledgments. The paper has been elaborated with financial support of TUL in the framework of Institutional Support for Longterm Development of Research Organizations.

References

1. Qiu, H. Wan, H.: Dynamic design of heald frame parts based on material properties. J. Silk **56**(7) (2019)
2. Lee, D.G., Lee, C.S., Oh, J.H., Jeon, H.S.: Composite heddle frame for high-speed looms. Compos. Struct. **47**(1–4), 507–517 (1999)
3. Xu, S., Feng, Z.: Modal analysis of heald frame and experimental verification. J. Soochow Univ. Eng. Sci. Edn. **28**(2), 34–36 (2008)
4. Bílek, M.: Analysis of the dynamic properties of the heald frame of a weaving loom. In: VIIth International Conference on the Theory of Machines and Mechanisms, TU Liberec (1996)
5. Bílek, M., Jágrová, J., Mrázek, M.: Calculation of the stiffness of the heald frame of a weaving loom. In: VIIIth International Conference on the Theory of Machines and Mechanisms, TU Liberec (2000)

6. Bílek, M., Kovář, Š.: Mathematical model of the heald shaft of the weaving loom. Buletinul institutului polytehnic din Iaşi. Tech. Univ. Iaşi **5**(1), 375–382 (2007)
7. Schwarzfischer, F., Kurtenbach, S., Onischke, J., Corves, B.: Design and development of a heddle shaft mechanism for air-jet weaving machines. In: Beran, J., Bílek, M., Žabka, P. (eds.) Advances in Mechanism Design II. MMS, vol. 44, pp. 23–28. Springer, Cham (2017). https://doi.org/10.1007/978-3-319-44087-3_3
8. Bílek, M., Skřivánek J.: Mathematical modeling of the system shedding motion – heald – warp. AUTEX Res. J. **14**(2), 42–46 (2013)
9. Gonda, J.: Bending Vibration of Beams. SAV, Bratislava (1977)

Measurement, Evaluation and Comparison of Behavior of Linear SGT Motor with Oscillating Mass

Zdeněk Braier(✉), Pavel Šidlof, Martin Pustka, and Pavel Fišer

VÚTS, a.s., Svárovská 619, Liberec XI-Růžodol I, 460 01 Liberec, Czech Republic
{zdenek.braier,pavel.sidlof,martin.pustka,pavel.fiser}@vuts.cz

Abstract. The designers and engineers design and operate many production lines and machines or tools with linear motor (sliders). The linear motion is used for positioning of tool or product to the exact position or combined with other precise machining during technology process. The article describes the measurement, evaluation and comparison of linear SGT motor behavior. The moving part (slider) of a real SGT motor prototype was completed with oscillating mass on springs. The system was operated by different stroke dependence. The measurement and evaluation were focused on eliminating or minimizing the undesirable oscillation of mass on springs a very short time. The acceleration sensors were mounted to the motor moving part (slider) and to the oscillating mass on springs. The moving part (slider) stroke within the exact positioning part was measured by RSLM linear scale system. The electrical quantities were measured by sensors in switching box. All electrical and mechanical quantities were synchronously measured and recorded for successive research and evaluation.

Keywords: Motor behavior · Linear SGT motor · Oscillating mass · Laboratory test device · Measurement and evaluation · Stroke dependencies

1 Introduction

The linear SGT motors with moving part (slider) are used in many production lines and machines. These motors are designed for exact positioning of part or tool during precise operations. The summary of design and dynamic properties with research and development of linear motor is given e.g. in [1–3]. The main topics and targets of this research and evaluation focus on linear motor possibilities (limits) and moving slider positioning precision at different machine settings. To study the influence of a dynamic load on the motor behavior, an oscillating mass on springs was mounted to motor moving part (slider) to excite auxiliary loading vibrations.

Figure 1a shows the detail of laboratory test device (equipment) with linear SGT motor [4, 5] moving (primary) part (slider), oscillation mass mounted on a pair of flat springs and acceleration sensors. Pos. 1 is the desk of moving (primary) part (slider). Pos. 2 is the metal cover of static (secondary) field with magnets. Pos. 3 is the oscillating loading mass with exactly defined weight, which is assembled to the moving slider by a pair

© The Author(s), under exclusive license to Springer Nature Switzerland AG 2022
J. Beran et al. (Eds.): TMM 2020, MMS 85, pp. 296–303, 2022.
https://doi.org/10.1007/978-3-030-83594-1_31

of flat springs (pos. 8). The total weight of parts (oscillating mass with brackets, springs, grips, screws, etc.) added to the desk of moving slider is 1.55 kg. The thermally insulated sensor Brüel&Kjaer 4383S (pos. 4, [6]) measures acceleration of oscillating mass and the insulated sensor Brüel&Kjaer 4383 (pos. 5) measures acceleration of motor moving slider. The thermal insulations were realized using foam for temperature drift elimination. Pos. 6 is the readhead SR030A of Renishaw linear system [7]. RSLM high accuracy stainless steel linear scale [8] is shown as pos. 7. Figure 1b presents the switching box (terminals) [9] with connection wires of servopack SGDV and single phase U, V, W wires. The regulation of switching frequency across the total power range was 10.668 kHz. Pos. 9 are the current probes U, V, W, pos. 10 are the clamps for U, V, W voltage measurement. The electrical quantities were measured on supply wires between the output of switching box (terminals) and the input connectors of linear SGT motor.

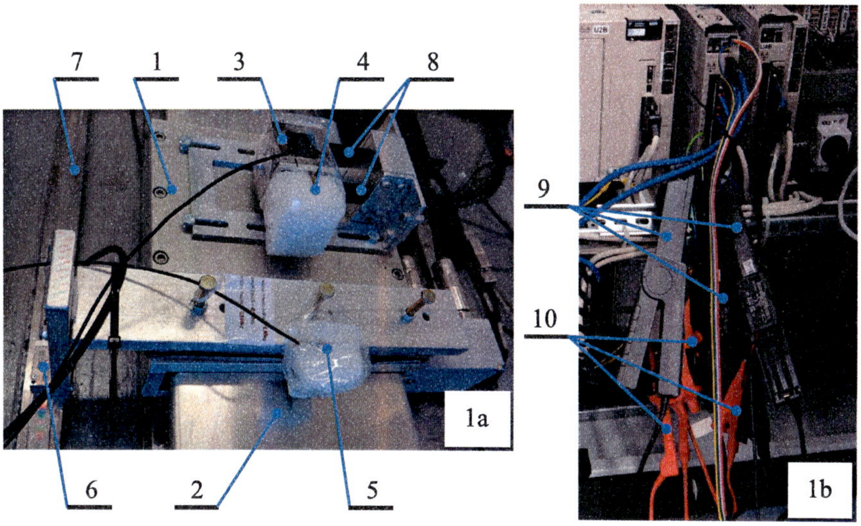

Fig. 1. A moving (primary) part (slider) with oscillation mass on springs, acceleration sensors, brackets, etc. (a), detail of electrical switching box (terminals) with servopack SGDV, current probes and voltage clamps (b)

Several different stroke dependencies were programed and prepared for simulation and evaluation of system behaviour and for measurement. The stroke dependencies were programed by means of block diagrams, software and simulation in MathWorks Matlab Simulink. Figure 2 shows the kinematic quantities of two types of stroke dependencies denoted by F9-1 and F9-2, i.e. stroke (green), velocity (blue) and acceleration (red). The horizontal axes are scaled in 0° to 360° equal to angle of virtual master axis, which was recalculated to time base of frequency 8 CPM (≈ cycles per minute) and stroke from 0 (≈ 1st limit, start position) to 140 mm (≈ 2nd limit, finish position). The total length 140 mm of stroke is located in the central part of the maximal possible stroke length of the linear motor. Equation (1) defines the filter that was used for calculation and simulation of stroke dependencies, i.e.

$$F_1 = 1/(1 + \tau s) \tag{1}$$

where F_1 is transfer function of filter, s is parameter of transfer function, τ is time parameter. The parameter τ was set to value 0.01 for the curve (stroke dependence) F9-1 and to value 0.1 for the curve (stroke dependence) F9-2. These simulation, evaluation and stroke dependencies are just a part of simulation and evaluation. Other stroke dependencies (functions) and linear SGT motor research, development, simulations, measurements, evaluations and results are summarized in [10–12].

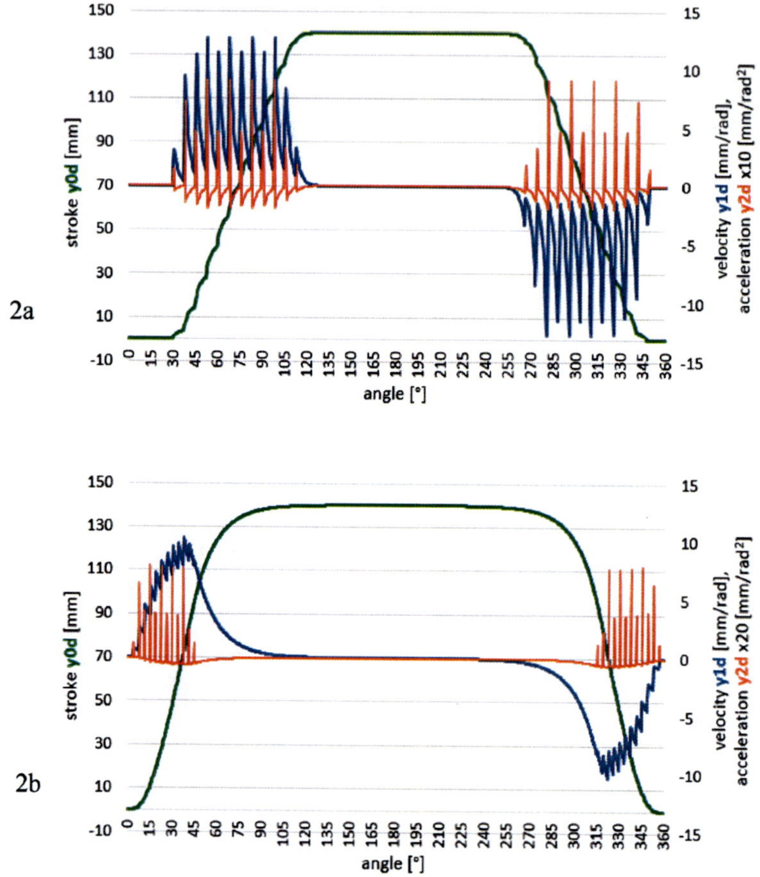

Fig. 2. Stroke dependence F9-1 (a, top) and stroke dependence F9-2 (b, bottom) with stroke, velocity and acceleration courses

2 Verification and Representation Method

All measurements and tests of linear motor behavior were prepared and realized by using the real laboratory test device, which included the linear SGT motor, the switching box, the dynamically (flexible) mounted oscillating mass on a flat springs and measurement equipment (sensors, probes, clamps, etc.). The variables and parameters were synchronously recorded by the portable modular All-in-One analyzer Dewetron DEWE-5000 [13]. The post-processing research, evaluation and comparison of the data with graphs were carried out by using Wolfram Mathematica software.

The linear SGT motor Yaskawa has the specification SGT-C-F35A120-0535-NA0020-EC, see Table 1.

Table 1. Yaskawa SGT linear motor specification.

Code	Specification
SGT	Sigma Trac Series Linear Slider
C	Not Specified
F	Type Iron Core
35	Magnet Height
A	Voltage 230 [VAC]
120	Length of Coil Assembly [mm]
0535	Stroke Length [mm]
N	Encoder Manufacturer (Numeric)
A	Incremental Encoder Type
0020	Encoder Pitch 20 [μm]
E	Not Specified (Home Sensor Option)
C	PNP Output Limit Switch

Table 2 Describes the Renishaw linear system specification and brüel&kjaer acceleration sensor sensitivity.

Table 2. Renishaw system and BK 4383 sensors specification

System, sensor	Specification	Value
Renishaw	Scale Length	1030 [mm]
RSLM-SS-20U3A-1030-A	Thermal Expansion	10.8 ± 0.2 [μm/m/°C]
	Accuracy grade	± 2.2 [μm]
	Error Span	± 1.16 [μm]
Brüel&Kjaer sensor 4383	Reference Sensitivity at 159.2 [Hz], 20 [ms^{-2}] RMS and 23 [°C]	3.101 [pC/ms^{-2}]
Brüel&Kjaer sensor 4383	Reference Sensitivity at 50 [Hz], 100 [ms^{-2}] and 23 [°C]	3.18 [pC/ms^{-2}]

3 Results

All time courses of electrical and mechanical quantities were straighten out to start of slider movement. The power supply voltage of wiring phases U, V, W of moving (primary) part (slider) is PWM modulated. There were measured the other voltage courses modulated to the original sinus courses during the slider movements between start and finish position. In the case when the moving part (slider) stops in the start or finish position, the original PWM non-modulated sinus voltage of phases U, V, W were measured. Figure 3 shows the measured time from −0.1 to 0.5 s of voltage phases U, V, W. The graphs show almost the similar courses and amplitudes of voltage for all phases and both stroke dependencies.

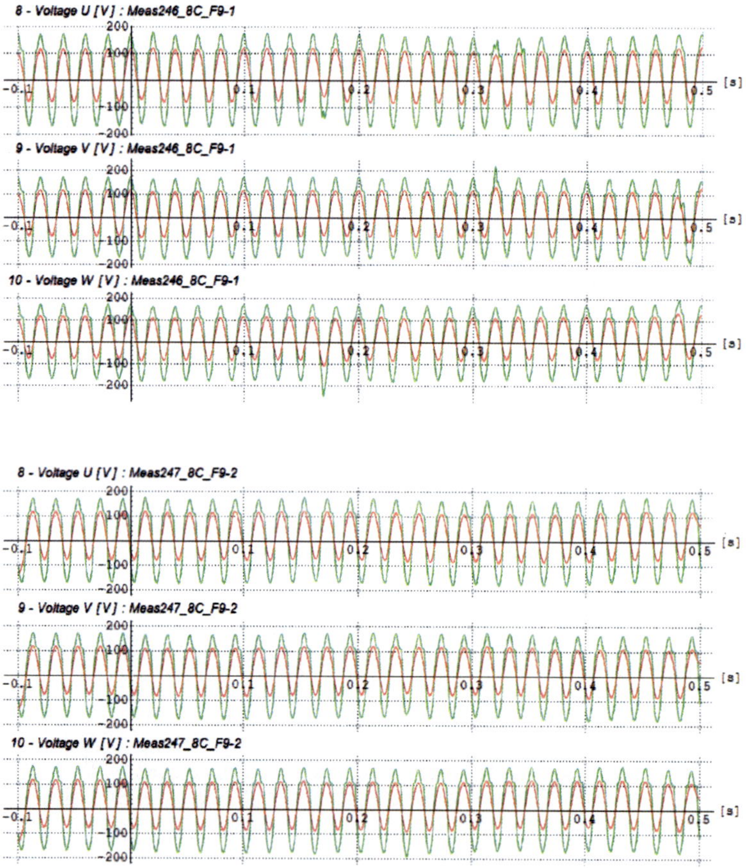

Fig. 3. Time courses of measured and evaluated the electrical quantities, voltage of phases U [V], V [V], W [V]. Stroke dependence F9-1 and F9-2, green - measured curve, red - curve filtered by a moving average, time range –0.1 to 0.5 s (correspond to the beginning of the movement from the zero state)

Figure 4 shows the time courses of current phases U, V, W. Comparing the graphs of single current phases for both strokes, it is possible to notice that the current amplitudes of stroke F9-2 are lower (approximately twofold) than of stroke F9-1. Also the time of power supply is shorter during moving part (slider) position change. This can be attributed to the setting of parameter τ in the filter equation.

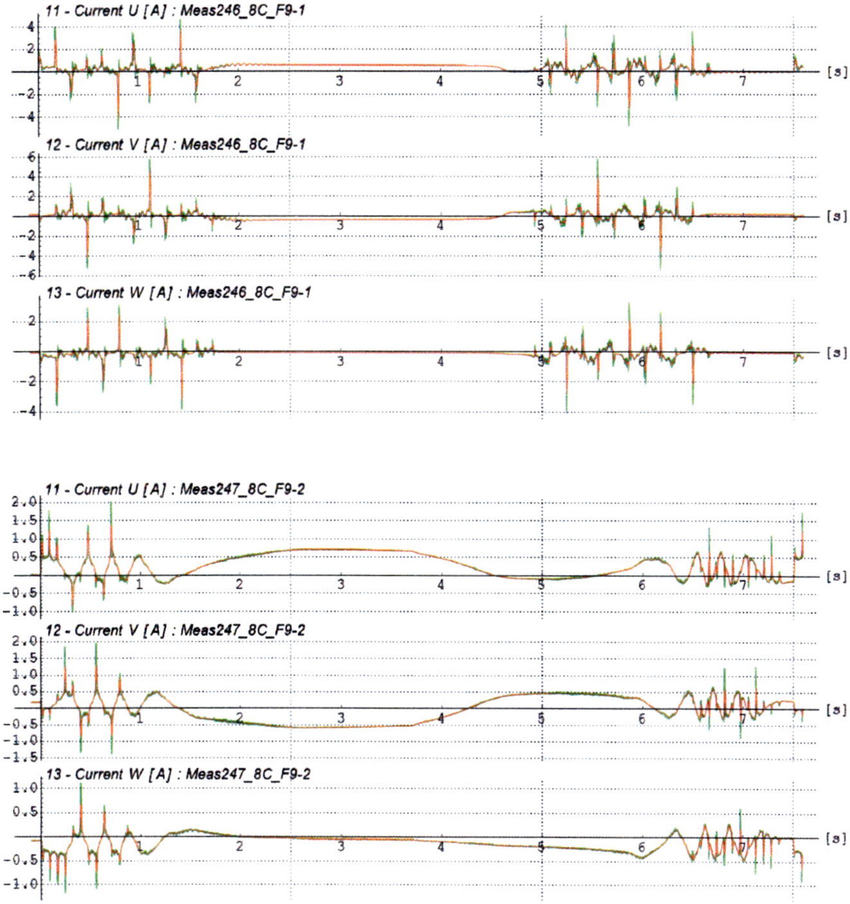

Fig. 4. Time course of measured and evaluated the electrical quantities, current of phases U [A], V [A], W [A]. Stroke dependence F9-1 and F9-2, green - measured curve, red-curve filtered by a moving, time range 0 to 7.5 s (equal to one complete stroke with length 140 mm, from start to finish position and backwards)

Figure 5 depicts the time courses of total power, motor stroke and relative oscillating mass stroke. Comparing the courses similarly as in previous paragraph, it is possible to analyze the energy savings within the motor stroke and also the reduction of mass oscillation. The energy saving of F9-2 is approximately fourfold compared to F9-1. The reduction of relative oscillating mass stroke is roughly tenfold. The time base of relative oscillating mass stroke for F9-2 is approximately twice shorter than time base for F9-1.

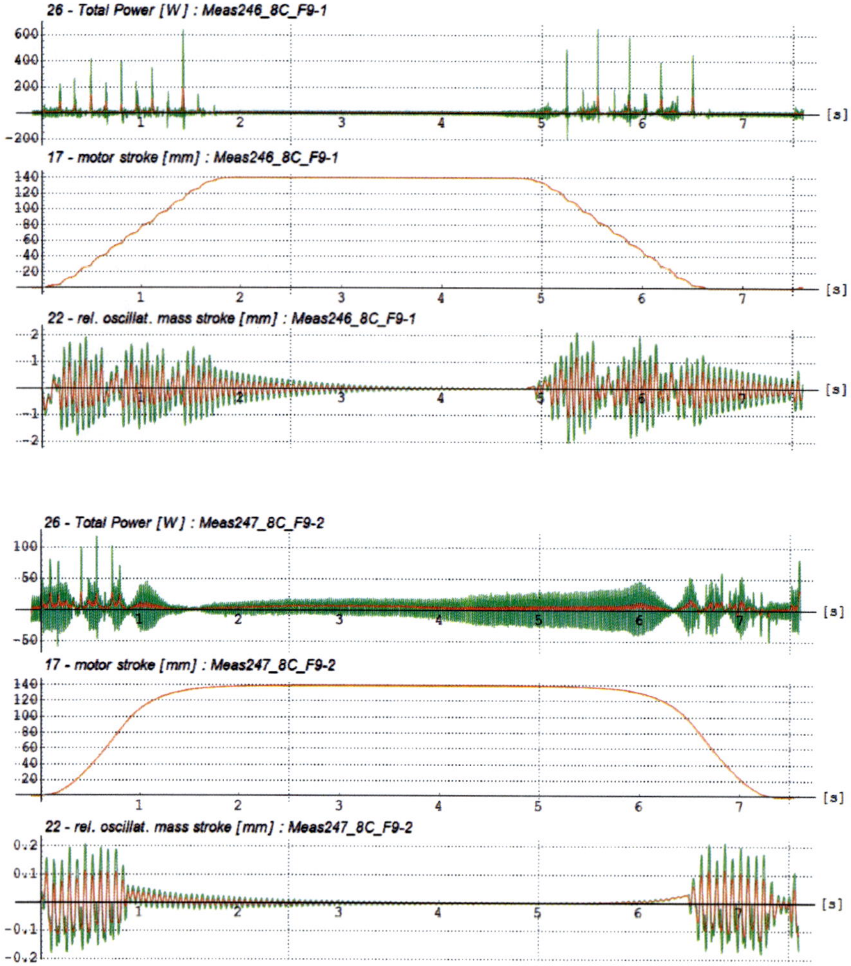

Fig. 5. Time courses of measured, evaluated and calculated electrical and mechanical quantities, total power [W], motor stroke [mm] and relative oscillating mass stroke [mm]. Stroke dependencies F9- and F9-2, green-measured curve, red-curve filtered by a moving average, time range 0 to 7.5 s (equal to one complete stroke with length 140 mm, from start to finish position and backwards)

4 Summary

The article deals with the dynamic behavior of the Linear SGT Motor with a flexible mounted loading mass. The main topic is focused on research, development and evaluation of linear motor settings and its impact on the response of the real laboratory test device (stend). The response of this device was measured and evaluated for two different stroke dependencies. The energy savings were achieved by the reduction of amplitude and by shortening the time of relative oscillating mass stroke. The results and graphs indicate the possible adjustment of linear motor settings and further:

- approx. fourfold energy savings (total power) calculated from the measured and evaluated three-phase currents and voltages between switching box (terminals) and connectors (input) of the linear motor,
- approx. tenfold reduction of the measured and evaluated amplitude of relative oscillating mass stroke, calculated from its acceleration,
- approx. twofold shorter time base of measured and evaluated relative oscillating mass stroke, calculated from its acceleration.

The presented savings and reductions are the results of research, development, measurement, evaluation and comparison of two different settings of stroke dependencies F9-1 (with filter parameter $\tau = 0.01$) and F9-2 (with $\tau = 0.1$), stroke amplitude (length of primary slider movement) 140 mm and frequency 8 CPM.

Acknowledgments. This work was supported by the Czech Ministry of Industry and Trade in the framework of the institutional support for long-term conceptual development of research organization – recipient VÚTS, a.s.

References

1. Souček, P.: Servomechanismy ve výrobních strojích, CTU Prague Publishing House, Prague (2004). ISBN: 80-01-02902-6
2. Jirásko, P.: a kolektiv autorů: Mechatronika pohonů pracovních členů mechanismů, VÚTS, a.s., Liberec (2015). ISBN: 978-80-87184-63-9
3. Hrabovcová, V., Rafajdus, P., Makyš, P.: Analýza elektrických strojov, Žilinská univerzita v Žilině, EDIS-vydavatel´ské centrum ŽU, Žilina (2017). ISBN: 978-80-554-1323-5
4. Linear slider datasheet Homepage. https://www.yaskawa.eu.com/products/motion-control/seriesdetail/serie/linear-servomotors_5919. Accessed 16 Apr 2021
5. Linear slider Homepage. https://www.yaskawa.com/products/motion/sigma-5-catalog. Accessed 16 Apr 2021
6. Brüel&Kjaer acceleration sensors Homepage. https://www.bksv.com/en/transducers/vibration/accelerometers/4383. Accessed 16 Apr 2021
7. Renishaw SIGNUM encoder system, datasheet L9517-9155-05-A. Renishaw plc, issued 0711 (2004–2011)
8. Renishaw RSLM high accuracy stainless steel scale, datasheet L-9517-9305—06-B, Renishaw plc, issued 1116 (2007–2016)
9. Crhák, V.: Stand Yaskawa v.2 – Manuál k rozvaděči ER1, VÚTS, a.s., Liberec (2015)
10. Braier, Z., Šidlof, P., Fišer, P.: Measurement, evaluation and comparison of positioning accuracy and other SGT linear motor quantities. In: 14th International Conference Mechatronic Systems and Materials MSM2018, AIP Conference Proceedings, vol. 2029, pp. 020008-1–020008-8, Opole University of Technology – Faculty of Mechanical Engineering, Zakopane, Poland (2018). ISBN: 978-0-7354-1451-9
11. Braier, Z.: Behavior analysis and measurement of linear motor with dynamically mounted oscillating mass on springs, Ph.D. Thesis, Technical University of Liberec, Faculty of Mechatronics, Informatics and Interdisciplinary Studies, Liberec (2019)
12. Pustka, M., Braier, Z.: Model of a linear motor mechanical load. ACC J. **XXVI**(A) (2020). https://doi.org/10.15240/tul/004/2020-1-004
13. Dewetron analyzer Homepage. https://www.dewetron.com/products/data-acquisition-hardware-chassis/all-in-one-chassis/. Accessed 16 Apr 2021

Check for updates

Verification of Mathematical Model of Mechanical System of Needle Bar Using Laser Doppler Vibrometer

J. Komárek(✉) and R. Nag

Technical University of Liberec, Liberec, Czech Republic
jiri.komarek@tul.cz

Abstract. This paper deals with verification of the accuracy of the mathematical model of the mechanical system of the needle bar. A special needle bar is part of the sewing machine, which creates a decorative stitch. In order to modify the selected mechanical system of the sewing machine for a purpose of reduce the vibrations and noise, a mathematical model was created. The parameters of the mathematical model were previously approximately determined. To refine these parameters, experimental measurements were performed using a Laser Doppler Vibrometer, thanks to which the kinematic quantities of selected parts of the mechanical system were obtained. The obtained curves were evaluated and compared with the results of the analysis of the mathematical model behaviour.

Keywords: Laser Doppler Vibrometer · Needle bar · Sewing machine · Mathematical model verification · Kinematic analysis · Experimental measurements

1 Introduction

The application of a higher degree of automation in production processes is a basic requirement across many industries, including the textile industry. There is, of course, great pressure to reduce production costs in the consumer industry. This trend is also appearing in the clothing industry, where there is an effort to at least partially automate the process of manufacturing clothing.

The subject of this research is a specialized sewing machine designed for sewing a decorative stitch. Replacing the original needle bars drive mechanism with a controlled servo drive should facilitate transition to automated production. Further requirements of the manufacturer are aimed at reducing vibrations and noise and increasing the operating speed of the machine at the same time, to which the proposed mechatronic system contributes. In order to maintain the reliability of the stitching process after increasing the speed, the needle bars must be also modified. For these purposes, a mathematical model was created, which allows to describe the behaviour of the investigated mechanism and facilitates the work associated with the design of needle bar modifications. A functional model of the needle bars drive mechanism with the controlled servo motors is prepared to verify the proposed modifications.

© The Author(s), under exclusive license to Springer Nature Switzerland AG 2022
J. Beran et al. (Eds.): TMM 2020, MMS 85, pp. 304–311, 2022.
https://doi.org/10.1007/978-3-030-83594-1_32

2 Description of Stitch-Forming Device

The examined sewing machine creates a stitch, which imitates the appearance of hand sewing as much as possible. These stitches are mainly used for decorative purposes and are used for edge stitching of pockets, coats and jackets, decorative sewing of all fabric uppers and leathers, but also for standard sewing of coats, jackets, collars, blouses and cuffs on shirts. To imitate a hand stitch, a so-called floating needle is used, which has a tip on both sides and an eye in the middle. By the use of two mechanical systems of the needle bar, one above and one below the work plate, a threaded needle is passed through the sewn material. The entire needle thus passes through the sewn material with each stitch. Of course, the sewing machine works with a limited supply of thread. The floating needle 1 held in the upper mechanical system of the needle bar is shown in Fig. 1a.

The needle bar (Fig. 1) performs a rectilinear reciprocating movement, according to the prescribed stroke function (Fig. 2), which is generated by the cam mechanism on the original sewing machine. The support shell of the needle bar 4 is driven by the carrier 12.

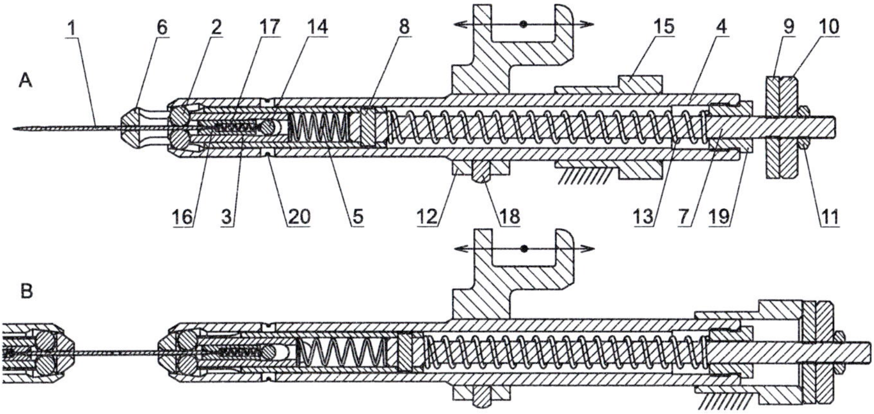

Fig. 1. Mechanical system of the needle bar, a) gripping the needle, b) handing over the needle.

The floating needle 1 is held in the needle bar by collets 2. The gripping and releasing of the needle are controlled by a control element which consists of parts 6, 7, 8, 9, 10, 11. The release of the needle is solved mechanically by the impact of the control element on the stopper 15 located on the machine frame, see the article [1]. This occurs during the first movement interval, when the needle bar reaches the needle interchange location. The condition of the needle bar at this location is shown in Fig. 1b.

The mechanical system of the needle bar is structurally designed for lower speed, and therefore at higher speed the inner parts of the needle bar oscillate, which is reflected in increased stress and endangering the sewing process. It has been found that at higher speeds, the inertia forces delay the control element.

For a detailed analysis of the behaviour of the mechanical system of the needle bar, a mathematical model was used, which is described in detail in [2]. The accuracy of the mathematical model was verified by measuring the position and velocity of the shell and the control element of the needle bar by using a Laser Doppler Vibrometer.

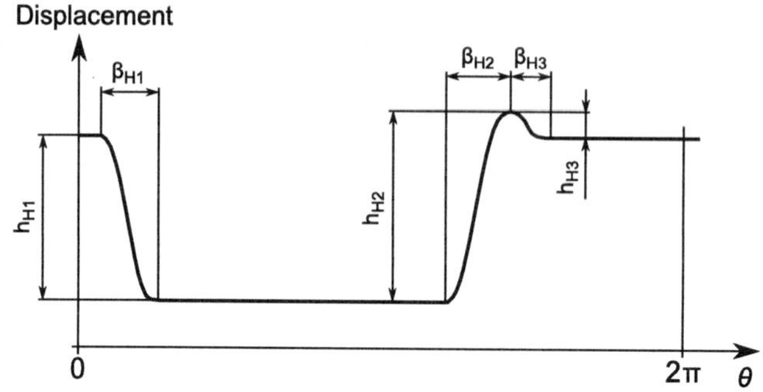

Fig. 2. Stroke function of the needle bar.

3 Principle of Laser Doppler Vibrometry

The Laser Doppler Vibrometer measures the frequency shift between a reference beam and a measurement beam, see Fig. 3. The laser beam, which is focused on the measured object, passes through a first beam splitter, where it is divided into the reference and the measurement beam. The reference beam is directed to the detector by means of a mirror. The measurement beam passes freely through a second beam splitter to the measured object, from which it is reflected. This reflected beam is now deflected on the second beam splitter and, on its subsequent path, blends with the reference beam on the detector. Due to the movement of the measured object, the length of the optical path of the measurement beam changes per unit time. This manifests itself as the Doppler frequency shift, which is added to the reflected beam [3].

The movement of an object away from the interferometer generates the same modulation pattern as the movement of the object towards the interferometer. For this reason, it is not possible to clearly determine in which direction the object is moving. In order to be able to detect not only the path length but also the direction of movement, a frequency shift of a known magnitude of 40 MHz is then added to the reference beam. This frequency shift creates an acousto-optic modulator (Bragg cell), which is placed in the path of the reference beam. This generates a typical interference pattern modulation frequency of 40 MHz when the measured object is at rest. If the object moves towards the interferometer, this modulation frequency increases and if it moves away from the interferometer, then the modulation frequency decreases [3]. Using a Laser Doppler Vibrometer, both the displacement and the velocity of the measured object can be measured directly. The advantage of the Laser Doppler Vibrometer, as follows from the description above, is the non-contact method of measuring the displacement and velocity of selected parts of the special needle bar, which, in the case of the contact measurement method, could lead to significant inaccuracies [4, 5].

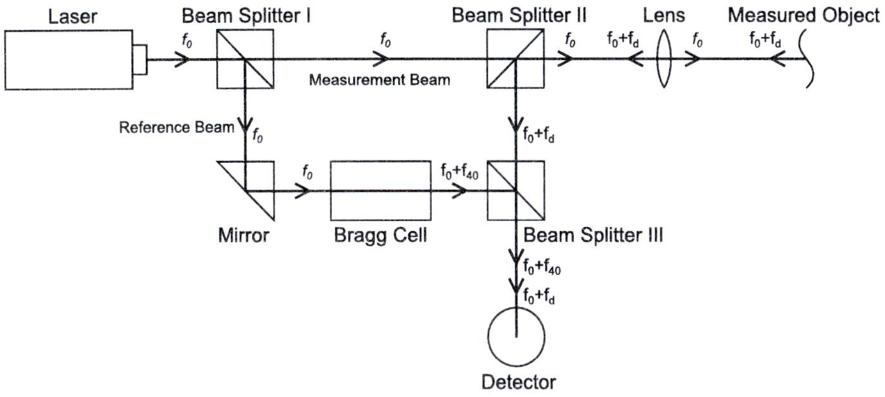

Fig. 3. Principle block diagram of Laser Doppler Vibrometer.

4 Measurement of the Position and Velocity of the Needle Bar

As already mentioned above, for a purpose of kinematic analysis, the mathematical model of the mechanical system of the needle bar is used. Its damping parameters were set to correspond to the behaviour of the real system, which was recorded using a high-speed camera [6]. From the video taken during the first working stroke, the value of the maximum change of position between the control cylinder 6 and the driven shell 4 and the maximum value of the bounce of the rubber pad 9 from the stopper 15 after their mutual collision were read. Now there is an opportunity to use a more advanced method to verify the mathematical model, which is non-contact measurement of the position and velocity of selected parts of the needle bar using the Laser Doppler Vibrometer.

The aim of the measurement was to obtain the actual position and velocity of the shell and the control element of the needle bar and to compare them with the curves obtained from the mathematical model. The measurement was performed on the functional model of the needle bars drive mechanism (Fig. 4), which was firmly attached to the worktop of the milling machine to prevent its movement. The movement of the needle bars is realized on the functional model by means of controlled servo drives. The Polytec PSV-400 Laser Scanning Vibrometer was used to measure velocity and position. The laser vibrometer was placed at such a distance from the measured object that it provided the highest signal level (maximum visibility). The maximum visibility is repeated every 204 mm, which corresponds to the length of the laser resonator. The head of the laser vibrometer was positioned so that the laser beam was directed to the axis of the needle bar. This is very important for obtaining accurate data. It is equally important that the beam be focused on a smooth surface perpendicular to the beam. The location of the device and the attachment of the functional model can be seen in Fig. 5. When measuring the needle bar shell, the beam was focused on the surface of the carrier 12, which is firmly connected to the needle bar shell. When measuring the control element of the needle bar, the beam was focused on the surface of the cylinder 10. The curves were measured for the operating speed of 250 stitches per minute. The measurement device was set to measure in the timeline. The scanning frequency was set to 8192 Hz. In this way, the position and velocity of the excitation part and subsequently also the response of the control element of the needle bar were measured.

Fig. 4. Location of the functional model.

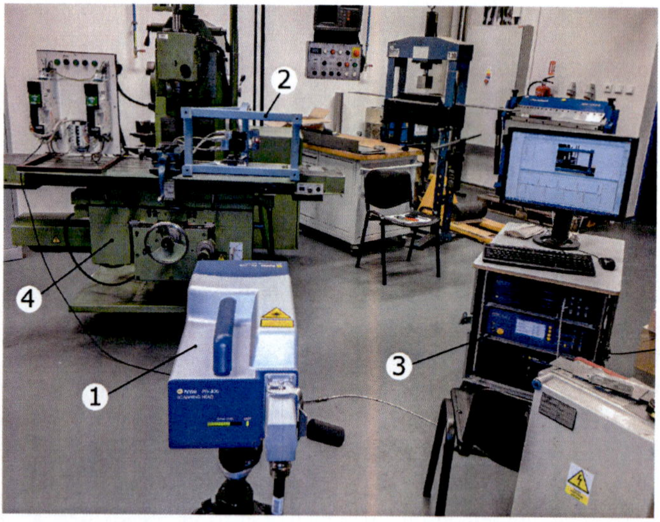

Fig. 5. Measurement: 1 - Laser Scanning Vibrometer, 2 - Functional model, 3 - Data acquisition unit, 4 - Milling machine.

5 Results

The aim was to verify the accuracy of the mathematical model of the mechanical system of the needle bar. The servo drive fulfils the required stroke with some accuracy, and our goal is to compare the response of the control element of the needle bar to the actual excitation. Therefore, instead of the theoretical stroke, the actual measured stroke of the needle bar shell was inserted into the mathematical model. Subsequently, an analysis was performed. Then the behaviour of the needle bar control element of the mathematical model was compared with the measured curves. The results are shown in Fig. 6 and Fig. 7.

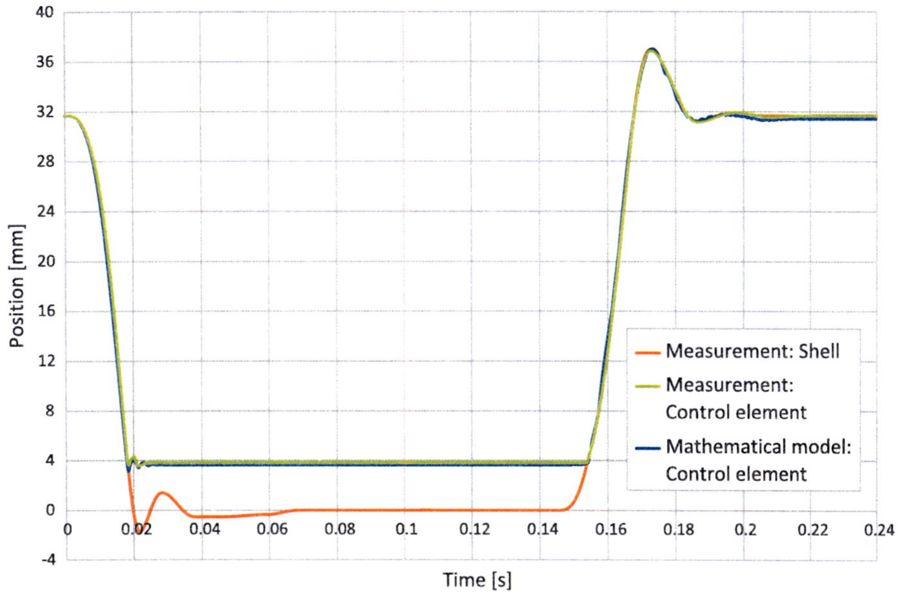

Fig. 6. Position of the shell and the control element of the needle bar - the whole stroke

Figure 6 shows the position versus time curves for one revolution of the virtual cam. The light grey curve describes the actual measured position of the needle bar shell, which was generated by the synchronous servomotor. The black curve describes the measured position of the needle bar control element and the grey curve describes the position of the needle bar control element from the mathematical model. Both positions of the control element seem to be almost identical.

Figure 7 shows the velocity versus time curves for the first movement interval of the stroke function. The light grey curve describes the actual measured velocity of the needle bar shell. The black curve describes the measured velocity of the needle bar control element and the grey curve describes the velocity of the needle bar control element from the mathematical model. In this movement interval, the needle is released due to the impact of the control element on the stopper. The impact occurs at a time of

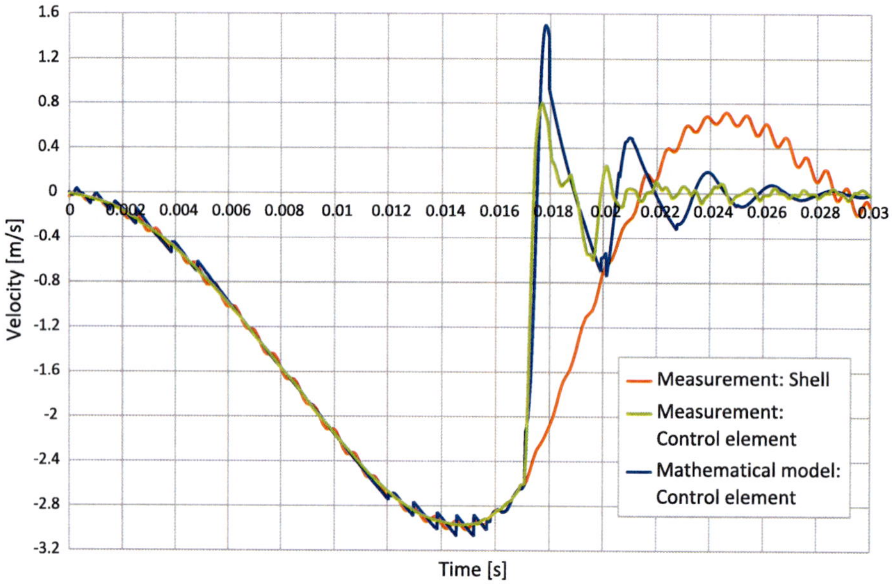

Fig. 7. Velocity of the shell and the control element of the needle bar - the first movement interval

approximately 0.017 s. After the control element bounces off the bearing surface, the system starts to oscillate.

6 Conclusions

Laser Doppler Vibrometer was used to measure the position and velocity of the selected parts of the needle bar. Its advantage is that the measurement is contactless. Therefore, the measured values are not affected by the mass of the sensor as with other contact measuring methods.

It is evident from the measured curves that there is no delay of the needle bar control element compared to the needle bar shell. Before the control element impact, the velocity of the shell and the control element is the same for the operating mode of 250 stitches per minute, which the mathematical model satisfies. However, when comparing the measured curves in detail with the behaviour of the mathematical model, is detected that on the mathematical model, the needle bar control element reaches a higher velocity after its impact than it actually is. The mathematical model also stabilizes more slowly after the impact. It follows that the real system is more damped. This can be caused by the low stiffness of the spring 13 or, on the contrary, by the high stiffness of the damping rubber pad 9. However, the stiffness of these parts was set in the mathematical model on the basis of experimental measurements of the force characteristic depending on the compression and at the moment there is no reason to change them. Other parameters that enter the mathematical model are the damping values, which were determined by monitoring the behaviour of the mechanical system of the needle bar, which was taken by a high-speed camera. There could certainly have been a significant error in reading the position between the control cylinder 6 and the driven shell 4 and the bounce of the rubber pad 9 mentioned in Chapter 4. It follows that finding suitable values for the mathematical model inputs, especially damping values, should be subjected to sensitivity analysis based on measured obtained with the Laser Doppler Vibrometer.

Acknowledgments. This work was partly supported by the Student Grant Competition of the Technical University of Liberec under the project No. SGS-2019–5055.

References

1. Komárek, J., Beran, J., Lima, M., Machado, J., Silva, J.: Finding the optimal setting of the sewing needle transfer mechanism using simulation software, In Proceedings of TRS 2012, The 41st Textile Research Symposium, 12–14 September 2012, Guimaraes, Portugal, pp. 401–404 (2012). ISBN: 978-972-8063-67-2
2. Komárek, J.: Dynamic model of the mechanical system of the needle bar, In: Proceedings of the XII International Conference on the Theory of Machines and Mechanisms, 6–8 September 2016, Liberec, Czech Republic, pp. 323–329 (2016). ISBN: 978-3-319-44086-6
3. Polytec. Laser Doppler vibrometry. In: Polytec [online]. https://www.polytec.com/eu/vibrometry/technology/laser-doppler-vibrometry. Accessed 25 May 2021
4. Kroschel, K. (ed.): Laser Doppler Vibrometry for Non-Contact Diagnostics Bioanalysis B, vol. 9. Springer, Cham (2020). https://doi.org/10.1007/978-3-030-46691-6
5. Nassif, H., Gindy, M., Davis, J.: Comparison of laser Doppler vibrometer with contact sensors for monitoring bridge deflection and vibration. NDT E Int. **38**(3), 213–218 (2005). https://doi.org/10.1016/j.ndteint.2004.06.012
6. Pejchar, K., Beran, J.: Optimation of the needle bar mechanism. In: Proceedings of Xth International Scientific Conference Transfer 2009, 17–18 November 2009, Trenčín, Slovak Republik, pp. 44–49 (2009). ISBN: 978-80-8075-414-3

Detection of Rabbit Skins for Robotic Handling

Š. Kovář(✉), M. Konečný, M. Bílek, P. Žabka, and J. Komárek

Department of Textile Machine Design, Technical University of Liberec, Studentská 2, 461 17 Liberec, Czech Republic
{simon.kovar,martin.konecny,martin.bilek,petr.zabka, jiri.komarek}@tul.cz

Abstract. Industrial automation is a major trend in today's light industry branches for handling textile products. The handling of natural material, in this case, rabbit skins, is also an industry requirement. This is a specific task where it is necessary to detect the fur side of the rabbit skin, which is characterized by a variety of shapes, colours and by its structure. This article deals with finding a simple, robust solution with minimal acquisition costs. To achieve this goal, different properties of the skin sides are utilized, specifically varying absorbency (reflectivity) of an ultrasonic beam.

Keywords: Ultrasonic sensor · Detection · Rabbit skin

1 Introduction

The article addresses issues related to the development of automation and robotics systems in textile processes having yet to be automated in greater measure. This is, for example, the process of preparing a semi-finished product for felt hats associated with the need to handle rabbit skins, including determining the desired orientation of the skin surface.

The processing technology compels the need to detect the side of the rabbit skin. If we want to automate the skin processing, we must be able to detect the front (fur) side from the back (flesh) side. The skins are always placed in storage rooms with the same sides together, thus eliminating contamination of the fur by grease exuded from the back side of the skin. Grease contamination leads to reduced functionality in subsequent operations and overall product quality. The need to replace the difficult, monotonous and dirty work of employees and unsatisfactory hygienic conditions forges the way to the automation of operation in the processing of rabbit skins. Here, the real possibility emerges from replacing their work using a robot to handle the skins instead.

2 Analysis of Side Detection Possibilities

The basic problem of detecting the side of rabbit skins is their wide variety of shapes, colours and surface structure. Current systems in the field of computer vision are difficult

© The Author(s), under exclusive license to Springer Nature Switzerland AG 2022
J. Beran et al. (Eds.): TMM 2020, MMS 85, pp. 312–318, 2022.
https://doi.org/10.1007/978-3-030-83594-1_33

to apply in terms of skin side recognition due to the variety of colours and their localization on the skin surface. This system might be used effectively in case of a well-defined colour difference on the fur side and back side of the skin, or possibly their contrast. However, this cannot be defined in the given application.

For this reason, it was necessary to seek out another solution. Options appear using varying properties of the individual skin sides. It can be assumed that the fur side has different absorption properties than those of the side without fur. This may concern the absorption of different types of radiation. For example, thermal radiation and the detection of different heat accumulation by the surface of the rabbit skin, as captured by the thermal camera, would be considered. The second option would be to exploit the varying absorption (reflectivity) of ultrasound. This option is interesting, given that these sensors are available in a wide range of designs [1].

3 Selection of the Ultrasonic Sensor

The ultrasonic sensor is designed for the detection of objects at a specified distance. The principle is to transmit and receive a reflected ultrasonic beam. This method has an advantage over optoelectronic sensors in the low sensitivity to colours, contaminated environments and surface quality. It can also detect various materials that are problematic for optoelectronic sensors, such as glass, certain polymer transparent materials, etc. Only materials that absorb the ultrasonic beam are problematic for ultrasonic detection. These materials include textile products such as carpets with shorn hair, certain felted textile structures, upholstery fabrics, as well as rabbit skins, or the hides of other animals naturally furry on the front side.

Based on the manufacturer's recommendation, several ultrasonic sensors were selected. Initial experiments have found that these sensors are able to reliably detect the back sides of the skins. In contrast, one of the sensors [4] had trouble detecting the front sides of the skins with fur. This behaviour was further exploited for skin side recognition. The problem only deepened using a so-called sonic tube, i.e., a plastic adapter (waveguide) of an ultrasonic sensor that directs an ultrasonic beam into a narrow space. This arrangement showed a significant error rate in the detection of the furry skin, which can be used to detect the skin's front side. Detecting the back side without fur showed a minimal error rate. The observed behaviour of the ultrasonic sensor could be used to detect the rabbit skin side as well as other flat materials having differences in the back and front side, and which demonstrate the ability of one side to absorb the detectable part of ultrasonic waves.

4 Laboratory Testing of the Sensor

The sensitivity of the sensor was examined on various materials. The goal of the testing was to determine the optimal settings. The sensor showed varying sensitivity for hard-to-detect materials, which depends on the ultrasound intensity and decreases with increased distance of the detected object from the sensor. Each sensor has a specified range of distance in which the object detection is guaranteed [2, 3]. On the contrary, in this application, we use the imperfections of objects sensing. Therefore, it is necessary to be in the area where the sensor is able to detect the back side of the skin, and while ensuring the same conditions, it must be incapable of detecting the front side of the rabbit skin. These signals can then be used when controlling a manipulation process so that the skin goes through the technological process properly oriented.

Testing of the sensors (Fig. 1) took place in motion. Measurements were performed during the vertical approach of sensors to the measured object successively in several different places. The movement of the sensors was controlled by the movement of the robotic arm. The distance between the sensor and the material varied in a range from 240 mm to 40 mm according to the sensing range of selected sensors (Fig. 2) and was verified by a laser scanner. For each specimen, 50 arrivals to the front side were made as well as 50 arrivals to the back side. On the basis of the obtained data, the empirical probability is evaluated of triggering the sensor depending on the distance from the measured object.

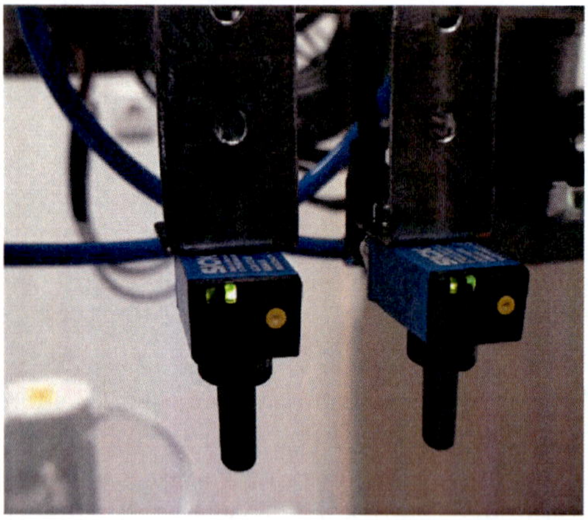

Fig. 1. Ultrasonic sensors UC4

Figures 3, 4, 5 show measured graphs of the empirical probability of triggering the sensor in relation to the distance from the measured object for varying specimens. The light-grey colour indicates the probability of triggering the sensor on the back side, while the dark colour indicates the probability of triggering the sensor on the front side.

It occurred in some cases that when approaching the skin, the already triggered sensor released for a brief moment. To achieve more reliable detection, only the first triggering of the sensor can be evaluated, and the subsequent "flicker" can be ignored. The values obtained by this method are indicated in the graphs by the hatched area.

Figure 3 shows two randomly selected specimens of rabbit skin, which represent the typical behaviour of the sensor. The specimen on which the most significant difference between the front side and the back side is achieved is shown in Fig. 4, Specimen C. Conversely, Specimen D in Fig. 4 shows the worst case detected, wherein side detection is the least reliable.

For reliable detection of the side, it is essential to find the distance at which the sensor is triggered for the back side, but also remains un-triggered for the front side. Measured

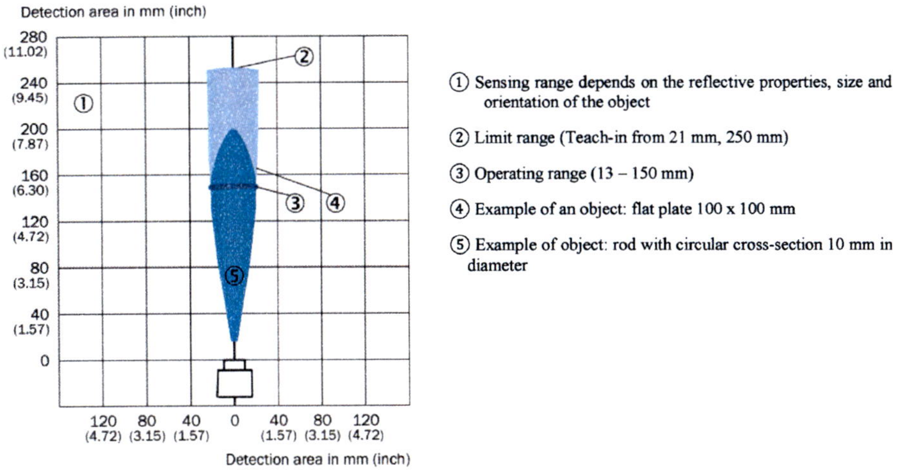

Fig. 2. Sensing range of ultrasonic sensors UC4-13 [4]

objects can be undulating to varying degrees, and the distance from the sensor to the object is not exactly guaranteed. It is thus necessary to expect that, at a given distance, the actual distance may differ slightly. Most specimens showed good readability in the range of 120–160 mm. Based on this analysis, the optimal distance was determined to be 140 mm.

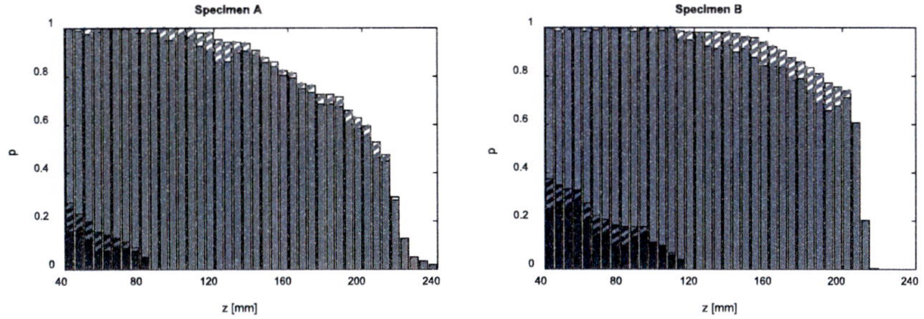

Fig. 3. Empirical probability of sensor triggering based on distance for two typical specimens

As the graphs indicate, in some instances (Specimen C), correct detection is practically guaranteed, whereas, for other specimens (Specimen D), it is not possible to achieve reliable identification. In this worst specimen, provided the optimum distance, the probability of correct identification is only 95%.

In addition to rabbit skin specimens, selected types of carpet were also examined (Specimens E and F in Fig. 5). Despite the fact that they also showed significantly varying absorption of ultrasonic waves for the back and front side, the detected difference is not large enough to allow the corresponding side to be reliably identified.

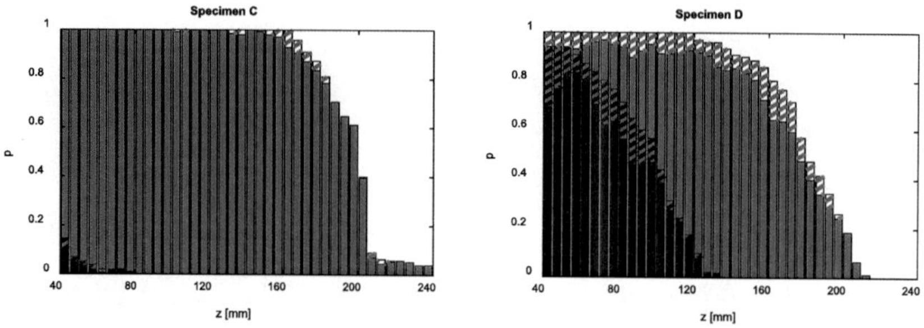

Fig. 4. Empirical probability of sensor triggering based on the distance for the best case (C) and the worst case (D)

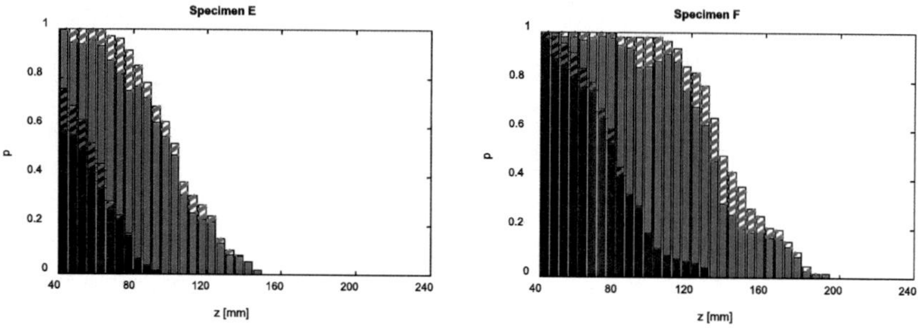

Fig. 5. Empirical probability of sensor triggering based on distance for carpet specimens

5 Draft Assessment Algorithm of the Skin Side

Laboratory testing of ultrasonic sensors has shown that the problem of detecting the side depends greatly on the type of test specimen. In rabbit skins, detection is not always unambiguous due to the great diversity of surface shape, colours, unevenness of the furry surface, which is given, among other things, by defects created during the processing of the raw semi-finished product, or possibly by the rabbit's own body proportions. For some materials, such as rabbit skins with white hair, it is easy to assess whether it is the back side or front side. In the case of other materials, it might be difficult to reliably detect which side of the skin (carpet) it concerns (e.g., different coloured rabbit skins, black rabbit skins, carpets, etc.). Even though the probability of triggering sensors when approaching different skin sides might show significant differences and allows side identification, there is still some level of uncertainty. Thus for proper detection, in order to decrease the error rate, one cannot rely on the signal from just one ultrasonic sensor, but it is necessary to evaluate multiple detected signals.

6 Laboratory Applications

To verify the detection of the side of a rabbit skin, a functional model of gripper (Fig. 6) was designed and built, and subsequently implemented on the end-member flange of a six-axis industrial robot. The robot's movements have also been programmed in such a way as to mirror the real handling of rabbit skins. Since, for technological reasons, these skins must always be stored with the same sides facing each other, the handling sequence must be equipped with skin-side recognition during the gripping process of the gripper. Testing was conducted by removing skins from a layered stack, where the skins were inserted randomly to verify their proper detection.

Fig. 6. Gripper with ultrasonic sensors for gripping skins

7 Conclusion

The proposed detection system of flat textile material is suitable for materials wherein one of its sides can absorb ultrasonic radiation. It is possible to detect the front and back side of these materials in a cost-effective manner. This can be applied in automation where these products must be handled while detecting the side of the processed material. The system was developed for the handling of rabbit skins, but it can also be used in other applications.

Acknowledgements. The paper has been elaborated with financial support of TUL in the framework of Institutional Support for Longterm Development of Research Organizations.

References

1. Martinek, R.: Senzory v průmyslové praxi. BEN-Technická literatura, Prague, p. 200. ISBN 80-7300-114-4 (2004)
2. Vojáček, A.: Ultrazvukové sensory přiblížení – funkce, provedení, použití. In: Automatizace.hw,cz. https://automatizace.hw.cz/ultrazvukove-senzory-priblizeni-funkce-provedeni-pouziti.html

3. Toa, M., Whitehead, A.: Ultrasonic sensing basic, application report. In: Texas Instruments, March 2020. https://www.tij.co.jp/jp/lit/an/slaa907c/slaa907c.pdf?ts=1620284205355&ref_url=https%253A%252F%252Fwww.google.com%252F
4. Sick: Product data sheet UC4-13341S01, Ultrasonic sensors. In: SICK. https://cdn.sick.com/media/pdf/6/46/346/dataSheet_UC4-13341S01_6049509_en.pdf

Simulation and Experimental Validation of Stable Multiple-Ballooning in Nanofibrous Yarn Production

Petr Žabka(✉), Ondřej Baťka, Josef Skřivánek, Jan Valtera, and Jaroslav Beran

Technical University of Liberec, Liberec, Czech Republic
{petr.zabka,ondrej.batka,josef.skrivanek,jan.valtera, jaroslav.beran}@tul.cz

Abstract. The article deals with the analysis of the process of yarn ballooning with the formation of multiple balloons. Ballooning plays a vital role in the production of composite nanofiber yarn. To achieve an even quality layer of nanofibers, a stable balloon shape is necessary to ensure. To do this, it is necessary to determine the influence of the individual production parameters and find their optimal values. Numerical simulations of balloon shape as a function of yarn tension were performed. The results were verified experimentally.

Keywords: Composite nanofibrous yarn · Electrospinning · Ballooning

1 Introduction

The production of nanofibers by spinning a polymer solution with an electric current is currently the most widely used method in the industry. The most common product is flat nanofiber membranes. Recently, attention is also paid to the research and development of technologies for the production of linear nanofibrous materials, such as nanofibrous yarns. At the Technical University of Liberec, a technology was developed to produce core nanoyarn [1], which uses AC-electrospinning [2]. A simplified diagram of a line for the production of core nanoyarn is in Fig. 1 [3]. The yarn moves at a set speed in the direction from the feeding device (1) to the winding device (8). At the beginning of the process, it passes through a twisting device (2) which rotates at a constant angular speed. The yarn emerges from the twirling device through a ceramic eyelet, which is positioned eccentrically, so that its axis does not coincide with the axis of rotation. As a result, the yarn forms a complex formation in the space, which is known from the processes of spinning or unwinding the yarn as a so-called balloon [4, 5]. It is characterized by its length and maximum radius, while with a large length, the balloon can be multiple [6, 7]. Ballooning is desirable in the production of a core nanoyarn, as it ensures the collecting of the nanofibers (6), which are by the electrode produced from a polymer solution (5). The nanofibers are thus deposited on the surface of the core, and a composite nanofibrous yarn is formed (Fig. 2, Fig. 3).

© The Author(s), under exclusive license to Springer Nature Switzerland AG 2022
J. Beran et al. (Eds.): TMM 2020, MMS 85, pp. 319–325, 2022.
https://doi.org/10.1007/978-3-030-83594-1_34

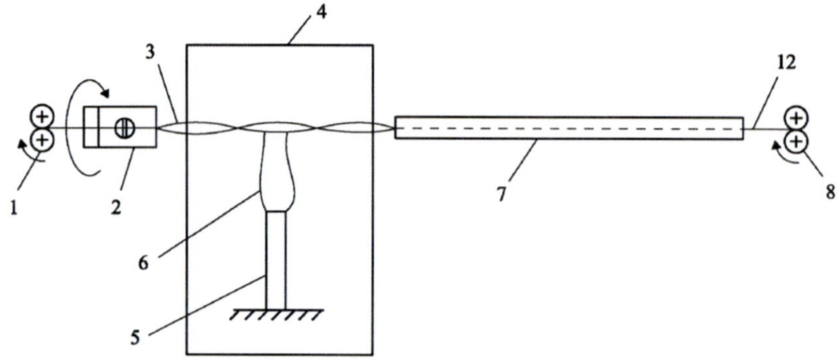

Fig. 1. Production line scheme [3]

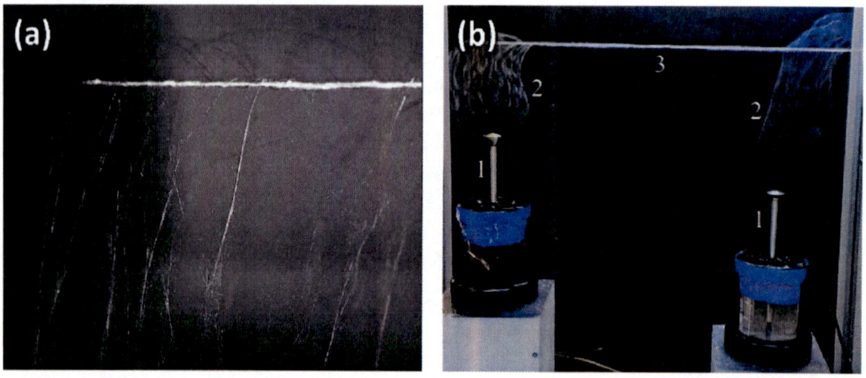

Fig. 2. (a) The yarn core axis and the axis of the nanofibrous plume emitted from the spinning electrode are almost perpendicular to one another. (b) Two screw pumps integrated with the spinning electrode with disc-like heads (1) and the polymer reservoir emitting nanofibrous plumes (2), which wrap around the ballooning yarn core (3) [1].

For correct deposition of nanofibers on the ballooning yarn, it is necessary to ensure that its radius has the required size at a given speed of the twisting device. It must be such that the circumferential speed of the yarn at the deposition place is higher than the production speed of the nanofiber plume, but it must not be high enough to cause it to break. Many parameters affect the rate of nanofiber production, such as the type of polymer solution to be spun, the humidity of the environment, the magnitude of the electrical voltage, and others. Typically, the production speed of nanofibers is between 30 and 60 m/min. To ensure stable deposition of fibres on the core, it is therefore essential to know the shape of the ballooning yarn, in particular the size of the maximum radius for various technological parameters, such as the speed of the twirling device and the tension in the yarn [1].

The paper deals with the simulation of the shape of a ballooning curve and the subsequent experimental verification. Particular attention is paid to the dependence of the maximum radius of the balloon on the tension in the yarn.

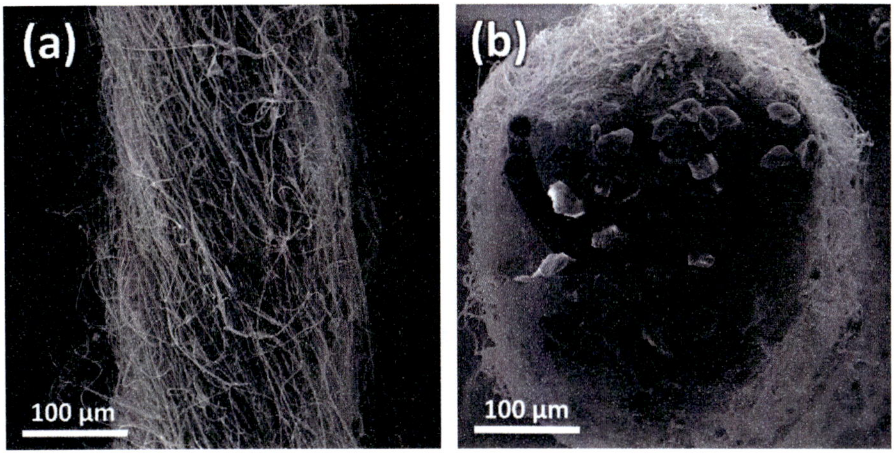

Fig. 3. (a) The nanofibre envelope with a twist value of about 103 m − 1. (b) A cross-section of a composite yarn with a polyester (PES) multifilament core with a linear weight of 330 dtex and a nanofibrous envelope made with polyamide 6 (PA6) [1].

2 Simulation

In order to determine the influence of individual parameters on the shape of the ballooning yarn, a number of simulations were performed. The mathematical model is based on the differential description according to Migushov [5–8] for a yarn that rotates at a constant speed, so that the shape of the balloon is stable. For the calculation, it is assumed that the yarn is uniform, perfectly bendable and unstretchable. The gravitational acceleration was neglected in the model. The simulations were performed for yarns with a fineness of 120 tex, speeds of 6000 RPM, 9000 RPM and 12000 RPM and for a spool diameter of 1 mm, an eccentricity of the rotating spool of 3 mm and a line length of 4 m. The simulation results are shown in Fig. 4.

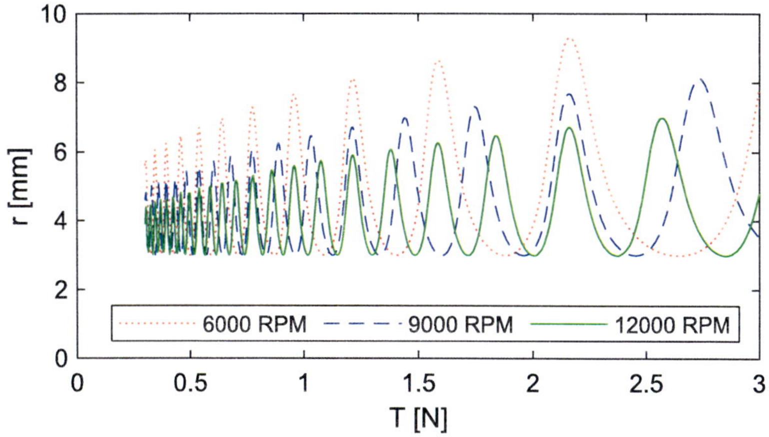

Fig. 4. Simulation results of maximum radius dependence on tension for several RPM

The simulation results show that the maximum radius of the balloon highly depends on the tension in the yarn. The diameter changes in waves, each time the multiplicity of the balloon changes. If the tension is small, the multiplicity of the balloon is greatest and frequently changes as the tension changes. If the tension is greater, the multiplicity of the balloon is more stable, and the regulation of the balloon shape is easier. However, the maximum tensile force is limited by the mechanical properties of the yarn.

3 Experiment

An experiment was performed to verify the obtained results. A diagram of an experimental apparatus used to observe the movement of a ballooning yarn is in Fig. 5. The yarn moves from the spool (1) and passes through a twirling device (Fig. 6) with an eccentric eyelet (3), which contributes significantly to the characteristic shape of the balloon (4). It then passes through a static eye (5), in which the ballooning ends. The yarn is then stored in the form of a spool by a winding system (7). Throughout the experiments, the tension in the yarn was measured by means of sensors (2) and (6). The tension in the yarn was controlled by changing the ratio of the feed and winding speeds of the yarn.

A video recording was made for each experiment, the analysis of which determined the maximum radius of the balloon depending on the tension in the yarn. The image captured by the camera is shown in Fig. 7. Due to the capabilities of the camera, only the area of nanofiber deposition was captured. The size of the view area was chosen to ensure that the location with the maximum radius of rotation is recorded, but at the same time to achieve the highest possible resolution when evaluating the radius of the balloon. The shutter speed was slower than the yarn rotation speed, so it is not the shape of the yarn that is recorded at one time but the envelope of the balloon.

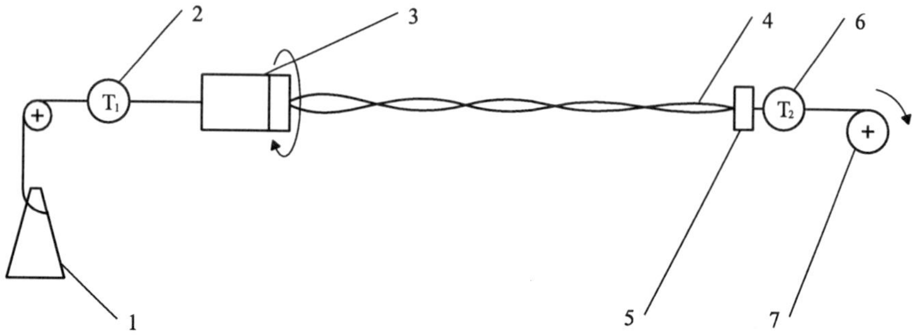

Fig. 5. Scheme of the experiment setup

Figure 6 shows a detail of the area between the tension sensor A and the point where the yarn leaves the twirling device B. The movement of the yarn through the twirling device is shown here. It can be seen from the figure that the yarn is rubbing against the metal part of the device, and thus the tension measured at point A is lower than the tension at point B, while the value of the tension in the simulation is being considered at the point B. It follows that the measured tension cannot be directly compared with the value used in the simulation, and it is necessary to make a correction.

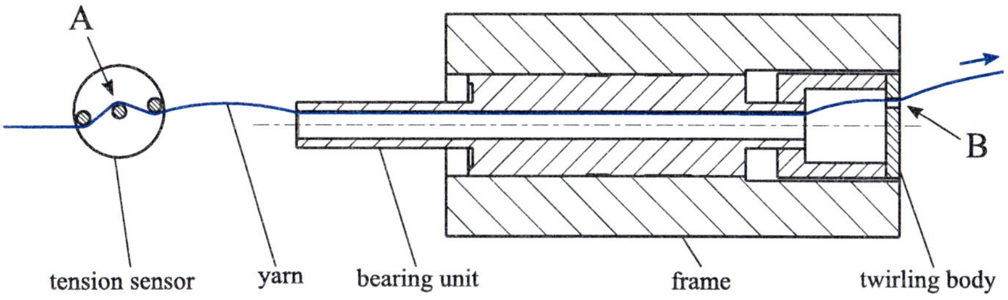

Fig. 6. Scheme of twirling device

Fig. 7. An image captured by a camera while measuring the shape of a balloon

To evaluate the shape of the balloon, the video recording (Fig. 7) was computer processed. The process of image processing is shown in Fig. 8. First, the amount of processed data was reduced by converting to grayscale and cropping the image (a). Subsequently, the background was subtracted from the image, which was obtained as the maximum brightness of the respective pixels during the selected part of the video. To reduce noise, the image was horizontally blurred. In the obtained image (b), the balloon envelope was detected by edge tracing. The starting point was always placed in the place with the smallest balloon diameter, where the automatic recognition of the balloon area is the most reliable. Subsequently, the upper and lower edges of the balloon envelope were detected. The maximum diameter of the balloon envelope was determined from the obtained balloon shape.

The experiments were performed with the same parameters as the simulation, i.e. for a yarn with a fineness of 120 TEX for speeds of 6000 RPM, 9000 RPM and 12000 RPM. Figure 9 shows a comparison of the simulation with the experiment. The experimentally obtained radius values were plotted as a function of the corrected tension, which is 1.3 times greater than the tension measured at the input sensor.

Fig. 8. Individual phases of computer image processing (cutout), the original image (a), processed image (b), threshold (c)

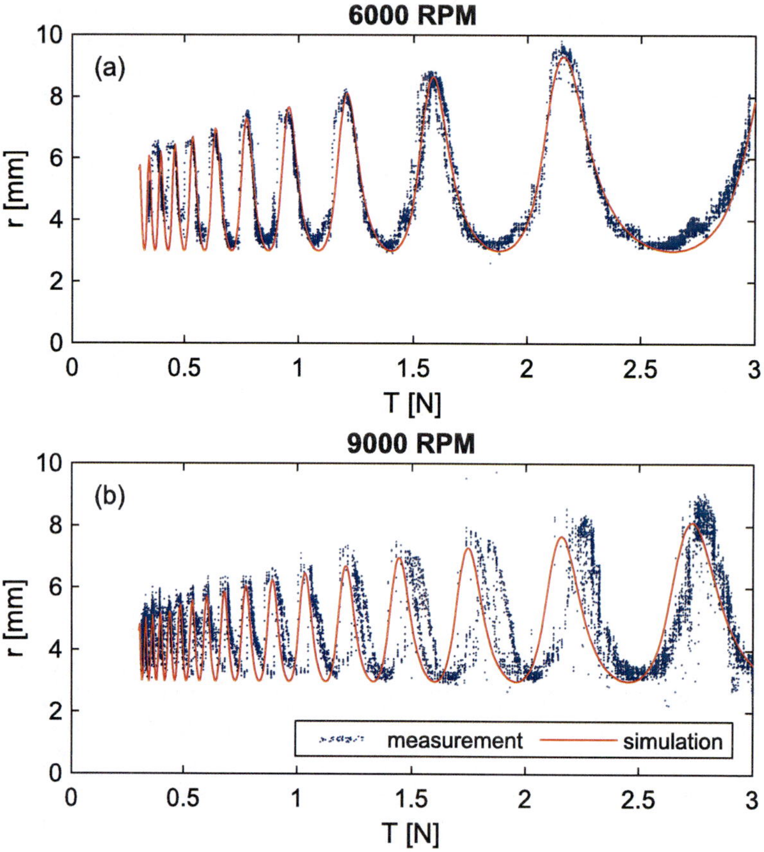

Fig. 9. Comparison of simulation results with experiment, a) 6000 RPM, b) 9000 RPM

4 Conclusion

The aim of the research was to analyse the process of depositing nanofibers to the core yarn. By ensuring the stability of this process, it is possible to improve the uniformity of the nanofiber coating and thus achieve a higher quality product. For even application of nanofibers to the core yarn, it is necessary to ensure a stable balloon shape. Above all, it is necessary to achieve a stable multiplicity of the balloon and to achieve an acceptable maximum diameter of the balloon. Based on the performed simulations, it was found that the shape of the balloon is very dependent on the tension in the yarn. During the production process, it is, therefore, necessary to precisely regulate the tension in the yarn. At higher values of tension, regulation is easier, but the maximum value is limited by the mechanical properties of the yarn. By precisely regulating the yarn tension, the optimal shape of the balloon can be achieved. The findings were experimentally verified.

Acknowledgements. This work was supported by the Student Grant Competition of the Technical University of Liberec under the project No. SGS-2021-5029.

References

1. Valtera, J., et al.: Fabrication of dual-functional composite yarns with a nanofibrous envelope using high throughput AC needleless and collectorless electrospinning. Sci. Rep. **9**(1) (2019). https://doi.org/10.1038/s41598-019-38557-z
2. Kocis, L., et al.: Method for production of polymeric nanofibers by spinning of solution or melt of polymer in electric field, and a linear formation from polymeric nanofibers prepared by this method [online]. EP2931951B1 (2019) https://patents.google.com/patent/EP2931951B1/en?oq=EP+2931951. Accessed 17 May 2021
3. Beran, J., et al.: A linear fibre formation with a case of polymeric nanofibres enveloping the supporting linear formation constituting the core, the method and equipment for its production [online]. CZ306428B6 (2017). https://patents.google.com/patent/CZ306428B6/en. Accessed 17 May 2021
4. Klein, W., Stalder, H.: A Practical Guide to Ring Spinning. B.m.: Textile Institute (1987). ISBN 978-0-900739-94-1
5. Beran, J.: Dynamic Analysis of the Ring Spinning Process, Engineering Mechanics, vol. 12, 2005, no. 6, pp. 393–401. ISSN 1805-4633
6. Batra, S. K., Ghosh, T. K., Zeng, Q., Robert, K. Q., Fraser, W.B.: An integrated approach to dynamic analysis of the ring spinning process: Part IV: inherent instability of the free balloon. Textile Res. J. **65**(7), 417–423 (1995). ISSN 0040-5175. https://doi.org/10.1177/004051759506500707
7. Tang, Z.X., Fraser, W.B., Wang, X.: Modelling yarn balloon motion in ring spinning. Appl. Math. Model. **31**(7), 1397–1410 (2007). ISSN 0307-904X. https://doi.org/10.1016/j.apm.2006.03.031
8. Migushov, I.I.: Mechanika tekstilnoj niti i tkani. B.m.: Legkaja Industrija (1980)

Optimization of the Batten of the Weaving Loom DIFA

Tomáš Koňařík[1], Radek Zbončák[1], and Josef Žák[1,2](✉)

[1] VÚTS, Svárovská 619, 460 01 Liberec, Czech Republic
josef.zak@tul.cz
[2] Technická Univerzita Liberec, FS KMP,
Studentská 1402/2, 461 17 Liberec, Czech Republic

Abstract. The weaving is considered as one of oldest human activities. It was also the first manufacture process where the modern industry principles were applied and hand-crafted production was replaced by the machines. Nevertheless the weaving machines remained long time merely more then a steam engine driven mechanical looms. It is only in the last 50 years or so when the looms became really automatized weaving machines. Invention of shuttle-less loom brought an enormous increase of productivity and at the same time automatisation of loom operation allowed production of very complex fabrics. Both these facts necessitated also a development of other mechanical parts of the loom. In this paper we present the way we upgraded the batten – the key machine part – used on a classical loom to be used on a weaving machine producing heavy 3D fabrics. While the methods of classical mechanics are sufficient to make the basic changes most recent methods such as FEM were used to finalize the design. Beside of modern computational tools we profited also of advances in the technology in the past 20 years since the original design has been first conceived.

Keywords: Reduction of inertia · Optimization of machine part · Wear reduction

1 Introduction to the Problematics

The batten (also slay) is the most dynamically loaded part of a weaving loom. It performs a rocking motion during the machine cycle with high angular acceleration. Therefore its inertia is the one of its properties that must be considered in the construction of a loom. At the same time, as the batten is that part which is forming the fabric, it is loaded by the forces generated during the beat up and thus it must sustain rather high loads. Therefore its stiffness is the crucial property in the final product forming. On a classical loom the highest beat up forces are partially compensated by the inertial forces and the above mentioned problem arises only for a high speed weaving machines if ever. The presented

© The Author(s), under exclusive license to Springer Nature Switzerland AG 2022
J. Beran et al. (Eds.): TMM 2020, MMS 85, pp. 326–336, 2022.
https://doi.org/10.1007/978-3-030-83594-1_35

work deals with a special 3D fabrics weaving loom which uses a non standard weaving sequence: it starts and stops several times a minute dependent on the pile yarns density. Each start and stop generates elevated loads on the cams and bearings so a minimum inertia moment of the batten is at most desirable. On the other hand its weaving speed is not sufficiently high to compensate the elevated loads emerging from the forming of the very heavy 3D fabrics. Considering these requirements a major redesign of the actual batten was needed when upgrading the VEGA loom to the DIFA distance fabrics weaving machine.

2 Analysis of Actual Design

Currently the batten is formed by a steel tube on which two laminated plates are stick using glue joint. The axis of the tube is also the axis of oscillation of the batten. The plates present a form of clamp in which the reed is fixed using screw and nut joints. The clamping force is given by the stiffness of the plates. The batten bears also other auxiliary devices such as relay nozzles or short pick detection device. The static imbalance caused by the reed and other devices must be compensated by means of separate masses placed inside of the tube. This design facilitates production, as the tube and the separate masses are easy to machine. Also the gluing of laminated plates is easy in this configuration. Its disadvantages are obvious: to achieve good static equilibrium rather heavy counterbalancing masses are needed. This latter increases unnecessarily the inertia moment.

Another source of concerns was the resistance of the batten to the elevated loads. On the DIFA loom the batten would have to sustain elevated weaving resistance, i.e. forces generated by the front of fabric during the binding points forming while the machine RPM are lower; the reserve of clamping force must be higher to prevent a wobbling out of the reed. Thus the reed must be fixed more firmly in the batten which required reworking of laminated plates concerning their stiffness.

The batten in current design when first used on the DIFA loom presented a misalignment which was not important on the VEGA loom. Contrary, due to the conception of the pile creation, this misalignment of batten – or more precisely that of the reed – was very embarrassing on the DIFA loom. A closer analysis of this problem had shown that a curved laminated plate can not be manufactured properly even when all rules of symmetric plies stacking are respected. Another reason of the laminated plates redesign arose, thus.

On the other hand the current design was found satisfactory concerning its torsional and bending stiffness; in fact its conception with double hollow section is the best that can be achieved in the given available space. With all this knowledge we proceeded to the major redesign of the batten.

3 Optimization of the New Design

3.1 Inertial Properties

In the new design of batten we abandoned the conception where the axis of oscillation coincides with the axis of central tube. We placed the centre of gravity of the tube outside of the oscillation axis and the tube – the heaviest part of batten – works as counter balance itself. In theory the best solution is to place very heavy balance mass near the axis of rotation:

$$J = J_t + m_t \cdot r_t^2 + J_b + m_b \cdot r_b^2 \tag{1}$$

where indexes t and b refer respectively to the dead load and counterbalancing masses. Position of a mass necessary to compensate m_t at r_t is $r_b = r_t \cdot \frac{m_t}{m_b}$ and (1) becomes thus:

$$J = J_t + m_t \cdot r_t^2 + J_b + \frac{m_t^2 \cdot r_t^2}{m_b} \tag{2}$$

For a mass in form of thin walled tube we get using simplified formulae:

$$J_b = \frac{D^2}{4} \cdot m_b = \frac{\pi}{4} \cdot D^3 \cdot t \cdot l \cdot \rho \tag{3}$$

where D is diameter of middle plane of tube wall of thickness t (assuming $t \ll D$), l is the length of the tube and ρ is the density. Then (1) using the expression $r_b = r_t \cdot \frac{m_t}{m_b}$ becomes:

$$J = J_t + m_t \cdot r_t^2 + \frac{\pi}{4} \cdot D^3 \cdot t \cdot l \cdot \rho + \frac{m_t^2 \cdot r_t^2}{\pi \cdot D \cdot t \cdot l \cdot \rho} \tag{4}$$

An unconstrained optimization with regard to D and t would not result to a solution as the J in form of (4) has no proper point with all derivatives vanishing (it has no local extreme with finite value). Thus we should use some functions to constrain the values of D and t.

First, t can not be greater then D and considering the assumption of thin walled tube, it should actually fall into interval $(0, \approx 0.1 \cdot D\rangle$. Such unsymmetric condition can be expressed as normalized penalization function:

$$C = \frac{1}{2} + \frac{1}{\pi} \cdot \arctan\left(\frac{1}{d} \cdot \left(q - \frac{D}{t_1}\right)\right) \tag{5}$$

where q is permissible value of ratio $\frac{D}{t}$ and d is a tolerance field of q (the d determines the *strongness* or intensity of the condition).

Secondly, we should take into account the fact that by minimizing the inertia moment we decrease also the torsional stiffness of the slay. This torsional stiffness is formed by the two hollow section:

$$(G \cdot J_k) = 4 \cdot \frac{U_{s,1}^2 \cdot \lambda_2 + \left(U_{s,1}^2 + 2 \cdot U_{s,1} \cdot U_{s,2} + U_{s,2}^2\right) \cdot \lambda_{12} + U_{s,2}^2 \cdot \lambda_1}{\lambda_1 \cdot \lambda_2 + \lambda_{12} \cdot (\lambda_1 + \lambda_2)} \tag{6}$$

where $U_1(D, r_b, r_t)$ and $U_2(D, r_b, r_t)$ are sectorial surfaces of hollows and $\lambda_i = \frac{s_i}{G_i \cdot t_i}$; s_i is length of wall section with thickness t_i and a mean shear modulus G_i (see Fig. 1). As the geometry of the section is rather complex we have opted for simplified expressions:

$$U_{s,1} = \frac{\pi \cdot D^2}{4} \qquad U_{s,2} = \frac{D}{2} \cdot (r_t + r_b) \qquad t_1 = t_{12}$$
$$s_1 = s_{12} = \frac{\pi \cdot D}{2} \quad s_2 = 2 \cdot \sqrt{\left(\frac{D}{2}\right)^2 + (r_t + r_b)^2}$$

Second condition expresses the requirement of conservation of torsional stiffness:

$$(G \cdot J_k) - (G \cdot J_k)_0 = 0 \tag{7}$$

where $(G \cdot J_k)_0$ is current value of torsional stiffness, determined by evaluating the relation (6) for current geometry values.

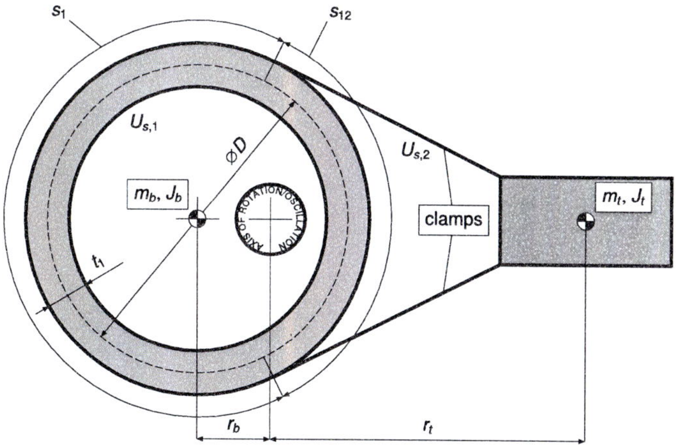

Fig. 1. Schema of batten configuration

The target function to minimize then takes the following form:

$$F(D, t_1, \lambda) = J \cdot (1 + C) + \lambda \cdot ((G \cdot J_k) - (G \cdot J_k)_0) \tag{8}$$

where J is given by (4), C is given by (5) and λ is Lagrange's multiplier, the other terms being expressed above.

Using conditions of stationarity of F defined by $\frac{\partial F}{\partial u_i} = 0$ with $u_i \in (D, t_1, \lambda)$ we get a non linear system of equations. Its solution has to be carried out using some non linear algorithm, e.g. Newton Raphson's method. To this purpose we have used the software of symbolic mathematics Maxima. Actually for the set of given construction values

$$\begin{array}{lll}
m_t = 8.723\,\mathrm{kg} & r_t = 0.151\,\mathrm{m} & J_t = 0.03273\,\mathrm{kg}\cdot\mathrm{m}^3 \\
\rho = 7800\,\frac{\mathrm{kg}}{\mathrm{m}^3} & G_1 = 81000\,\mathrm{MPa} & G_2 = 15000\,\mathrm{MPa} \\
l = 2.5\,\mathrm{m} & t_2 = 4\,\mathrm{mm} & \\
q = 7.5 & d = 0.001 &
\end{array}$$

and for the set of initial – current – values (which are required for evaluation of current torsional stiffness):

$$[D = 107.5\,\text{mm}\ t_1 = t_{12} = 7.5\,\text{mm}\ r_b = 0\,\text{mm}]_0$$

we have got:

$$D = 94.5\,\text{mm}\ t_1 = 12.5\,\text{mm}\ r_b = 18.3\,\text{mm}$$

By evaluating the value of J following the relation (4) we get approximately the same value for the new configuration $J \approx 0.416\,\text{kg} \cdot \text{m}^2$ as we get for the old one *without* the counterbalancing masses. The final value of $\frac{D}{t} = 7.6$ close to the accepted minimum of 7.5 suggests that by increasing the tube wall thickness and decreasing its diameter we could further reduce the inertia moment.

Of course, this theoretical configuration had to be revised while finishing the actual design with respect to available tube. The values of D and t_1 were selected in the available tube range and the value of r_b had to be determined consecutively. Comparison of the old configuration and that one finally chosen is on the Fig. 2.

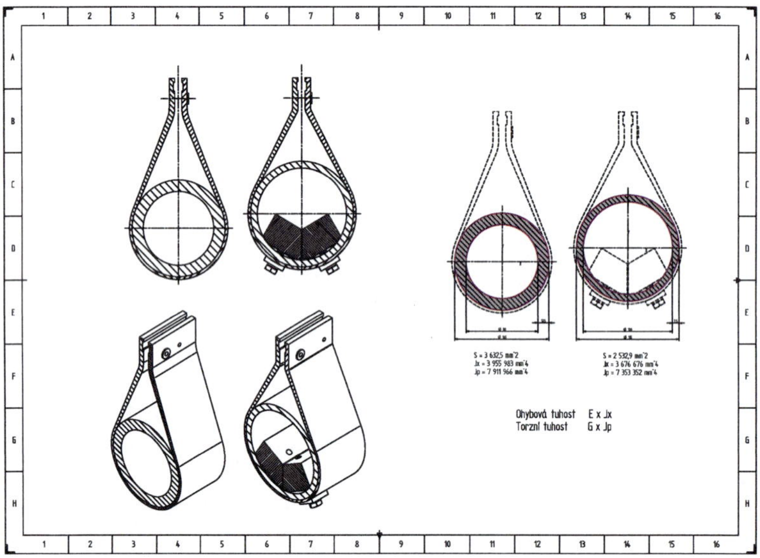

Fig. 2. Comparison of current and new configuration

3.2 Design of Laminated Composite Clamps

When first used on the 3D fabric weaving machine DIFA the batten went almost unchanged from its original configuration as used on the VEGA loom, only with minimal necessary modifications. Soon it appeared that a misalignment of the reed had an adverse effect on the function of the loom during the insertion of the pulling bar (basic part of 3D fabric creating mechanism). The first solution used a changement of the batten bearing position but this could be a source of other problems. A study in depth was thus carried out and a viable solution was found.

Stacking Sequence. The stacking sequence was determined using simplified relations for flexion of flat plates. Schema of batten is on the Fig. 3. The thickness

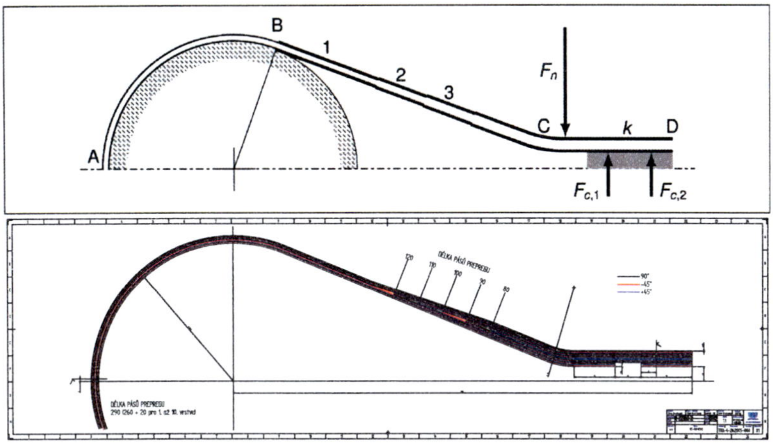

Fig. 3. Schema of the clamp (top) and example of actual design (bottom)

in individual sections and corresponding stacking were determined using classical laminating theory which implies the use of the cylindrical flexion of thin flat plates theory (thus we are neglecting effects of shearing stresses):

$$M = -(E \cdot J) \cdot \frac{d^2u}{dx^2} \tag{9}$$

where M is bending moment over a length unit, u is flexion and value of $E \cdot J$ is corresponding bending stiffness; yet again, this value is related to a length unit. For a laminated plate, i.e. a plate which is composed of several plies with different mechanical properties, the value of $E \cdot J$ is given by:

$$E \cdot J = \sum_{i=1}^{N} E_i \cdot \frac{z_i^3 - z_{i-1}^3}{3} \tag{10}$$

where E_i is Young modulus of i-th ply in direction of bending, z_{i-1} and z_i are respectively its bottom and top distance from the neutral surface. We will return on the latter relation in the next subsection more closely, for the purpose of this section we will retain this form.

By examining closely the final form of plates it is obvious that several sections have a stacking sequence that does not respect the mirror symmetry. This fact can make difficult a precise fabrication of the clamps due to their thermic distortion during the curing (or more precisely during their cooling down after the curing). Although the mirror symmetry is considered as almost imperative in fabrication of laminated parts we were aware of another issue in the design of the clamps so no particular measures were taken to resolve this shortcoming.

Curved Laminated Plates. The current design of clamping plates respects the above said mirror symmetry of ply stacking. However a non negligible distortion appears on the final laminated parts. A study in depth using classical theory of laminates [1] was carried out and soon it appeared that a phenomenon joint to the curved form of clamps would practically prevent the solution of this issue.

Let us return on the relation (10). It is only part of a more complex relation between loads and deformations of a thin plate (we are using annotation as used by [2]):

$$\begin{vmatrix} \mathbf{N} \\ \mathbf{M} \end{vmatrix} = \begin{vmatrix} \mathbf{A} & \mathbf{B} \\ \mathbf{B'} & \mathbf{C} \end{vmatrix} \cdot \begin{vmatrix} \varepsilon_\mathbf{0} \\ \kappa_\mathbf{0} \end{vmatrix} - \Delta T \cdot \begin{vmatrix} \langle \alpha \cdot \mathbf{E} \cdot \mathbf{h} \rangle \\ \langle \alpha \cdot \mathbf{E} \cdot \mathbf{h^2} \rangle \end{vmatrix} \tag{11}$$

where $\varepsilon_\mathbf{0} = [\varepsilon_{0,x}, \varepsilon_{0,y}, \gamma_{0,xy}]'$, $\kappa_\mathbf{0} = [\kappa_x, \kappa_y, \kappa_{xy}]'$, $\langle \alpha \cdot E \cdot h \rangle$ and $\langle \alpha \cdot E \cdot h^2 \rangle$ being the terms of thermal expansion. In this relation the index 0 refers to the deformations of neutral surface. For a flat plate the terms of sub matrices **A**, **B** and **C** are defined using the above mentioned classical theory as follows (see [2]); per example

$$\begin{aligned} A_{11} &= \sum_{k=1}^{K} \int_{h_{k-1}}^{h_k} \overline{E_{11k}} \cdot dz \quad B_{11} = -\sum_{k=1}^{K} \int_{h_{k-1}}^{h_k} \overline{E_{11k}} \cdot z \cdot dz \\ (B'_{11}) &= B_{11} \qquad\qquad\qquad C_{11} = \sum_{k=1}^{K} \int_{h_{k-1}}^{h_k} \overline{E_{11k}} \cdot z^2 \cdot dz \end{aligned} \tag{12}$$

where K is total number of plies and z is distance from the neutral surface as given by the definition of ε by Kirchhoff's theory. The terms representing the thermal expansion are defined by:

$$\langle \alpha \cdot E \cdot h \rangle_x = \sum_k \overline{\alpha \cdot E_{1,k}} \cdot (h_k - h_{k-1}) \text{ etc.}$$

For a flat laminated plate respecting mirror symmetry of ply stacking this surface lies in the plane of symmetry and the terms of B vanish per definition. This implies that there is no coupling between behavior in membrane and in bending and thus there is no flexion/torsion caused by thermic expansion. Actually if we let vanish the left hand side matrix of loads in relation (11) the solution of resulting algebraic system for unknown deformations can be decoupled into two separated subsystems of 3 unknowns each, the deformations in membrane $\varepsilon_{0,i}$, and curvatures $\kappa_{0,i} = -\frac{\partial^2 u_{0,i}}{\partial x_i^2}$.

Let us assume flexion of a general curved and twisted plate. Due to non symmetry of the field of strain (see Fig. 4) the neutral surface is displaced by $e > 0$ to the interior of curved plate:

$$\varepsilon_x(z) = \frac{z-e}{R_x - z} \cdot \frac{d\varphi}{d\psi} = \frac{z-e}{1-\frac{z}{R_x}} \cdot \frac{d\varphi}{ds} \tag{13}$$

Replacing the value of strain in the middle plane $\varepsilon_{0,x} = -e \cdot \frac{d\varphi}{ds}$ in (13) we get:

$$\varepsilon_x(z) = \frac{\varepsilon_{0,x}}{1-\frac{z}{R_x}} + \frac{z}{1-\frac{z}{R_x}} \cdot \frac{d\varphi}{ds} = \frac{1}{1-\frac{z}{R_x}} \cdot \varepsilon_{0,x} + \frac{z}{1-\frac{z}{R_x}} \cdot \kappa_x \tag{14}$$

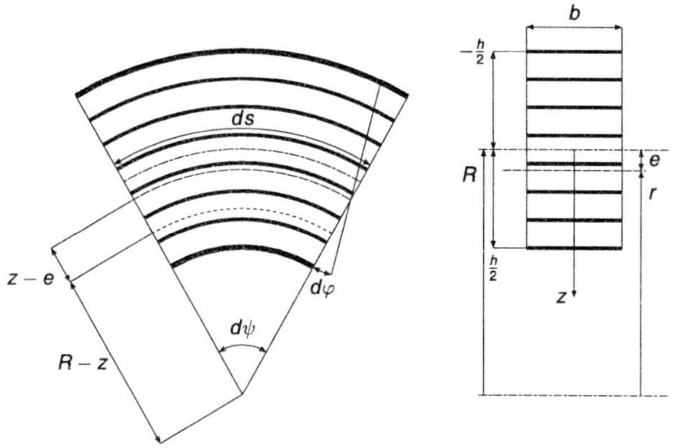

Fig. 4. Kinematic schema of a bent curved plate

It is obvious that for $R_x \to \infty$ we get the expression for flat plates:

$$\varepsilon_x(z) = \varepsilon_{0,x} + z \cdot \kappa_x$$

Expressions for ε_y and γ_{xy} analogous to (14) can be easily found. Let's express the field of stresses in the pli k (Hook's law) and put the resulting relations for σ into equilibrium equation. We get for membrane forces in laminated plate composed of K plies relations in the following form (only x direction is presented):

$$N_x = \sum_{k=1}^{K} \left(\int_{h_{k-1}}^{h_k} \left(\overline{E_{11k}} \cdot \varepsilon_x(z) + \overline{E_{12k}} \cdot \varepsilon_y(z) + \overline{E_{13k}} \cdot \gamma_{xy}(z) \right) \cdot dz \right)$$

and for the bending moment of a laminated plate composed of K plies:

$$M_x = \sum_{k=1}^{K} \left(\int_{h_{k-1}}^{h_k} \left(\overline{E_{11k}} \cdot \varepsilon_x(z) + \overline{E_{12k}} \cdot \varepsilon_y(z) + \overline{E_{13k}} \cdot \gamma_{xy}(z) \right) \cdot z \cdot dz \right)$$

We can rewrite the previous relations in form of (11), where the terms of sub-matrices can be obtained by integration analogous to (12) and by using (14); again only selected matrix members are presented for example:

$$A_{11} = \sum_{k=1}^{K} \int_{h_{k-1}}^{h_k} \overline{E_{11k}} \cdot \frac{dz}{1-\frac{z}{R_x}} \quad B_{11} = -\sum_{k=1}^{K} \int_{h_{k-1}}^{h_k} \overline{E_{11k}} \cdot z \cdot \frac{dz}{1-\frac{z}{R_x}}$$
$$(B'_{11}) = B_{11} \qquad C_{11} = \sum_{k=1}^{K} \int_{h_{k-1}}^{h_k} \overline{E_{11k}} \cdot z^2 \cdot \frac{dz}{1-\frac{z}{R_x}} \tag{15}$$

It follows from the form of the integrand in (15) that for $R_i < \infty$ the members of **B** will generally not vanish even for a symmetric stacking order of plies. The consequence of this is the fact that for zero values of **N** and **M** (i.e. unloaded plate) and non-zero ΔT we must get non-zero values of ε_0 and κ_0.

Coupling Flexion-Torsion of Symmetric Laminates. By using the integration in form either of (12) either of (15) we get the members of stiffness matrix C_{31} and C_{32} zero only for a flat plate and, moreover, only in the case where the directions of orthotropy of *all* plies lie in the plane of flexion. It follows that if there are unidirectional – even symmetric – plies under ±45° to the plane of flexion the above mentioned non zero members of **C** will cause a coupling between flexion of plates due to clamping force and their deformation in torsion.

3.3 Final Design of Batten

Due to all the phenomena described above, we chose the following construction of the beam. In contrast to the current practice, two *mirror* versions of the clamps will be produced, *right* and *left* or *even* and *odd*, which will be combined on one batten, see Fig. 5. The two versions will differ in their stacking sequence where the plies under 45° in the *right* plate will be placed under −45° in the *left* one and vice versa. In such manner we suppose minimize the final misalignment of the reed and this at the price of non constant clamping force along the batten.

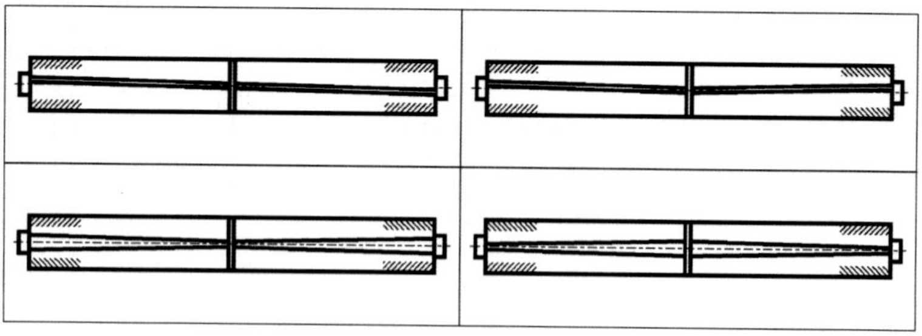

Fig. 5. Top view of batten in various possible configurations; the current configuration of clamps is top left; the preferred one is the bottom right

4 Conclusions

By using methods of classical mechanics we succeeded in optimisation of the batten in order to minimise the loads of cams. The calculations were verified using CAD software during final design of the batten, see Fig. 6.

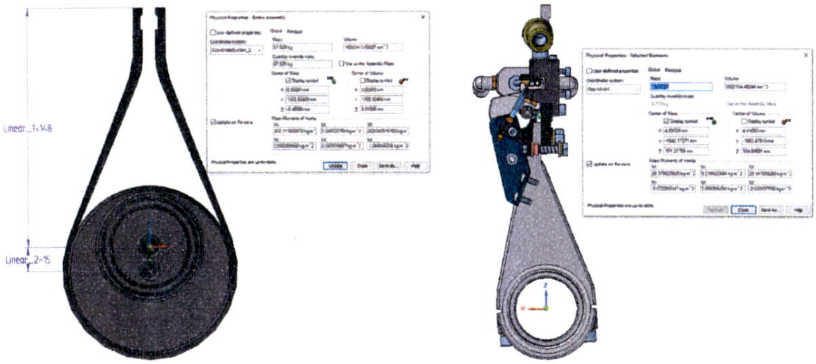

Fig. 6. Evaluation of various parts of batten using CAD system Solid Edge

Another evaluation of analytical calculation was carried out on the deformation caused by thermal expansion. Its results are presented on the Fig. 7. It is obvious that a combination of flexure and twist occurs during cooling down of the cured laminated composite.

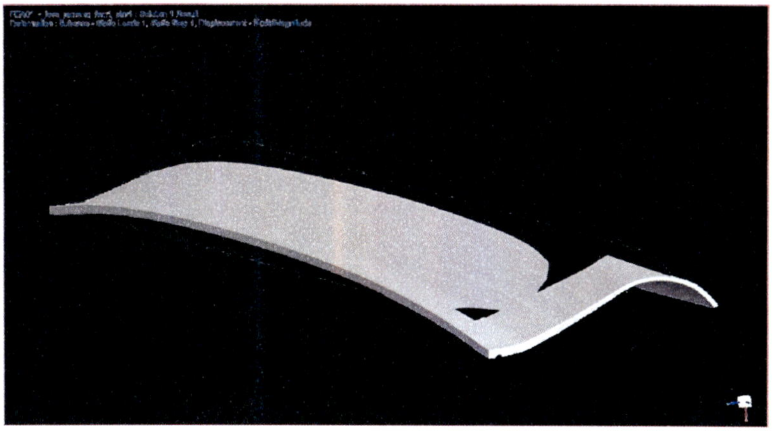

Fig. 7. Deformation due to thermal expansion calculated by using FEM model

In conclusion, we have succeeded in our efforts to increase the utility value of the batten and thus we have increased also the utility value of the entire machine. Both the optimisation of inertia properties and the evaluation of elastic properties were carried out by using in parallel the methods of classical mechanics and modern computational tools such as FEM and CAD which proves that classical methods may be still useful.

References

1. Berthelot, J.-M.: Matériaux composites. MASSON Editeur, Paris (1992)
2. Gay, D.: Matériaux composites. 3e éd. HERMES, Paris (1991)

Author Index

© The Editor(s) (if applicable) and The Author(s), under exclusive license
to Springer Nature Switzerland AG 2022

J. Beran et al. (Eds.): TMM 2020, MMS 85, pp. 337–338, 2022.
https://doi.org/10.1007/978-3-030-83594-1